Soil and Groundwater Quality and Resources Assessment, 2nd Edition

Soil and Groundwater Quality and Resources Assessment, 2nd Edition

Guest Editors

Wanjun Jiang
Yizhi Sheng
Hairu Mao

MDPI

Basel • Beijing • Wuhan • Barcelona • Belgrade • Novi Sad • Cluj • Manchester

Guest Editors
Wanjun Jiang
Tianjin Center
China Geological Survey
Tianjin
China

Yizhi Sheng
State Key Laboratory of
Biogeology and
Environmental Geology
China University of
Geosciences
Beijing
China

Hairu Mao
College of Water Conservancy
North China University of
Water Resources and Electric
Power
Zhengzhou
China

Editorial Office
MDPI AG
Grosspeteranlage 5
4052 Basel, Switzerland

This is a reprint of the Special Issue, published open access by the journal *Water* (ISSN 2073-4441), freely accessible at: https://www.mdpi.com/journal/water/special_issues/58XF43RSR9.

For citation purposes, cite each article independently as indicated on the article page online and as indicated below:

Lastname, A.A.; Lastname, B.B. Article Title. *Journal Name* **Year**, *Volume Number*, Page Range.

ISBN 978-3-7258-6650-2 (Hbk)
ISBN 978-3-7258-6651-9 (PDF)
https://doi.org/10.3390/books978-3-7258-6651-9

Contents

About the Editors

Wanjun Jiang

Wanjun Jiang is an Associate Researcher at the Tianjin Center of China Geological Survey, holding a Ph.D. in Hydrogeology from the China University of Geosciences (Beijing). His research focuses on isotope hydrogeochemistry applied to groundwater environments, regional groundwater circulation and evolution, the genesis of deep formation water, paleoclimate change, and paleo-groundwater recharge. He serves as a Youth Editorial Board Member for *Hydrology* and a Guest Editor for several SCI journals (e.g., *Water, Sustainability*). With over 20 publications in journals such as *Science of the Total Environment* and *Journal of Hydrology*, his work advances the understanding of groundwater evolution, environmental tracers, and ancient recharge processes.

Yizhi Sheng

Yizhi Sheng is a Professor and Doctoral Supervisor at the China University of Geosciences (Beijing). He earned his Ph.D. through a joint program with Pennsylvania State University (USA) and conducted postdoctoral research at Tsinghua University and the University of Miami. Prof. Sheng serves as Associate Editor for *Geo-Bio Interfaces* and as an Editorial Board Member for journals including *Earth Science Frontiers*. His research centers on biogeochemical cycles and soil/groundwater pollution remediation. He has published over 60 papers in leading journals such as *Nature Reviews Earth & Environment* and *Environmental Science & Technology*, contributing to microbial–environment interactions and sustainable remediation.

Hairu Mao

Hairu Mao is an Associate Researcher at the North China University of Water Resources and Electric Power, with a Ph.D. in Hydrogeology from the China University of Geosciences (Beijing). Her research specializes in isotope hydrogeochemistry, regional groundwater circulation, and groundwater environments. She serves as a Guest Editor for *Water* and *Water Environment Research*. She has published over 20 papers in journals including *Water Resources Research*, *Journal of Hydrology*, and *Water Research*.

Preface

This volume, titled "Soil and Groundwater Quality and Resources Assessment, 2nd Edition", is dedicated to a critical and timely subject in environmental earth sciences. Its scope encompasses the multifaceted challenges facing soil–groundwater systems under the dual pressure of climate change and intensive anthropogenic activities, including contamination, resource depletion, and ecological degradation. The work aims to present a cohesive collection of advanced research that integrates innovative methodologies—spanning numerical modeling, hydrochemical and isotopic tracing, geospatial analysis, and data-driven risk assessment—to elucidate complex subsurface processes. Its fundamental purpose is to bridge the gap between mechanistic scientific understanding and actionable management strategies, thereby providing a robust knowledge base for the accurate assessment and sustainable governance of these indispensable resources in diverse geographical and climatic settings.

The primary motivation for compiling this scientific work stems from the urgent need to address the escalating environmental threats to our terrestrial and subsurface environments. Groundwater, often the sole source of potable water in arid regions, and soil, the foundation of agriculture and ecosystems, require enhanced protective measures informed by cutting-edge science. This volume is foremost addressed to researchers, hydrogeologists, and environmental scientists engaged in subsurface process studies and pollution mitigation. It is equally pertinent to policymakers, resource managers, and engineers who are tasked with designing and implementing sustainable practices, risk mitigation frameworks, and climate-resilient water governance policies, offering them both theoretical insights and practical tools derived from contemporary case studies and analytical innovations.

Wanjun Jiang, Yizhi Sheng, and Hairu Mao
Guest Editors

water

MDPI

Editorial

Challenges and Advances in Soil and Groundwater Quality and Resources Assessment

Wanjun Jiang [1,2], Yizhi Sheng [3,*] and Hairu Mao [4]

1 Tianjin Center, China Geological Survey, Tianjin 300170, China; jiangwanjun0718@163.com
2 Tianjin Key Laboratory of Coast Geological Processes and Environmental Safety, Tianjin 300170, China
3 State Key Laboratory of Geomicrobiology and Environmental Changes, China University of Geosciences, Beijing 100083, China
4 College of Water Conservancy, North China University of Water Resources and Electric Power, Zhengzhou 450046, China; maohairu@163.com
* Correspondence: shengyz@cugb.edu.cn

1. Introduction

Groundwater and soil systems represent indispensable yet vulnerable components of the Earth's critical zone, underpinning drinking water supply, agricultural production, ecosystem stability, and sustainable socio-economic development [1–7]. However, these vital resources face unprecedented pressures from the synergistic effects of climate change and intensive anthropogenic activities, including agricultural expansion, industrial discharge, rapid urbanization, and unsustainable resource extraction [6,8–11]. These compound stressors have significantly altered hydrological cycles and biogeochemical processes, leading to widespread contamination, resource depletion, and ecological degradation [7,8,12,13]. Consequently, the accurate assessment and sustainable management of soil-groundwater systems have emerged as a paramount challenge for both scientific research and environmental governance.

The complexity of pollutant behavior within the interconnected soil-groundwater continuum is a central concern [14–16]. Contaminants, encompassing both legacy and emerging inorganic (e.g., heavy metals, nutrients, and geogenic arsenic/fluoride) and organic (e.g., pesticides, industrial solvents, pharmaceuticals, perfluorinated compounds (PFCs), and microplastics) compounds, originate from diverse point and non-point sources [17–24]. Their mobilization, speciation, and ultimate fate are governed by a complex interplay of physical transport (advection, dispersion), geochemical reactions (adsorption–desorption, dissolution-precipitation, and redox transformations), and microbiological processes [20,25]. These dynamics are further modulated by site-specific factors such as lithology, hydraulic properties, and the frequency of surface water-groundwater exchange. Climate change intensifies these challenges by altering precipitation patterns and freeze–thaw cycles, which can accelerate contaminant leaching and remobilize stored pollutants. Moreover, elemental biogeochemical cycling has been shown to be critical factors controlling the transformation and fate of materials in these belowground environments [26].

Conventional assessment frameworks exhibit significant limitations in characterizing the spatial heterogeneity of contaminants and quantifying uncertainties in long-term risks under compound stress. This limitation arises primarily from inadequate representation of coupled multi-scale processes and oversimplified descriptions of complex transport-transformation behaviors in heterogeneous media. Therefore, developing an interdisciplinary framework integrating high-resolution in situ monitoring, process-based mechanistic modeling, multi-isotopic tracing (e.g., stable and non-traditional isotopes), and data-driven analytics is imperative [5,6,12,27–35]. Synthesizing these complementary approaches enables systematic elucidation of contaminant dynamics in multiphase media, reduces predictive uncertainty, and establishes a scientific foundation for precise environmental risk assessment and sustainable management strategies.

This Special Issue, entitled "Soil and Groundwater Quality and Resources Assessment," brings together nine research articles that address these challenges through innovative methodologies and case studies across diverse geographical and climatic settings. Collectively, these contributions advance our understanding of contaminant transport mechanisms, groundwater dynamics, hydrogeochemical evolution, and integrated assessment frameworks under varying natural and anthropogenic stressors. This editorial synthesizes key findings from the published studies, highlights methodological innovations, and identifies emerging research directions for the scientific and managerial communities.

2. Modeling Transport Processes and Environmental Geological Effects

Several studies emphasize the power of numerical modeling in disentangling complex subsurface processes. An integrated numerical framework combining the Richards equation and advection–dispersion modeling was developed to simulate glyphosate mobility in agricultural soils (Contribution 1). Through inverse calibration, they revealed that the governing role of soil texture and clay content in contaminant mobility, providing a predictive tool for assessing groundwater vulnerability under long-term agrochemical practices. At the urban and engineering scale, Li et al. (Contribution 2) investigated the mechanical response of buildings to dynamically rising groundwater levels using FLAC3D simulations. Their findings identified phased settlement patterns and tensile stress concentrations at slab-wall junctions, which heighten cracking and seepage risks. The findings underscore the necessity of integrating dynamic groundwater monitoring and differentiated foundation design into structural risk management. Additionally, Zheng et al. (Contribution 3) integrated MIKE URBAN modeling with a genetic algorithm to optimize drainage network layout for annual cost minimization. By linking hydraulic performance with an annual cost system, their approach demonstrates how simulation-based optimization can enhance drainage efficiency while achieving substantial economic savings, providing a practical pathway for climate-resilient agricultural water management.

3. Hydrochemical Processes and Source Identification

Understanding groundwater chemistry evolution under natural and anthropogenic forcing is a recurrent theme across this Special Issue. For example, Ouedraogo et al. (Contribution 4) combined hydrochemistry, stable isotopes (δ18O, δ2H), GIS and multivariate statistics to elucidate surface water–groundwater interactions, the key hydrogeochemical processes and anthropogenic pollution in the Kou basin. Their findings confirm strong hydraulic connectivity between aquifer layers and nitrate contamination from anthropogenic activities, providing a baseline for integrated water resource management in water-scarce regions. In arid endorheic systems, Guo et al. (Contribution 5) integrated inverse hydrochemical modeling with isotopic evidence to quantify mineral dissolution, evaporation concentration, and cation exchange processes along groundwater flow paths in the Qaidam Basin. The work provides a mechanistic understanding of salinity evolution and highlights hydrochemical processes as key drivers of mineral transport and enrichment in arid basins. Zhang et al. (Contribution 6) further advanced the analysis of groundwater dynamics by developing a GIS-based multi-factor weighted comprehensive evaluation system to classify groundwater response types under compound stress. Their results demonstrate how precipitation variability, groundwater extraction, and land use jointly shape spatial het-

https://doi.org/10.3390/w18020160

erogeneity in groundwater behavior, offering a framework for differentiated groundwater management strategies.

4. Risk, Eco-Environmental Assessment and Sustainable Management

Several studies address groundwater quality degradation and its implications for human health and ecosystem security. Li Jiani et al. (Contribution 7) proposed an improved DRASTICL model and a kernel density estimation–trapezoidal fuzzy number–Monte Carlo simulation (KDE-TFN-MCSS) model to assess the vulnerability index and health risk in the Songnen Plain, respectively. Furthermore, groundwater quality index (E-GQI) and self-organizing maps (SOM) were employed for water quality assessment and contaminant source apportionment. Applied to a cold-region aquifer system, this method reduces uncertainty in risk estimation and reveals persistent non-carcinogenic and carcinogenic risks associated with NH4+ and As, particularly under freeze–thaw and urbanization influences. Focusing on industrial regions, Du et al. (Contribution 8) applied spatial autocorrelation analysis and SOM coupled with K-means clustering to investigate the spatial distribution and key influencing factors of nine heavy metals (Cr, Fe, Mn, Ni, Cu, Zn, As, Ba, and Pb) in southern China. Their results highlight the intertwined impacts of industrial pollution, domestic sewage, agricultural activities, and natural mineral dissolution, demonstrating the necessity of integrated source-oriented management in industrialized landscapes. At the watershed scale, Hao et al. (Contribution 9) constructed an ecological–environmental geological quality evaluation system for the Kundulun River Basin, emphasizing the dominant role of land use, precipitation, and groundwater depth in shaping ecological risk patterns. Their zonal management recommendations provide actionable guidance for balancing resource exploitation and ecological protection in arid and semi-arid regions.

5. Synthesis and Future Perspectives

Collectively, the studies in this Special Issue underscore several key insights. First, groundwater systems respond nonlinearly to environmental stresses, with physical, chemical, and anthropogenic processes interacting across scales. Second, integrating numerical models with hydrochemical, isotopic, and data-driven approaches is essential for capturing system complexity and reducing uncertainty. Third, management-oriented frameworks, linking scientific understanding with engineering design and policy needs, are increasingly critical under accelerating climate change and human pressure.

Future research should prioritize (1) coupling groundwater models with climate projections and land-use change scenarios, (2) advancing uncertainty-aware risk assessment tools for vulnerable populations and ecosystems, (3) strengthening cross-scale integration from pore-scale processes to basin-scale management, and (4) integrating the hydro-biogeochemical processes. By doing so, groundwater science can more effectively support sustainable water governance and resilience under compound environmental stress.

The contributions assembled in this Special Issue provide both methodological innovations and practical insights, collectively advancing the frontiers of groundwater and soil-water system research. We hope that this collection will stimulate further interdisciplinary collaboration and inform adaptive management strategies in diverse hydrogeological settings worldwide.

Author Contributions: Conceptualization, W.J. and Y.S.; writing—original draft preparation, W.J.; writing—review and editing, Y.S. and H.M. All authors have read and agreed to the published version of the manuscript.

Funding: This research was funded by the Tianjin Natural Science Foundation Project (25JCYBJC00670).

https://doi.org/10.3390/w18020160

Acknowledgments: We would like to thank all the authors who contributed to this Special Issue. We would also like to thank all the reviewers for providing their valuable comments, which have greatly improved the quality of the published papers.

Conflicts of Interest: The authors declare no conflicts of interest.

List of Contributions:

1. García-Gallego, J.; Fuentes, S.; Mercado-Fernández, T.; Ventura-Ramos, E.; Treviño-Reséndez, J.; García-Espinoza, J.D.; Fuentes, C.; Chávez, C. Integrated Numerical Approach to Glyphosate Transport in Soil Profiles Under Farming Conditions. *Water* **2025**, *17*, 3569.
2. Li, H.; Gu, M.; Zhang, M.; Ma, B.; Zhu, X.; Gu, L.; Tai, J.; Chen, L. Research on the Influence of Groundwater Level Dynamic Rising Process on Buildings Based on Numerical Simulation. *Water* **2025**, *17*, 3014.
3. Zheng, Z.; Li, M.; Wang, T.; Ren, H. Study on the Factors Affecting the Drainage Efficiency of New Integrated Irrigation and Drainage Networks and Network Optimization Based on Annual Cost System. *Water* **2025**, *17*, 1201.
4. Ouedraogo, I.; Vanclooster, M.; Huneau, F.; Vystavna, Y.; Kebede, S.; Koussoubé, Y. Surface Water–Groundwater Interactions in a Sahelian Catchment: Exploring Hydrochemistry and Isotopes and Implications for Water Quality Management. *Water* **2025**, *17*, 2756.
5. Guo, L.; Ding, Y.; Fang, H.; An, C.; Jiang, W.; Yang, N. Integrating Inverse Modeling to Investigate Hydrochemical Evolution in Arid Endorheic Watersheds: A Case Study from the Qaidam Basin, Northwestern China. *Water* **2025**, *17*, 2074.
6. Zhang, Q.; Zhang, M.; Jiang, W.; Hao, Y.; Chen, F.; Zhang, M. Characterization of Groundwater Dynamics and Their Response Mechanisms to Different Types of Compound Stress in a Typical Hilly Plain Area. *Water* **2025**, *17*, 1846.
7. Li, J.; Wang, Y.; Bian, J.; Sun, X.; Feng, X. Health Risk Assessment of Groundwater in Cold Regions Based on Kernel Density Estimation–Trapezoidal Fuzzy Number–Monte Carlo Simulation Model: A Case Study of the Black Soil Region in Central Songnen Plain. *Water* **2025**, *17*, 2984.
8. Du, J.; Liao, F.; Zhang, Z.; Du, A.; Qian, J. Spatial Heterogeneity and Controlling Factors of Heavy Metals in Groundwater in a Typical Industrial Area in Southern China. *Water* **2025**, *17*, 2012.
9. Hao, Y.; Wan, J.; Xin, Y.; Zhou, W.; Li, Y.; Mao, L.; Li, X.; Mo, L.; Li, R. Research on Ecological–Environmental Geological Survey and Evaluation Methods for the Kundulun River Basin in Baotou City. *Water* **2025**, *17*, 1926.

References

1. Cuthbert, M.O.; Gleeson, T.; Moosdorf, N.; Befus, K.M.; Schneider, A.; Hartmann, J.; Lehner, B. Global patterns and dynamics of climate-groundwater interactions. *Nat. Clim. Change* **2019**, *9*, 137–141.
2. Hou, D.Y.; Jia, X.Y.; Wang, L.W.; McGrath, S.; Zhu, Y.G.; Hu, Q.; Zhao, F.J.; Bank, M.S.; O'Connor, D.; Nriagu, J. Global soil pollution by toxic metals threatens agriculture and human health. *Science* **2025**, *388*, 316–321. [CrossRef]
3. Hou, D.; O'Connor, D.; Igalavithana, A.D.; Alessi, D.S.; Luo, J.; Tsang, D.C.W.; Sparks, D.L.; Yamauchi, Y.; Rinklebe, J.; Ok, Y.S. Metal contamination and bioremediation of agricultural soils for food safety and sustainability. *Nat. Rev. Earth Environ.* **2020**, *1*, 366–381. [CrossRef]
4. Rohde, M.M.; Albano, C.M.; Huggins, X.; Klausmeyer, K.R.; Morton, C.; Sharman, A.; Zaveri, E.; Saito, L.; Freed, Z.; Howard, J.K.; et al. Groundwater-dependent ecosystem map exposes global dryland protection needs. *Nature* **2024**, *632*, 101–107. [CrossRef]
5. Jasechko, S.; Seybold, H.; Perrone, D.; Fan, Y.; Shamsudduha, M.; Taylor, R.G.; Fallatah, O.; Kirchner, J.W. Rapid groundwater decline and some cases of recovery in aquifers globally. *Nature* **2024**, *625*, 715–721. [CrossRef]
6. Kuang, X.X.; Liu, J.G.; Scanlon, B.R.; Jiao, J.J.; Jasechko, S.; Lancia, M.; Biskaborn, B.K.; Wada, Y.; Li, H.L.; Zeng, Z.Z.; et al. The changing nature of groundwater in the global water cycle. *Science* **2024**, *383*, eadf0630. [CrossRef]
7. Jiang, W.J.; Meng, L.S.; Liu, F.T.; Sheng, Y.Z.; Chen, S.M.; Yang, J.L.; Mao, H.R.; Zhang, J.; Zhang, Z.; Ning, H. Distribution, source investigation, and risk assessment of topsoil heavy metals in areas with intensive anthropogenic activities using the positive matrix factorization (PMF) model coupled with self-organizing map (SOM). *Environ. Geochem. Health* **2023**, *45*, 6353–6370.
8. Niazi, H.; Wild, T.B.; Turner, S.W.D.; Graham, N.T.; Hejazi, M.; Msangi, S.; Kim, S.; Lamontagne, J.R.; Zhao, M.Q. Global peak water limit of future groundwater withdrawals. *Nat. Sustain.* **2024**, *7*, 413–422. [CrossRef]

https://doi.org/10.3390/w18020160

9. Long, D.; Yang, W.; Scanlon, B.R.; Zhao, J.S.; Liu, D.; Burek, P.; Pan, Y.; You, L.Z.; Wada, Y. South-to-North Water Diversion stabilizing Beijing's groundwater levels. *Nat. Commun.* **2020**, *11*, 3665. [CrossRef]
10. Abascal, E.; Gómez-Coma, L.; Ortiz, I.; Ortiz, A. Global diagnosis of nitrate pollution in groundwater and review of removal technologies. *Sci. Total Environ.* **2022**, *810*, 152233.
11. Kurwadkar, S.; Kanel, S.R.; Nakarmi, A. Groundwater pollution, occurrence, detection, and remediation of organic and inorganic pollutants. *Water Environ. Res.* **2020**, *92*, 1659–1668. [CrossRef]
12. Scanlon, B.R.; Fakhreddine, S.; Rateb, A.; de Graa, I.; Famiglietti, J.; Gleeson, T.; Grafton, R.Q.; Jobbagy, E.; Kebede, S.; Kolusu, S.R.; et al. Global water resources and the role of groundwater in a resilient water future. *Nat. Rev. Earth Environ.* **2023**, *4*, 87–101. [CrossRef]
13. Sheng, Y.Z.; Jiang, J.W.; Zhang, M. Mobilization, Speciation, and Transformation of Organic and Inorganic Contaminants in Soil-Groundwater Ecosystems. *Appl. Sci.* **2023**, *13*, 11454. [CrossRef]
14. Aullón, A.A.; Schulz, C.; Bundschuh, J.; Jacks, G.; Thunvik, R.; Gustafsson, J.P.; Mörth, C.M.; Sraceke, O.; Ahmad, A.; Bhattacharya, P. Hydrogeochemical controls on the mobility of arsenic, fluoride and other geogenic co-contaminants in the shallow aquifers of northeastern LaPampa Province in Argentina. *Sci. Total Environ.* **2020**, *715*, 136671.
15. Wei, Y.Q.; Chen, Y.L.; Cao, X.D.; Yeh, T.C.J.; Zhang, J.; Zhan, Z.; Cui, Y.D.; Li, H. Modeling of Microplastics Migration in Soil and Groundwater: Insights into Dispersion and Particle Property Effects. *Environ. Sci. Technol.* **2024**, *58*, 15224–15235. [CrossRef]
16. Doyi, I.; Essumang, D.; Gbeddy, G.; Dampare, S.; Kumassah, E.; Saka, D. Spatial distribution, accumulation and human health risk assessment of heavy metals in soil and groundwater of the Tano Basin, Ghana. *Ecotoxicol. Environ. Saf.* **2018**, *165*, 540–546. [CrossRef]
17. Gandhi, T.P.; Sampath, P.V.; Maliyekkal, S.M. A critical review of uranium contamination in groundwater, treatment and sludge disposal. *Sci. Total Environ.* **2022**, *825*, 153947. [CrossRef] [PubMed]
18. Wang, F.; Xiang, L.; Leung, K. Emerging contaminants: A One Health perspective. *Innovation* **2024**, *5*, 100612. [PubMed]
19. Wang, Y.X.; Li, J.X.; Ma, T.; Xie, X.J.; Deng, Y.M.; Gan, Y.Q. Genesis of geogenic contaminated groundwater: As, F and I. *Crit. Rev. Environ. Sci. Technol.* **2020**, *51*, 2895–2933. [CrossRef]
20. Amarasiri, M.; Sano, D.; Suzuki, S. Understanding human health risks caused by antibiotic resistant bacteria (ARB) and antibiotic resistance genes (ARG) in water environments, Current knowledge and questions to be answered. *Crit. Rev. Environ. Sci. Technol.* **2020**, *50*, 2016–2059. [CrossRef]
21. Picetti, R.; Deeney, M.; Pastorino, S.; Miller, M.R.; Shah, A.; Leon, D.A.; Dangour, A.D.; Green, R. Nitrate and nitrite contamination in drinking water and cancer risk, A systematic review with meta-analysis. *Environ. Res.* **2022**, *210*, 112988. [CrossRef]
22. Zainab, S.M.; Junaid, M.; Xu, N.; Malik, R.N. Antibiotics and antibiotic resistant genes (ARGs) in groundwater, A global review on dissemination, sources, interactions, environmental and human health risks. *Water Res.* **2020**, *187*, 116455. [CrossRef]
23. Liu, Y.; Wang, P.; Gojenko, B.; Yu, J.J.; Wei, L.Z.; Luo, D.G.; Xiao, T.F. A review of water pollution arising from agriculture and mining activities in Central Asia, Facts, causes and effects. *Environ. Pollut.* **2021**, *291*, 118209. [CrossRef] [PubMed]
24. Dong, Y.; Jiang, Z.; Hu, Y.; Jiang, Y.; Tong, L.; Yu, Y.; Cheng, J.; He, Y.; Shi, J.; Wang, Y. Pathogen contamination of groundwater systems and health risks. *Crit. Rev. Environ. Sci. Technol.* **2023**, *54*, 267–289. [CrossRef]
25. Sheng, Y.; Dong, H.; Coffin, E.; Myrold, D.; Kleber, M. The important role of enzyme adsorbing capacity of soil minerals in regulating β-glucosidase activity. *Geophys. Res. Lett.* **2022**, *49*, e2021GL097556.
26. Sheng, Y.; Baars, O.; Guo, D.; Whitham, J.; Srivastava, S.; Dong, H. Mineral-bound trace metals as cofactors for anaerobic biological nitrogen fixation. *Environ. Sci. Technol.* **2023**, *57*, 7206–7216. [CrossRef] [PubMed]
27. Xie, J.; Liu, X.; Jasechko, S.; Berghuijs, W.; Wang, K.; Liu, C.; Reichstein, M.; Jung, M.; Koirala, S. Majority of global river flow sustained by groundwater. *Nat. Geosci.* **2024**, *17*, 770–777. [CrossRef]
28. Jiang, W.J.; Wang, G.C.; Sheng, Y.Z.; Shi, X.M.; Zhang, H. Isotopes in groundwater (^{2}H, ^{18}O, ^{14}C) revealed the climate and groundwater re-charge in the Northern China. *Sci. Total Environ.* **2019**, *666*, 298–307. [CrossRef] [PubMed]
29. Jiang, W.J.; Sheng, Y.Z.; Wang, G.C.; Shi, Z.M.; Liu, F.T.; Zhang, J.; Chen, D.L. Cl, Br, B, Li, and noble gases isotopes to study the origin and evolution of deep groundwater in sedimentary basins, a review. *Environ. Chem. Lett.* **2022**, *20*, 1497–1528. [CrossRef]
30. Cao, W.G.; Zhang, Z.; Fu, Y.; Zhao, L.H.; Ren, Y.; Nan, T.; Guo, H.M. Prediction of arsenic and fluoride in groundwater of the North China Plain using enhanced stacking ensemble learning. *Water Res.* **2024**, *259*, 121848. [CrossRef]
31. Rateb, A.; Scanlon, B.R.; Pool, D.R.; Sun, A.; Zhang, Z.Z.; Chen, J.L.; Clark, B.; Faunt, C.C.; Haugh, C.J.; Hill, M.; et al. Comparison of Groundwater Storage Changes From GRACE Satellites With Monitoring and Modeling of Major U.S. Aquifers. *Water Resour. Res.* **2020**, *56*, e2020WR027556. [CrossRef]
32. Yang, J.; Pan, Y.; Zhang, C.; Gong, H.; Xu, L.; Huang, Z.; Lu, S. Comparison of groundwater storage changes over losing and gaining aquifers of China using GRACE satellites, modeling and in-situ observations. *Sci. Total Environ.* **2024**, *938*, 173514. [CrossRef] [PubMed]

33. Mukherjee, A.; Coomar, P.; Sarkar, S.; Johannesson, K.H.; Fryar, A.E.; Schreiber, M.E.; Ahmed, K.M.; Alam, M.A.; Bhattacharya, P.; Bundschuh, J.; et al. Arsenic and other geogenic contaminants in global groundwater. *Nat. Rev. Earth Environ.* **2024**, *5*, 312–328. [CrossRef]
34. Podgorski, J.; Berg, M. Global analysis and prediction of fluoride in groundwater. *Nat. Commun.* **2022**, *13*, 4232. [CrossRef]
35. Liu, R.N.; Li, Z.C.; Xie, X.J.; Hou, Q.X.; Han, D.Y.; Yuan, Z.T.; Huang, G.X. A multi-method for identifying and spatial modeling of groundwater nitrate natural background levels in an urbanized delta in China: Insight from Cl/Br ratios combining with KNN algorithm and EBK. *Water Res.* **2026**, *289*, 124953. [PubMed]

https://doi.org/10.3390/w18020160

water

MDPI

Article

Integrated Numerical Approach to Glyphosate Transport in Soil Profiles Under Farming Conditions

Jesús García-Gallego [1], Sebastian Fuentes [1], Teobaldis Mercado-Fernández [2], Eusebio Ventura-Ramos [1], José Treviño-Reséndez [3], Josué D. García-Espinoza [3], Carlos Fuentes [4,*] and Carlos Chávez [1,*]

1 Water Research Center, Department of Irrigation and Drainage Engineering, Autonomous University of Queretaro, Cerro de las Campanas SN, Col. Las Campanas, Santiago de Querétaro 76010, Queretaro, Mexico; jgarcia341@alumnos.uaq.mx (J.G.-G.); sebastian.fuentes@uaq.mx (S.F.); eusebio.ventura@uaq.mx (E.V.-R.)

2 Facultad de Ciencias Agrícolas, Universidad de Córdoba, Carrera 6 No. 77-305, Montería 230002, Córdoba, Colombia; tmercado@correo.unicordoba.edu.co

3 Centro de Investigación en Química para la Economía Circular—CIQEC, Facultad de Química, Universidad Autónoma de Querétaro, Cerro de las Campanas SN, Col. Las Campanas, Santiago de Querétaro 76010, Queretaro, Mexico; jose.trevino@uaq.edu.mx (J.T.-R.); josue.daniel.garcia@uaq.mx (J.D.G.-E.)

4 Mexican Institute of Water Technology, Paseo Cuauhnáhuac Núm. 8532, Jiutepec 62550, Morelos, Mexico

* Correspondence: cfuentes@tlaloc.imta.mx (C.F.); chagcarlos@uaq.mx (C.C.)

Abstract

Glyphosate is the most widely used herbicide in the world for weed control; however, due to lixiviation, wind and runoff effects, an important fraction can reach the soil, aquifers and surface waters, affecting environmental and human health. The behavior of glyphosate in two agricultural soils (C1: silty clay texture, and C2: silty loam texture) was analyzed in this study using a laboratory-scale model. Water transfer was modeled with the Richards equation, while glyphosate transport was modeled using the advection–dispersion equation, with both solved using finite difference methods. The glyphosate dispersion coefficient was obtained from laboratory concentration data derived from the soil profile via inverse modeling using a non-linear optimization algorithm. The goals of this study were to (i) quantify glyphosate retention in soils with different physical and chemical properties, (ii) calibrate a numerical model for the estimation of dispersivity and simulation of short- and long-term scenarios, and (iii) assess vulnerability to groundwater contamination. The results showed that C1 retained a greater amount of glyphosate in the soil profile, while C2 was considered more vulnerable as it liberated the contaminant more easily. The model accurately reproduced the measured concentrations, as evidenced by the RMSE and R^2 statistics, thus supporting further scenario simulations allowing for prediction of the fate of the herbicide in soils. The approach utilized in this study may be useful as a tool for authorities in environmental fields, enabling better control and monitoring of soil contamination. These findings highlight potential risks of contamination and reinforce the importance of agricultural management strategies.

Keywords: contamination; pesticides; advection–dispersion equation; inverse model

1. Introduction

Agricultural activities are among the most significant sources of diffuse contamination in soils and aquifers, given the intensive use of pesticides for the control of pests in crops. These sources represent a significant risk affecting the supply of water for human consumption, as well as aquatic ecosystems in recharge zones, due to their toxicity as contaminants [1]. The soil acts as the main recipient of these chemicals, and their fate in the environment depends on the physicochemical interactions between soil components, water and the applied contaminant [2]. These interactions determine the sorption and desorption degree of the pesticide in the soil, which influences both its availability for the control of weeds and its environmental impacts. When sorption is low, most of the pesticide remains in the soil solution, allowing for its absorption by weed seeds and seedlings and improving their control [3,4].

Glyphosate [N-(phosphonomethyl) glycine] is a non-selective post-emergent systemic herbicide, often sprayed on the leaves of weeds to control their growth. It has been proven that a significant amount of this product can reach the soil, as well as surface and ground waters, negatively impacting ecosystems and being transported throughout the trophic chain [5].

Despite the fact that soil serves as a storage reservoir for glyphosate, due to its high affinity to agrochemicals, the risks related to contamination and affectation of surface and groundwater resources are noticeable. This is because small concentrations of this contaminant are accumulated in animals and humans; therefore, it can be amplified through the food chain, resulting in problems not only for the environment but also human health [6,7].

In Mexico, glyphosate is found as an active ingredient in many commercial herbicides, and it is used mainly in crops such as soybean, wheat, cotton and corn [8]. The United Nations Food and Agricultural Organization (FAO) reported that 4.55 tons of pesticides per 1000 hectares were applied in Mexico between 2009 and 2010. The situation did not improve by the year 2013, when 31,195 tons of herbicides were used [9]. At present, farmers and Mexican authorities are legally battling due to the significant impacts of this chemical compound on ecosystems and public health. In addition, some carcinogenic pesticides—such as dichlorodiphenyltrichloroethane (DDT), lindane, heptachlorine, endosulfan, aldrin, endrin and dieldrin (organochlorine)—are still used, which also have significant impacts on human and environmental health [10].

At present, studies focused on the resistance of various plant species to the action of glyphosate are more abundant than those assessing the behaviors of this contaminant in the environment using numerical tools [11,12]. Therefore, it is important to perform risk studies related to the management, transport and accumulation of glyphosate in the soil, to develop tools that allow for prediction of its fate, especially in soils with different textures.

Villholth et al. [1] studied the particle-facilitated transport of procholaz in soil with subsurface drainage and a sandy loam texture in Denmark, and found that the lost fraction of pesticides in the drainage was 0.2% of the applied mass, 6% of which was associated with the solid particle phase. While studying the behaviors and chemical properties of pesticides in soils cultivated with rice, Zhou et al. [2] observed that the population and activity of methanogens increased slightly with the addition of carbofuran or butachlor in the soil; however, this activity was significantly inhibited when additional butachlor was added. Similarly, Sun et al. [3] explored how environmental variables and levels of contamination due to organochlorine pesticide (OCP) influence the structure of bacterial communities in an abandoned pesticide plant site in China, observing that the soil microbial activity was significantly influenced by the level of OCP (%).

Similarly, da Silva et al. [4] examined the dynamics of hexazinone in three types of Brazilian agricultural soils with different textures: Arenosol (sandy loam), Cambisol (sandy clay) and Ferrasol (sandy loam). The authors found that Cambisols showed a greater lixiviation potential, resulting in an Aquifer Vulnerability Index (GUS) of 3.9, when compared with the Arenosols and Ferrasols, which presented GUS values of 3.6 and 3.2, respectively. Bernasconi et al. [5] analyzed the spatial–temporal characterizations of 19 pesticides in soils (included glyphosate and AMPA), and reported the presence of three or more pesticides in 93% of samples obtained from agrochemical production systems

(CBA) and 32% of those from agricultural systems based on agroecological practices (ABA). Both glyphosate and AMPA showed a higher detection frequency, accounting for 90% of the total charge of pesticides in CBA and ABA. The risk of aquifer contamination was studied by Hintze et al. [6]. They found that, for soils where the last application of chloridazon occurred from 5 to 10 years ago, concentrations of its metabolites were 10 times (desphenyl-chloridazon: 0.22–7.4 μg/kg) and 6 times (methyl-desphenyl-chloridazon: 0.12–3.1 μg/kg) greater in comparison with that of chloridazon (<0.050–1.0 μg/kg).

The use of numerical models to investigate the transport of pesticides in soils has been the object of many investigations. For example, Diamantopoulos et al. [7] compared the results of HYDRUS (2D/3D) modeling with those obtained using the unidimensional models PEARL and PELMO, commonly used for the assessment of pesticides. They concluded that HYDRUS (2D/3D) is a viable alternative for the evaluation of contaminants, as the results were consistent with those obtained by PEARL and PELMO, while providing the advantage of allowing simulations in two and three dimensions. Márquez et al. [8] examined the transport of the nematicide DiTera in soils using HYDRUS 1D modeling. The predictive model was found to precisely represent the behavior of DiTera under controlled conditions, demonstrating its tendency to accumulate in the first 20 cm of the soil.

Ona [9] carried out HYDRUS 1D modeling to analyze the transport of pesticides in disturbed and undisturbed soil columns. They found that the patterns of lixiviation were influenced mainly by the relative extension of preferential flow and the sorption capacity of the compounds present in different types of soil. In the study performed by Villarreal et al. [10], HYDRUS 1D modeling was performed to predict the dynamics of glyphosate and AMPA (aminomethylphosphonic acid)—its main degradation product—considering constant and variable hydraulic properties during the studied crop cycle.

Hmimou et al. [11] analyzed the transport of carbofuran in sandy soils from Mnasra, Morocco, using a model based on the Richards and advection–diffusion equations. They found that despite the limited irrigation and rainfall regime, carborufan was lixiviated almost completely beyond the root zone; furthermore, with a more intensive irrigation regime, this compound could reach a depth of up to 100 cm.

Lei et al. [12] investigated the transport behavior of 3,5,6-trichloro-2-pyridinol (TCP)—one of the most well-known products of pesticide degradation—by performing experiments and simulations in agricultural soils. They found that the transport of TCP in soil follows complex patterns that are influenced significantly by chemical and physical instabilities and identified a greater risk of contamination in water bodies near to rice and canola fields due to a more pronounced preferential flow. On the other hand, Dusek et al. [13] used a numerical 1D model based on the Richards equation to find the flow of water, while the advection–dispersion equation was used to simulate the movement and transport of five pesticides (atrazine, imazaquin, sulfometuron methyl, S-metolachlor and imidachloprid).

Padilla and Selim [14] studied the sorption in plots as well as miscible displacement experiments to analyze the mobility of glyphosate in agricultural soils with two different textures. The miscible displacement experiments showed that the mobility of glyphosate is extremely limited in both soils, with only 3% and 2% of the mass of herbicide applied recovered in the effluents of silty loam and clay soils, respectively. Nahar and Niven [15] experimentally and numerically evaluated the behavior of glyphosate using data from soil lixiviation columns and the HYDRUS 1D model, considering flow rates representative of field conditions and using bromide as a non-reactive tracer. After three applications of glyphosate, the columns with coarse textured soils showed the highest lixiviation rates, which were closely related to their macropore structure.

It is possible to observe that in research focused on the transport and distribution of glyphosate in the soil, analyses of the contaminant in soils with different characteristics or under different crop conditions have been reported. On the other hand, researchers have also performed numerical modeling to study the behaviors of the compound in soil columns in detail. Although previous studies have modeled solute transport in soils, most have focused on pesticides transported under surface irrigation or short duration tracer tests, and have not combined experimental infiltration testing, soil-specific dispersivity estimates, and long-term scenario simulations. This study addresses this gap by integrating laboratory evidence with a numerical model calibrated for each soil type, providing a robust tool for predicting the mobility of glyphosate and the associated contamination risks.

With the objective of examining the vulnerability of soils under the application of glyphosate, the movement and fate of this herbicide were analyzed in two types of agricultural soils through experimentation and numerical modeling. For this purpose, a silty clay soil from the middle valley of the Sinú River in Montería, Córdoba, Colombia, and a silty loam soil from the San Juan River Valley, municipality of San Juan del Río, Querétaro, México, were used. The experiments involved an infiltration test, in which a solution of glyphosate was applied in soil columns. The infiltration depth of glyphosate was obtained, and the saturated hydraulic conductivity and characteristic pressure were optimized. Similarly, the concentrations of glyphosate were analyzed in the laboratory before and after the test, to optimize these coefficients.

2. Materials and Methods

2.1. Study Area

This study considers soils obtained from Colombia and Mexico. The Colombian soil (Case 1, C1) corresponds to a silty clay soil at an experimental agricultural plot of the University of Cordoba Central Campus at Montería, Colombia. This site is located at the coordinates 8°47′25″ N and 75°51′31″ W (Figure 1) and forms part of the Caribbean region of Colombia. Soils in this zone are on an alluvial natural plane relief; are formed from fine and medium size sediments deposited during the Quaternary; tend to be moderately deep, with medium fine and very fine textures; are susceptible to flooding, with occasional ponding; and are classified as Inceptisols according to the Soils Chart for the Department of Cordoba [16,17].

The Mexican soil (Case 2, C2) is located at the coordinates 20°27′12.2″ N and 99°59′10.8″ W and corresponds to agricultural soil of silty loam texture from the municipality of San Juan del Río, State of Queretaro, Mexico (Figure 1). Soils in this area occur on a flooding plain relief; are formed from quaternary deposits on volcanic materials from the Upper Tertiary; tend to be shallow in depth; are fertile; and are classified in the order of Vertisols, according to the Soils Chart for the State of Queretaro [18,19].

2.2. Soil Sampling

According to the Official Mexican Norm NOM-021-RECNAT-2000 [20], soil sampling was performed for the purpose of soil classification. To define the number of sampling points in each soil type, the procedures established by the Mexican Norm NMX-AA-132-SCFI-2016 [21] were observed. In accordance with this procedure, a preliminary field survey and soil profile description were performed to identify horizon differentiation, texture changes, and structure discontinuities, following FAO and IGAC criteria. Detailed soil sampling over a regular area of 1000 m^2 was considered, taking one sample in each corner (4) and one more in the center (1), for a total of 5 sub-samples. A composed sample was then obtained with the objective of obtaining a representative sample; this approach is recommended for agricultural soils, where short-scale variability in clay content and organic matter may strongly influence chemical retention and hydraulic behaviors. Soil sampling was performed in 2024.

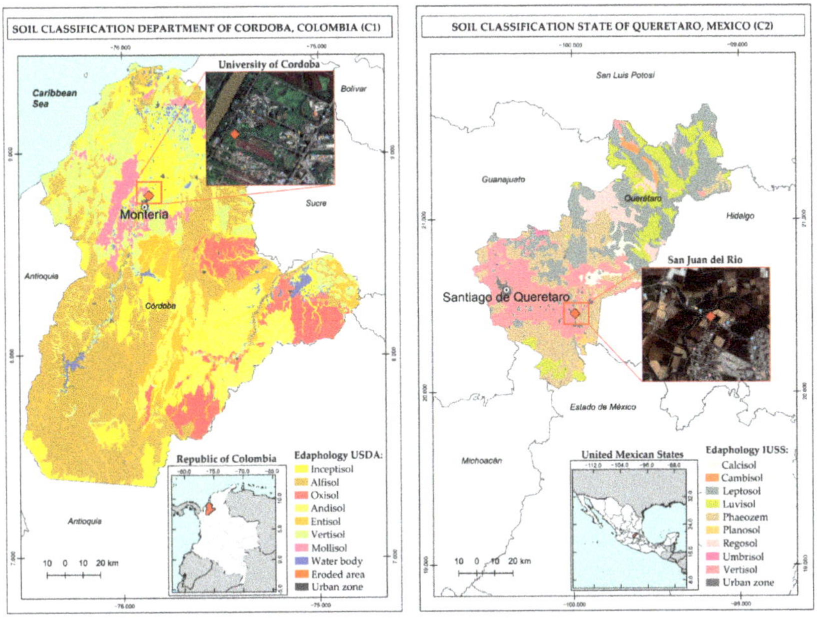

Figure 1. Study areas: studied soils from Colombia (C1) and Mexico (C2).

2.3. *Laboratory Analysis*

2.3.1. Soil Properties

The water content, bulk density and particle density of the soil were analyzed in the laboratory using the methods described in Table 1. C1 samples were analyzed in the Soil and Water Laboratory at the University of Cordoba (Colombia), while the C2 soil samples were analyzed in the Soil Mechanics Laboratory at the Autonomous University of Queretaro (Mexico).

Table 1. Laboratory methods used for determining soil properties.

Parameter	Method	Guide
Soil moisture	Gravimetry	Analytical methods of the soil laboratory of the Agustín Codazzi Geographic Institute ([22]).
Particle density	Alcohol	
Bulk density	Paraffin	

2.3.2. Soil Granulometry

Soil texture was determined following the procedures described in norm ASTM D 422-63 ([23]) of the American Society for Testing and Materials. After drying and sieving using a 2 mm mesh, the particle size distribution was assessed using sieves, according to Cabrera et al. [24], and a Bouyoucos Hydrometer. A particle size distribution curve was generated and the corresponding percentages of sand, silt and clay were determined to obtain the textural class of each soil [24].

2.3.3. Glyphosate Analysis in Soil

Analysis of glyphosate concentrations in the soil and leachate for the C1 soil sample was performed using gas chromatography in the Laboratory of Environmental Services at the University of Cartagena, Colombia; meanwhile, C2 samples were analyzed using spectrophotometry in the Center of Chemical Research for Circular Economy (CIQEC) at the

Autonomous University of Queretaro, Mexico. In both cases, quantification was based on calibration curves, ensuring consistency and comparability of the reported concentrations despite the use of different analytical techniques. Extraction of glyphosate from the soil was performed following the technique recommended by Druart et al. [25]; in brief, a 5 g soil sample was taken and dissolved in 40 mL of distilled water, followed by agitation for 1 h, resting for 1.5 h, and agitation for another 1 h. Finally, the sample was centrifuged for 20 min at 3000 rpm. After that, following the method of Bhaskara and Nagaraja [26], 0.5 mL of the supernatant was reacted in a double boiler with 1 mL of Ninhidrine solution at 3% and 1 mL of sodium molybdate at 3%, at a temperature between 90 and 100 °C. Spectrophotometry was performed at 570 nm, and the concentration of glyphosate was obtained from the calibration curve using the following equation:

$$\text{Concentration} = \frac{0.0168 + \text{absorbance}}{0.0065} \tag{1}$$

2.3.4. Glyphosate Infiltration Model

The infiltration model consisted of 10 cm diameter acrylic columns, with a depth of 50 cm in the case of C1 samples and 8 cm for C2 samples. The objective of the experiment was not to compare identical geometries, but to reproduce the real hydraulic behavior of each soil. Comparability between soils does not rely on column depth, but on soil-specific hydrodynamic parameters (K_s, ψ_d and dispersivity), which were calibrated experimentally and used in the numerical model to normalize the effects of geometry. The insides of the columns were partially treated with wax to create roughness between the soil sample and the acrylic column, thus avoiding preferential flows. A lid with orifices was placed at the lower end of the column to allow lixiviation. The setup of the columns can be observed in Figure 2.

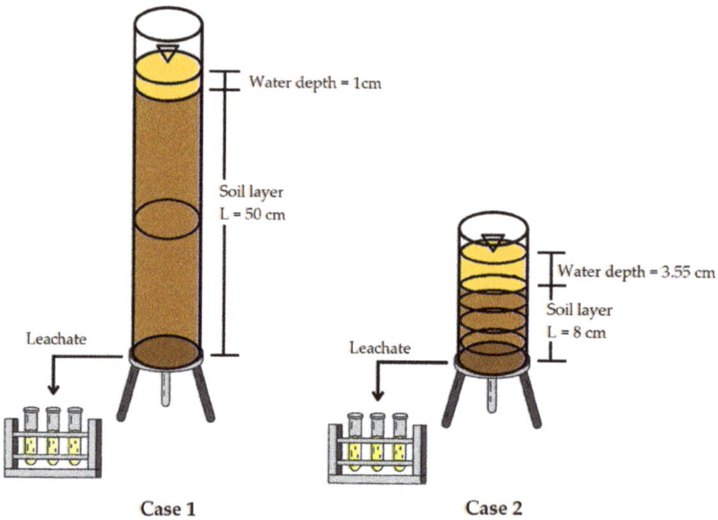

Figure 2. Soil column infiltration experiment setup.

Glyphosate at a concentration of 40.33 g/L was applied to the columns, which was selected to ensure detectable glyphosate levels throughout the soil column, according to the detection limits of the analytical method. Following the procedure described by Chávez [27], constant and variable water heads were used during the trials. Initially, with a constant water head, the volume of water needed for soil saturation (maximum infiltrated water depth) was calculated from the bulk density of soil and the initial water content, which was applied until the first leachate occurred. The time for lixiviation initiation was registered and, subsequently, the water head was considered variable (only for C2), and the time and leachate volume were recorded every time the water head decreased by 0.5 cm. At the end of the trial, soil samples were extracted from the columns and taken to the laboratory for determination of glyphosate concentrations.

2.4. Numerical Modeling

A numerical one-dimensional model with finite differences coupling the Richards equation [28] and the advection–dispersion equation [29] was implemented, as given by Equations (2) and (3):

$$C(\psi)\frac{\partial\psi}{\partial t} = \frac{\partial}{\partial z}\left[K(\psi)\left(\frac{\partial\psi}{\partial z}+1\right)\right] - Y \tag{2}$$

$$\frac{\partial(\theta C)}{\partial t} + \frac{\partial(qC)}{\partial z} = \frac{\partial}{\partial z}\left(\theta Da\frac{\partial C}{\partial z}\right) + R_s \tag{3}$$

where $C(\psi)$ corresponds to the specific capacity; ψ is the water pressure potential in the soil; t denotes time; $K(\psi)$ represents the hydraulic conductivity as a function of water pressure in the soil; z is the vertical coordinate; Υ is a sink or source term, which represents the inputs or outputs of water in the system due to external processes; $\theta(\psi)$ is the volumetric water content; C is the solute concentration in the soil; Da is the water dispersion coefficient; and R_s is the term related to the addition or removal of solute due to its extraction by plants.

To find the numerical solution of the Richards equation using the finite difference method, iterative schemes based on the local mass balance can be employed. These include explicit, implicit, and intermediate approaches. Among them, the Laasonen method has been highlighted as a particular case of the fully implicit scheme [30].

Temporal discretization was performed using a weighted implicit formulation, where variables at an intermediate time level $k+\omega$ are computed as a function of values at times k and $K+1$. The time step Δt_k was adaptively adjusted according to the model's convergence, ensuring mass balance closure in each iteration.

The calibration of the coupled flow–transport model was performed through an inverse modeling procedure. First, the hydraulic parameters required by the Richards equation—saturated hydraulic conductivity (K_s) and air-entry/pressure parameter (ψ_d)—were estimated using the Levenberg–Marquardt non-linear optimization algorithm, minimizing the error between simulated and measured infiltration curves obtained from the column experiment. Once the hydraulic solution was stable, the advection–dispersion equation was solved using the same hydraulic state, and the soil dispersivity (Da) was adjusted by fitting the simulated glyphosate concentration profiles to those measured at the end of the experiment. The boundary conditions followed the experimental setup: (i) the upper boundary was defined by the applied concentration (Dirichlet), and (ii) the lower boundary was defined as a zero-concentration-gradient (C1) or a time-dependent Neumann flux derived from leachate measurements (C2). Model performance was quantified using RMSE and the coefficient of determination (R^2), demonstrating that the simulated profiles reproduced experimental trends and retained the dominant transport mechanisms (advection and dispersion).

The advection term of the advection–dispersion equation was solved using a finite-difference formulation as the model solves for flow and transport simultaneously on the same computational grid, preserving local mass balance at each node. In problems where the velocity field is continuously updated using the Richards equation, the method of characteristics requires characteristic tracking and interpolation between hydraulic and solute trajectories, producing numerical dispersion and loss of mass conservation.

For this reason, a fully implicit finite-difference scheme was adopted, which ensures numerical stability both during long simulations and under the strong non-linearity of the soil hydraulic properties.

The spatial domain of the column was discretized into one-dimensional control volumes, with refinement near the inlet and outlet boundaries to resolve sharp concentration and pressure gradients while maintaining a coarser spacing in the interior to reduce computational cost. The time step was adaptively adjusted based on the convergence of the non-linear iterations of the Richards equation: if mass balance was not met during an iteration, the time step was automatically reduced. Convergence testing was conducted by refining both the grid spacing (Δz) and time step (Δt). Convergence was reached when further refinement produced no appreciable change in breakthrough curves or glyphosate concentration profiles, nor improvement in RMSE/R^2. The numerical model was implemented using an in-house computational code, which allowed for explicit control of discretization, convergence criteria and parameter calibration.

2.4.1. Hydraulic Soil Water Properties

To solve the Richards equation, the van Genuchten model [31] was used to represent the water retention curve, as shown in Equation (4). Meanwhile, for the hydraulic conductivity, two of the fractal models proposed by Fuentes et al. [32] were used: the neutral pore model and big pore model, given in Equations (5) and (6), respectively. These models allow for description of the complexity and heterogeneity associated with porous media in a more precise way than traditional approaches, improving the estimation accuracy when integrating information on the variability of the size distribution of particles, pores and aggregates, which is important for the transfer of water and transport of solutes [33].

$$\theta(\psi) = \theta_r + (\theta_s - \theta_r)\left[1 + (\psi/\psi_d)^n\right]^{-m} \quad (4)$$

$$K(\Theta) = K_s\left[1 - \left(1 - \Theta^{\frac{1}{m}}\right)^{2sm}\right],\ 0 < 2sm = 1 - \frac{4s}{n} < 1 \quad (5)$$

$$K(\Theta) = K_s\Theta^s\left[1 - \left(1 - \Theta^{1/m}\right)^{sm}\right],\ 0 < sm = 1 - \frac{4s}{n} < 1 \quad (6)$$

where θ_r is the residual volumetric water content; θ_s is the volumetric water content at saturation; ψ_d is the characteristic soil pressure; $s = D/E$ (D is the soils fractal dimension, and E is the Euclidean space dimension) is obtained from Equation (7) as a function of soil porosity, considering the relationship between total bulk density of soil (ρ_t) and density of solids (ρ_s), $\phi = 1 - (-\rho_t/\rho_s)$; $\Theta = (\theta - \theta_r)/(\theta_s - \theta_r)$ is the effective degree of saturation; and m and n are dimensionless parameters.

$$(1 - \phi)^s + \phi^{2s} = 1 \quad (7)$$

The estimation of m and n was performed using the granulometric curve, following the methodology proposed by Fuentes [34].

Optimization of K_s and ψ_d in the Richards equation, as well as the Dispersivity (Da) value, was performed using the Levenberg–Marquardt algorithm [35].

2.4.2. Initial and Boundary Conditions

The experiment with the C1 sample was performed at constant head, while that with the C2 sample was performed at both constant and variable head; the latter started when lixiviation initiated at the bottom of the column.

The glyphosate concentrations in the soils previous to the trials were 4.68 g/kg for C1 and 15.57 g/kg for C2. Notably, the experiment was performed with a glyphosate solution at a concentration of 40.33 g/L.

The glyphosate solution was applied under a constant head at the upper boundary until the first leachate occurred, which defines the effective duration of the input: C1 = 109 h 46 min 28 s (head = 1 cm); C2 = 5.77 h (head = 3.55 cm). After the first leachate in C2, the test continued under variable head, recording the time and leachate volume for each 0.5 cm decrease. The lower boundary was free drainage (zero pressure gradient, $\partial\psi/\partial z = 0$). To determine the initial conditions for the Richards equation, the initial volumetric water content prior to applying the solution was used.

2.4.3. Simulation Scenarios

Once the solute transport model was optimized for each case, simulation scenarios were applied to evaluate the behaviors and predict the possible fates of glyphosate. The scenarios included the application of herbicide at specific timings, taking as reference the research of Su et al. [36]. The evaluated scenarios are listed in Table 2.

Table 2. Details of simulation scenarios.

	Scenario	Change
No.1	Continuity in the application of glyphosate	The application of glyphosate continuously in both cases until 1000 h after the test was completed.
No.2	Soil washing scenario	The same concentration of glyphosate is supplied, following which water is added to wash the soil and reduce the concentration.

3. Results

3.1. Soil Properties

The values of physical soil characteristics and glyphosate concentrations, measured in the laboratory, are reported in Table 3. The soil texture for C1 corresponded to a silty clay class, with its clay content is almost twice that in C2, which corresponded to a silty loam texture class. C2 had a significant percentage of silt and more than double the content of sand with respect to C1.

Table 3. Results of soil properties and initial glyphosate concentration.

Parameter	C1	C2
Sand (%)	6.05	15.23
Silt (%)	51.35	61.43
Clay (%)	42.50	23.34
Initial soil moisture (cm^3/cm^3)	0.071	0.046
Particle density (g/cm^3)	2.41	2.32
Bulk density (g/cm^3)	1.23	1.35
Glyphosate (g/kg)	4.68	17.57

Note: Glyphosate (g/kg) corresponds to the total extractable concentration in soil obtained after the complete extraction procedure described in Section 2.3.

Concentrations of glyphosate in the C2 sample were significantly greater than those found in the C1 sample. This could be because the C1 samples were taken from an experimental plot at the Faculty of Agricultural Sciences of the University of Cordoba, while the C2 samples were obtained from an irrigated intensive agricultural area, where more herbicides are used. In both cases, the glyphosate concentrations were very high, when compared with the permitted maximum limits in soils suggested by the Canadian Council of Ministers of the Environment (CCME). The council establishes that, for agricultural and residential land uses, the level must not be greater than the upper limit of 0.0007 g/kg to prevent environmental and human health risks [37,38].

3.2. Hydrodynamic Characteristics of Soils

Hydrodynamic characterization of soils was performed for each type of soil to estimate the values for the water retention and hydraulic conductivity curves. It was assumed that the residual water content was equal to zero (θ_r = 0 cm^3/cm^3) [39], and that the water content was related to soil porosity ($\theta_s = \phi$). The obtained results for the particle size distribution in each soil (m, n) were used to optimize the pore models proposed by Fuentes et al. [40]. The big pore model showed a better fit for the C1 sample, while the neutral pore model provided a more accurate adjustment for C2, according to the respective RMSE values. The results are shown in Table 4.

Table 4. Values obtained during the optimization process.

Parameter	C1	C2
m	0.19724	0.18609
n	3.8826	3.1784
RMSE	0.0175	0.0166
K_s (cm/h)	1.3002	1.4689
ψ_d (cm)	−54.6826	−12.4318
RMSE	0.5356	0.4595

3.3. Dispersion Coefficient of Glyphosate in Agricultural Soils: Modeling via the Advection–Dispersion Equation

Glyphosate concentrations at different depths were obtained by sectioning the soil column used for the infiltration test and taking these samples to the laboratory. Two samples were taken for the C2 soil (one at a depth of 12.5 cm and the other at 37.5 cm), while four samples were taken for the C2 soil (at 1, 3, 5 and 7 cm depths). The results obtained for each soil are shown in Figure 3.

The optimized glyphosate dispersion coefficients are given in Table 5. The glyphosate concentration in C1 soil had a significantly smaller value than in C2 soil (Table 5), suggesting that soil heterogeneities, such as porosity and structure, reduce glyphosate dispersion in the C1 soil. Therefore, its movement is slower and less extensive in comparison with that in the C2 soil, which has a smaller dispersion coefficient; thus, enabling faster dissemination (Figure 3).

Table 5. Dispersion coefficient values.

Parameter	C1	C2
Dispersion coefficient Da (m^2/s)	2.7×10^{-6}	2.21×10^{-5}
RMSE	0.223	12.077
R^2	1.00	0.8472

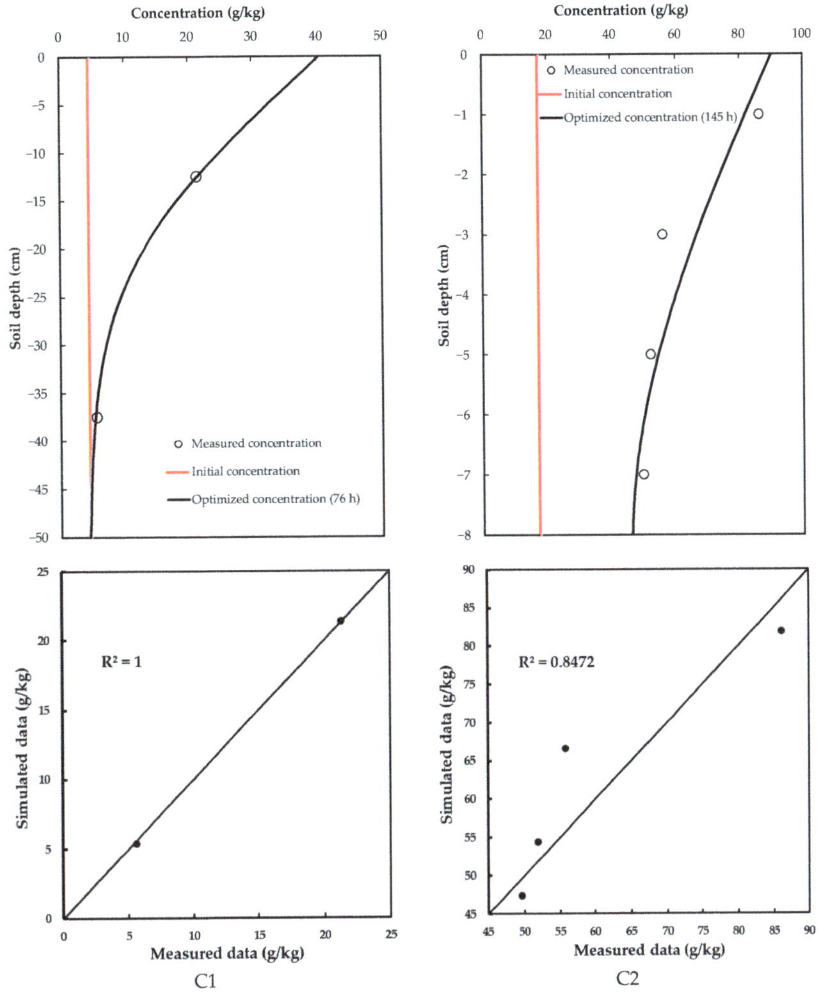

Figure 3. Optimized concentration of glyphosate in the soil profiles.

3.4. Simulation Scenarios

3.4.1. Scenario 1: Continuous Application of Glyphosate

In Scenario 1, continuous application of glyphosate for up to 1000 h during the test was considered, allowing for a direct assessment of the dynamics, movement and degradation of the contaminant in the soil, given that considerably high concentrations would be reached. This ensured that the soil was contaminated, and thus, the risk of this chemical compound of reaching the aquifers could be determined.

For this scenario, the results indicated a direct increase in glyphosate levels in the soil, with no significant decrease in concentration with depth. The C1 samples showed a slight decrease in glyphosate concentration in the soil column at the final time; for C2 samples, this decrease was practically null, indicating that the initial (input) and final (output) concentrations of glyphosate were practically balanced (Figure 4).

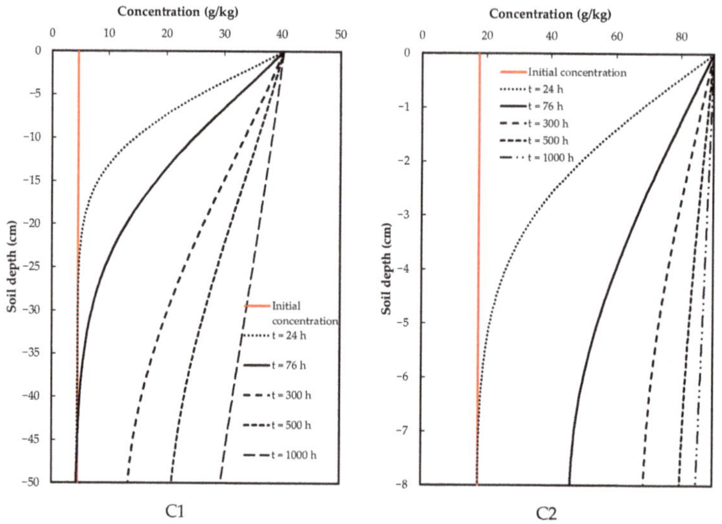

Figure 4. Results of scenario 1: continuous application of glyphosate.

3.4.2. Scenario 2: Soil Washing

In this scenario, glyphosate was only applied at a specific time, followed by the addition of water only (i.e., washing). This considerably reduced the concentration of the contaminant in both cases. In Figure 5, $t = 0$ refers to the final time of the test [41,42] in both cases, where a given concentration of glyphosate was applied, followed by the application of clean water. Considering the times at which the concentration values decrease below those at the reference time ($t = 0$), it is possible to observe a noticeable difference between the C1 and C2 samples. A slower process occurs in the C1 sample, where soil washing increased concentrations in the deeper soil layers, approaching the reference value ($t = 0$); meanwhile, concentrations at the same depth were greater and were reached faster for the C2 sample, indicating greater vulnerability.

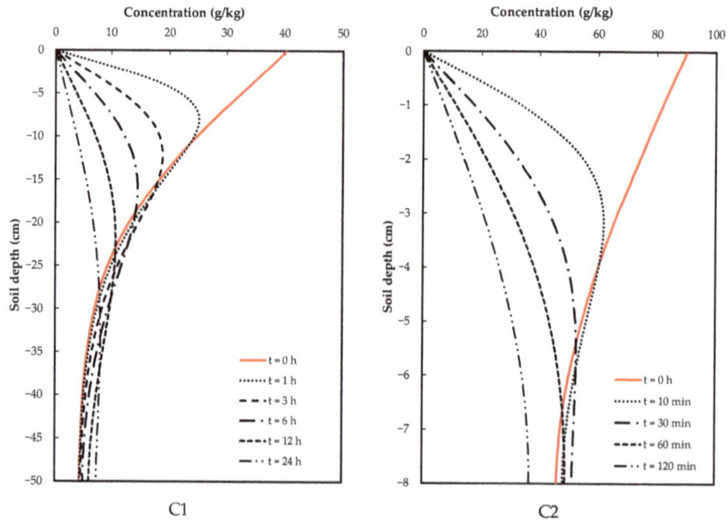

Figure 5. Glyphosate concentrations at different soil depths in two different soils for the soil washing scenario.

4. Discussion

Soil texture characteristics are related to water transfer and solute transport behaviors in the soil, as indicated by the results obtained in this study for two types of soil. As soil pores are the conduits through which water moves and soil texture conditions porosity, it is possible to infer that there is predominance of microporosity in the C2 soil, while microporosity is dominant in the C1 soil [43]. The soil bulk density results support this finding, as the corresponding value for the C1 soil was smaller than that for C2 soil sample. The pore space tends to be reduced when the presence of finer soil particles dominate [44]. On the other hand, the initial concentrations of glyphosate in both soils is of concern, as they were both above the established upper limit values for agricultural soils indicated by the CCME. The values found in this study are similar to those previously reported in agricultural soils by Gomaracsca et al. [45], Mencaroni et al. [46], and Nahar and Niven [15].

The infiltration test using glyphosate solution in the soil allowed, through the use of an inverse modeling method, the saturated hydraulic conductivity (K_s) and characteristic pressure (ψ_d) values to be obtained for each case. These variables are required to solve the Richards equation numerically and reflect the hydrodynamic characteristics of the soil. The accuracy of these values was very good, according to the RMSE. A higher value of K_s was found in C2 soil, reflecting its textural characteristics; in particular, it has bigger and more interconnected pores, facilitating the movement of water.

Taking into consideration the concentrations of glyphosate measured before and after the infiltration test, it is possible to observe that values registered for C1 in the deepest layer were relatively close to the initial concentration. This was not the case for C2, where the concentrations of glyphosate in the deepest layer were similar to the initial value (higher concentration), indicating a more critical situation.

These values allowed the glyphosate dispersion coefficient to be obtained for each case. According to the RMSE criterion, the fit was "very good" for the C1 case (0.223 g/kg). Meanwhile, C2 presented a value of 12.077 g/kg—although this seems high, it is congruent with the context of the problem. The simulated curves adequately replicated the trends observed in the experimental data, reflecting the dominant physical processes of advection and dispersion.

In the experiment, a range of glyphosate concentrations were considered (up to 100 g/kg). The value of RMSE represented approximately 12% of the total range, which is acceptable in environmental studies—especially when uncertainties inherent to experimental measurements and soil heterogeneity are considered—without affecting the capacity of the model to describe the behavior of glyphosate in a useful and realistic way [47,48].

The dispersion coefficients obtained in this study (2.7×10^{-6} m^2/s for C1 and 2.21×10^{-5} m^2/s for C2) are consistent with values reported in the literature for soils of similar texture under unsaturated conditions. For instance, Padilla and Selim [14] found hydrodynamic dispersion coefficients on the order of 10^{-7} m^2/s in silt loam and clay columns under saturated flow, whereas Nahar and Niven [15] observed dispersion coefficients of approximately 10^{-8} m^2/s in unsaturated leaching experiments on agricultural soils. It is also recognized that unsaturated field conditions and soil heterogeneity can lead to higher dispersivity; indeed, studies using non-reactive tracers have reported dispersion coefficients approaching 10^{-6}–10^{-5} m^2/s in structured or macroporous soils.

Thus, the values calibrated for glyphosate in our silty clay (C1) and silty loam (C2) soils are in line with these reported ranges, indicating that the model's dispersion parameterization reflects realistic physical transport processes in such soils. This agreement with published values reinforces the credibility of the models in capturing the distribution of glyphosate in the soil profiles.

The contrasting behaviors observed between C1 and C2 are in agreement with previous studies reporting that clay- and oxide-rich soils enhance herbicide sorption, while sandy or shallow soils increase leaching vulnerability. Similar behaviors have been documented regarding the movement of glyphosate and AMPA toward shallow aquifers in agricultural regions.

The simulation scenarios allowed for analyses of soil vulnerability due to the action of glyphosate. In the case of scenario 1 (continuous application), C1 soil showed slower transport when compared with C2 soil. In the latter, more rapid transport and higher concentrations in the lower layer were observed, with similar concentration values to those applied at the top of the soil column indicating a loss in the soil's capacity to degrade the contaminant. Similar results have been reported by Su et al. [36] who, by means of simulations, showed that pollutants present a decreasing tendency if it reaches a greater depth in the soil profile and, when the application is greater than the assimilation capacity of the soil, it may lead to severe consequences for ecosystems.

For scenario 2 (soil washing), there was a noticeable difference between soils C1 and C2. A slower process occurred in C1, with soil washing causing concentrations to move closer to the reference value ($t = 0$) in the deeper soil layers; meanwhile, the concentrations in the same layers ware greater and reached in shorter times in C2, indicating a greater vulnerability when compared with C1. This implies that, in scenario 2, the implications of applying glyphosate to the soil would be more significant due to a considerable decrease in pollutant load (understood as the product of concentration and incoming flow during a contamination event). Consequently, the risk of contamination would be significantly reduced for both the soil and its underlying aquifers.

The higher glyphosate retention observed in C1 was considered to be strongly associated with its finer texture and higher clay fraction (42.5%) relative to C2 (23.3%). Clay minerals—particularly those containing Fe and Al oxides—provide abundant surface sites capable of ligand exchange reactions with the phosphonate group of glyphosates, promoting inner-sphere complex formation and thus stronger sorption. This mechanism has been widely reported for Fe/Al oxide-rich soils under near-neutral pH conditions, which favor deprotonation of surface hydroxyl groups and enhance electrostatic attraction [14]. Additionally, although the organic matter content was slightly lower in C1 (2.35%) than in C2 (2.61%), its humic constituents contribute carboxylic and phenolic functional groups that can complex cations and forming ternary complexes with glyphosate, further increasing sorptive capacity. The near-neutral pH (6.8) of C1 also reduces glyphosate ionization, enhancing its affinity for oxide and clay surfaces. In contrast, the more acidic pH of C2 (6.05) and its higher sand content (15.2%) limit the number of reactive sites, thus increasing mobility and leaching potential. Therefore, the combined effect of higher clay and oxide content, together with favorable pH conditions, explains the enhanced glyphosate retention and lower dispersion coefficient measured in C1 compared with C2.

5. Conclusions

The initial concentration of glyphosate in the soil was found to be relatively high, especially when compared with the maximum limit established by the CCME for soils; this was even the case for the C1 soil, which corresponds to a non-agricultural experimental lot. This indicates the occurrence of contamination processes in the soils and, hence, a risk to groundwater quality. This is especially true in the case of C2 soil, located on the San Juan del Rio Valley Aquifer in the State of Queretaro, from which water is extracted for human consumption.

The initial and experimental glyphosate concentrations in the soil columns for both samples allowed for adjustment and optimization of the parameters of the advection–dispersion equation, achieved by means of calculating the coefficient of dispersion. Consequently, the behaviors of the contaminant in the considered soils could be accurately modeled.

The implemented simulation scenarios allowed for evaluation of the vulnerability of each soil under different glyphosate application scenarios. In the scenario involving continuous application of glyphosate, it was observed that the C2 soil presented significantly higher concentrations, representing a severe risk for groundwater contamination. A similar tendency was observed in the case of the C1 soil, but with lower values and requiring a longer time to reach higher concentrations. On the other hand, the soil-washing scenario, in which glyphosate was applied only for a certain time, could represent a positive alternative if the idea is to reduce contamination; in particular, it was observed that soil washing after the application of glyphosate can lead to lower concentration of the contaminant in the soil after a given period of time. However, at longer time scales, this process could represent a greater risk as the herbicide may reach the underlying aquifers.

This study demonstrated that the mobility of glyphosate in the soil profile is strongly influenced by the soil's texture, clay content, pH and reactive oxides, which control sorption and retention processes. Although the present results were obtained under controlled laboratory conditions and do not represent ecological impact by themselves, they reveal that soils with lower clay content and higher sand proportion (e.g., C2) exhibit a greater risk of downward glyphosate transport and potential migration toward groundwater. Conversely, soils with higher clay and oxide content (e.g., C1) show stronger retention capacity, thus reducing vertical mobility. These findings highlight the potential risk of contamination, rather than a deterministic evaluation of environmental decline, and support the need for improved agricultural management practices—such as reducing application frequency, implementing buffer zones, and adopting precision dosing—to minimize the transport of glyphosate from agricultural areas to surrounding ecosystems. Figure 6 illustrates the workflow summarizing the methodology adopted in this study and the main results obtained.

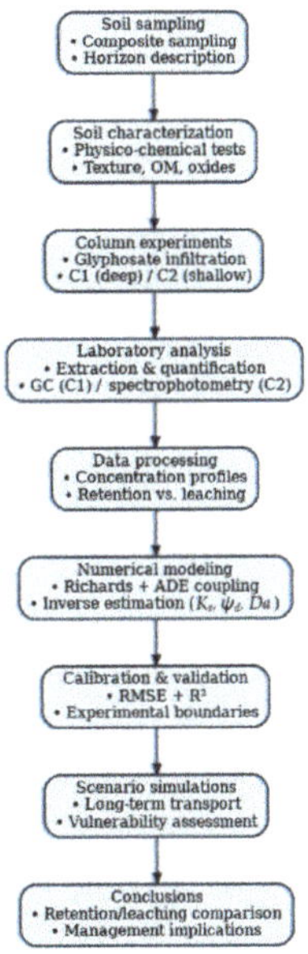

Figure 6. Workflow diagram summarizing the experimental–numerical methodology used in this study, including soil sampling, column experiments, glyphosate quantification, inverse calibration of hydraulic and transport parameters, numerical modeling and scenario simulations.

Author Contributions: Methodology, J.G.-G., T.M.-F., J.T.-R. and J.D.G.-E.; software, C.F., S.F. and C.C.; validation, J.G.-G., T.M.-F., J.T.-R. and J.D.G.-E.; investigation, J.G.-G., T.M.-F., E.V.-R., C.F. and S.F.; writing—original draft preparation, J.G.-G., C.F., S.F. and C.C.; writing—review and editing, E.V.-R., S.F. and C.C. All authors have read and agreed to the published version of the manuscript.

Funding: This research received no external funding.

Data Availability Statement: The original contributions presented in this study are included in the article. Further inquiries can be directed to the corresponding authors.

Acknowledgments: The first author is grateful to SECIHTI for funding his postgraduate studies (scholarship number 1183723), which enabled the development of this research.

Conflicts of Interest: The authors declare no conflicts of interest.

References

1. Villholth, K.G.; Jarvis, N.J.; Jacobsen, O.H.; de Jonge, H. Field Investigations and Modeling of Particle-Facilitated Pesticide Transport in Macroporous Soil. *J. Environ. Qual.* **2000**, *29*, 1298–1309. [CrossRef]
2. Zhou, X.; Shi, X.; Zhang, L.; Zhou, Y. Effects of Pesticide-Contamination on Population and Activity of Bacteria in Purple Paddy Soil. *Energy Procedia* **2012**, *16*, 284–289. [CrossRef]
3. Sun, G.; Du, Y.; Yin, J.X.; Jiang, Y.Z.; Zhang, D.; Jiang, B.; Li, G.; Wang, H.; Kong, F.; Su, L.; et al. Response of Microbial Communities to Different Organochlorine Pesticides (OCPs) Contamination Levels in Contaminated Soils. *Chemosphere* **2019**, *215*, 461–469. [CrossRef]
4. Da Silva, C.C.; Souza, M.d.F.; Passos, A.B.R.d.J.; Silva, T.S.; Borges, M.P.d.S.; dos Santos, M.S.; Silva, D.V. Risk of Environmental Contamination Due to the Hexazinone Application in Agricultural Soils in Northeastern Brazil. *Geoderma Reg.* **2022**, *28*, e00481. [CrossRef]
5. Bernasconi, C.; Demetrio, P.M.; Alonso, L.L.; Mac Loughlin, T.M.; Cerdá, E.; Sarandón, S.J.; Marino, D.J. Evidence for Soil Pesticide Contamination of an Agroecological Farm from a Neighboring Chemical-Based Production System. *Agric. Ecosyst. Environ.* **2021**, *313*, 107341. [CrossRef]
6. Hintze, S.; Cochand, F.; Glauser, G.; Hunkeler, D. Soil and Unsaturated Zone as a Long-Term Source for Pesticide Metabolites in Groundwater. *Water Res.* **2024**, *261*, 121901. [CrossRef]
7. Diamantopoulos, E.; Šimůnek, J.; Oberdörster, C.; Hammel, K.; Jene, B.; Schröder, T.; Harter, T. Assessing the Potential Exposure of Groundwater to Pesticides: A Model Comparison. *Vadose Zone J.* **2017**, *16*, 1–13. [CrossRef]
8. Márquez, D.; Faúndez, C.; Aballay, E.; Haberland, J.; Kremer, C. Assesing the Vertical Movement of a Nematicide in a Sandy Loam Soil and Its Correspondence Using a Numerical Model (HYDRUS 1D). *J. Soil Sci. Plant Nutr.* **2017**, *17*, 167–179. [CrossRef]
9. Ona, J. Leaching Risk Assessment of Pesticides from Soil as Function of Soil Properties, Pesticide Properties and Rainfall Characteristics Using HYDRUS 1D Numerical Model. *Poljopr. Tehnol.* **2022**, *47*, 67–81. [CrossRef]
10. Villarreal, R.; Soracco, C.G.; Salazar, M.P.; Bellora, G.L.; Valdés-Abellán, J.; Lozano, L.A. Glyphosate Dynamics Prediction in a Soil under Conventional and No-Tillage Systems during the Crop Cycle. *Rev. Bras. Cienc. Solo* **2020**, *44*, e0190130. [CrossRef]
11. Hmimou, A.; Maslouhi, A.; Tamoh, K.; Candela, L. Experimental Monitoring and Numerical Study of Pesticide (Carbofuran) Transfer in an Agricultural Soil at a Field Site. *Comptes Rendus Geosci.* **2014**, *346*, 255–261. [CrossRef]
12. Lei, W.; Tang, X.; Zhou, X. Transport of 3,5,6-Trichloro-2-Pyrdionl (a Main Pesticide Degradation Product) in Purple Soil: Experimental and Modeling. *Appl. Geochem.* **2018**, *88*, 179–187. [CrossRef]
13. Dusek, J.; Dohnal, M.; Snehota, M.; Sobotkova, M.; Ray, C.; Vogel, T. Transport of Bromide and Pesticides through an Undisturbed Soil Column: A Modeling Study with Global Optimization Analysis. *J. Contam. Hydrol.* **2015**, *175*, 1–16. [CrossRef] [PubMed]
14. Padilla, J.T.; Selim, H.M. Glyphosate Transport in Two Louisiana Agricultural Soils: Miscible Displacement Studies and Numerical Modeling. *Soil Syst.* **2018**, *2*, 53. [CrossRef]
15. Nahar, K.; Niven, R.K. An Analysis of Miscible Displacement and Numerical Modelling of Glyphosate Transport in Three Different Agricultural Soils. *Agronomy* **2023**, *13*, 2539. [CrossRef]
16. Gobernación de Córdoba. *Capítulo 1 Generalidades del Departamento de Córdoba*; Gobernación de Córdoba: Montería, Colombia, 2010. Available online: https://repositoriocdim.esap.edu.co/bitstream/handle/20.500.14471/24787/Capitulo1GeneralidadesDelDepartamentoDeCordoba-1.pdf?sequence=1&isAllowed=y (accessed on 23 July 2024).
17. IGAC. *Estudio General de Suelos y Zonificación de Tierras Del Departamento de Córdoba*; Instituto Geográfico Agustín Codazzi: Bogotá, Colombia, 2009. Available online: https://metadatos.icde.gov.co/geonetwork/srv/api/records/14138630 (accessed on 11 December 2025).

18. INEGI. *Aspectos Geográficos: Querétaro*; Santiago de Querétaro; Instituto Nacional de Estadística y Geografía: Aguascalientes, Mexico, 2021; Volume 60. Available online: https://www.inegi.org.mx/contenidos/productos/prod_serv/contenidos/espanol/bvinegi/productos/nueva_estruc/889463913993.pdf (accessed on 22 November 2025).
19. Concyteq. *Uso Actual y Potencial del Suelo en los Municipios Conurbados de Querétaro*; Consejo de Ciencia y Tecnología del Estado de Querétaro: Santiago de Queretaro, Mexico, 2002; ISBN 9786077710165. Available online: https://koha.cidesi.mx/cgi-bin/koha/opac-detail.pl?biblionumber=2585 (accessed on 22 November 2025).
20. SEMARNAT. Establishing the Specifications of Fertility, Salinity and Soil Classification. Studies, Sampling and Analysis—NOM-021-RECNAT-2000; Ciudad de México, México. 2000. Available online: https://www.ordenjuridico.gob.mx/Documentos/Federal/wo69255.pdf (accessed on 22 November 2025).
21. Secretaría de Economía. *Norma Mexicana NMX-AA-132-SCFI-2016*; Secretaría de Economía: Ciudad de México, México, 2016. Available online: http://www.economia-nmx.gob.mx/normas/nmx/2010/nmx-aa-132-scfi-2016.pdf (accessed on 22 November 2025).
22. IGAC. *Métodos Analíticos del Laboratorio de Suelos*, 6a ed.; Instituto Geográfico Agustín Codazzi: Bogotá D.C., Colombia, 2006. Available online: https://www.car.gov.co/uploads/files/654c1bdeed14c.pdf (accessed on 22 November 2025).
23. *ASTM D422*; ASTM D-422 Standard Test Method for Particle Size Analysis. ASTM International: West Conshohocken, PA, USA, 2007; Volume 63. Available online: https://pdfcoffee.com/download/astm-d422-1963-2007-pdf-free.html (accessed on 23 November 2025).
24. Cabrera, A.; Bustos, M.; Rosas, M.; Romero, M.; Díaz, L. *Manual Para el Curso de Edafología Laboratorio y Campo*; Universidad Michoacana de San Nicolás de Hidalgo: Morelia, Michoacán, 2019.
25. Druart, C.; Delhomme, O.; De Vaufleury, A.; Ntcho, E.; Millet, M. Optimization of Extraction Procedure and Chromatographic Separation of Glyphosate, Glufosinate and Aminomethylphosphonic Acid in Soil. *Anal. Bioanal. Chem.* **2011**, *399*, 1725–1732. [CrossRef] [PubMed]
26. Bhaskara, B.L.; Nagaraja, P. Direct Sensitive Spectrophotometric Determination of Glyphosate by Using Ninhydrin as a Chromogenic Reagent in Formulations and Environmental Water Samples. *Helv. Chim. Acta* **2006**, *89*, 2686–2693. [CrossRef]
27. Chávez, G.C.A. *Solución Numérica de las Ecuaciones de Transferencia de Agua y Solutos en Riego y Drenaje*; Universidad Autónoma de Querétaro: Santiago de Querétaro, Mexico, 2010.
28. Richards, L.A. Capillary Conduction of Liquids through Porous Mediums. *J. Appl. Phys.* **1931**, *1*, 318–333. [CrossRef]
29. Liu, K.; Huang, G.; Xu, X.; Xiong, Y.; Huang, Q.; Šimůnek, J. A Coupled Model for Simulating Water Flow and Solute Transport in Furrow Irrigation. *Agric. Water Manag.* **2019**, *213*, 792–802. [CrossRef]
30. Fuentes, S.; Trejo-Alonso, J.; Quevedo, A.; Fuentes, C.; Chávez, C. Modeling Soilwater Redistribution under Gravity Irrigation with the Richards Equation. *Mathematics* **2020**, *8*, 1581. [CrossRef]
31. Van Genuchten, M.T. A Closed-Form Equation for Predicting the Hydraulic Conductivity of Unsaturated Soils. *Soil Sci. Soc. Am. J.* **1980**, *44*, 892–898. [CrossRef]
32. Fuentes, C.; Chávez, C.; Brambila, F. Relating Hydraulic Conductivity Curve to Soil-Water Retention Curve Using a Fractal Model. *Mathematics* **2020**, *8*, 2201. [CrossRef]
33. Ming, F.; Zhang, M.; Pei, W.; Chen, L. A New Hydraulic Conductivity Model of Frozen Soil Considering the Hysteresis Effect Based on Fractal Theory. *Geoderma* **2024**, *442*, 116790. [CrossRef]
34. Fuentes-Ruiz, C. *Approche Fractale des Transferts Hydriques dans les dols Non-Saturés*; Université Joseph Fourier: Grenoble, France, 1992.
35. Moré, J.J. The Levenberg-Marquardt Algorithm: Implementation and Theory. In *Numerical Analysis*; Springer: Berlin/Heidelberg, Germany, 1978; pp. 105–116.
36. Su, C.; Yang, Y.; Jia, M.; Yan, Y. Integrated Framework to Assess Soil Potentially Toxic Element Contamination through 3D Pollution Analysis in a Typical Mining City. *Chemosphere* **2024**, *359*, 142378. [CrossRef] [PubMed]
37. Rocha, B.P.; Giacheti, H.L. Site Characterization of a Tropical Soil by in Situ Tests. *DYNA* **2018**, *85*, 211–219. [CrossRef]
38. Ferreira, N.G.C.; da Silva, K.A.; Guimarães, A.T.B.; de Oliveira, C.M.R. Hotspots of Soil Pollution: Possible Glyphosate and Aminomethylphosphonic Acid Risks on Terrestrial Ecosystems and Human Health. *Environ. Int.* **2023**, *179*, 108135. [CrossRef]
39. Haverkamp, R.; Leij, F.J.; Fuentes, C.; Sciortino, A.; Ross, P.J. Soil Water Retention: I. Introduction of a Shape Index. *Soil Sci. Soc. Am. J.* **2005**, *69*, 1881–1890. [CrossRef]
40. Fuentes, C.; Antonino, A.C.D.; Sepúlveda, J.; Zataráin, F.; de León, B. Predicción de La Conductividad Hidráulica Relativa de Los Suelos Con Modelos Fractales. *Ing. Hidraul. Mex.* **2003**, *18*, 31–40. Available online: https://www.academia.edu/41913545/Predicci%C3%B3n_de_la_conductividad_hidr%C3%A1ulica_relativa_de_los_suelos_con_modelos_fractales (accessed on 22 November 2025).
41. Bautista, E.; Schlegel, J.L. Modeling Solute Transport in the WinSRFR Surface Irrigation Software. *J. Irrig. Drain Eng.* **2020**, *146*, 4020035. [CrossRef]
42. Ojaghlou, H.; Sohrabi, T.; Abbasi, F.; Javani, H. Development and Evaluation of a Water Flow and Solute Transport Model for Furrow Fertigation with Surge Flow. *Irrig. Drain.* **2020**, *69*, 682–695. [CrossRef]

43. FAO. *Guía Para la Descripción de Suelos*; Food and Agriculture Organization (FAO): Rome, Italy, 2009; Volume 3. Available online: https://openknowledge.fao.org/server/api/core/bitstreams/b54d0348-dfce-413c-bd5d-142b3a14a049/content (accessed on 22 November 2025).
44. IGAC. *Descripción y Muestreo de Suelos*; Instituto Geográfico Agustín Codazzi: Bogotá, Colombia, 2021. Available online: https://www.igac.gov.co/sites/default/files/listadomaestro/in-gag-pc05-05_descripcion_y_muestreo_de_suelos.pdf (accessed on 22 November 2025).
45. Gomarasca, S.; Stefani, F.; Fasola, E.; La Porta, C.A.; Bocchi, S. Regional Evaluation of Glyphosate Pollution in the Minor Irrigation Network. *Chemosphere* **2024**, *355*, 141679. [CrossRef]
46. Mencaroni, M.; Longo, M.; Cardinali, A.; Lazzaro, B.; Zanin, G.; Dal Ferro, N.; Morari, F. Glyphosate and AMPA Dynamics during the Transition towards Conservation Agriculture: Drivers under Shallow Groundwater Conditions. *Soil Tillage Res.* **2023**, *229*, 105659. [CrossRef]
47. Willmott, C.J.; Matsuura, K. Advantages of the Mean Absolute Error (MAE) over the Root Mean Square Error (RMSE) in Assessing Average Model Performance. *Clim. Res.* **2005**, *30*, 79–82. [CrossRef]
48. Kuriakose, S.L.; Devkota, S.; Rossiter, D.G.; Jetten, V.G. Prediction of Soil Depth Using Environmental Variables in an Anthropogenic Landscape, a Case Study in the Western Ghats of Kerala, India. *Catena* **2009**, *79*, 27–38. [CrossRef]

water

Article

Research on the Influence of Groundwater Level Dynamic Rising Process on Buildings Based on Numerical Simulation

Hongzhao Li [1,2,3], Mingxu Gu [1,2,*], Ming Zhang [4], Baiheng Ma [1,2], Xiaolong Zhu [1,2], Liangyu Gu [1,2], Jiaoyang Tai [1,2] and Lili Chen [1,2]

1 Hebei Provincial Institute of Geological Environment Monitoring, Shijiazhuang 050021, China; hjjcylhz@163.com (H.L.)
2 Hebei Key Laboratory of Geological Resources and Environment Monitoring and Protection, Shijiazhuang 050021, China
3 Postdoctoral Research Workstation of Hebei Bureau of Geology and Mineral Resources Exploration, Shijiazhuang 050081, China
4 School of Engineering, China University of Geosciences, Wuhan 430074, China
* Correspondence: gumingxu1981@126.com

Abstract

In the North China region, measures such as restricting groundwater extraction and promoting cross-basin water diversion have effectively alleviated the problem of excessive groundwater exploitation. Nevertheless, the continuous rise in groundwater levels may alter the mechanical properties of foundation soil layers, potentially leading to geotechnical hazards such as foundation instability and the uneven settlement of structures. This study employs FLAC3D software to simulate the displacement, deformation, and stress–strain behavior of buildings and their surrounding strata during the dynamic recovery of groundwater levels, aiming to assess the impact of this process on structural integrity. Research findings indicate that the maximum building settlement within the study area reaches 54.8 mm, with a maximum inter-column differential settlement of 8.9 mm and a peak settlement rate of 0.16 mm/day. In regions where differential settlement aligns with the interface between the floor slab and walls, tensile stress concentrations are observed. The maximum tensile stress in these zones increases progressively from 1.8 MPa to 2.19 MPa, suggesting a potential risk of tensile cracking in the concrete structures. The influence of groundwater level recovery on buildings exhibits distinct phase characteristics, and the response mechanisms of different lithological strata vary significantly. Therefore, particular attention should be given to the physical properties and mechanical behavior of strata that are highly sensitive to variations in moisture content. These findings hold significant reference value for the sustainable development and utilization of underground space in the North China region.

Keywords: dynamic rise in groundwater level; numerical simulation; unsaturated–saturated; displacement deformation; stress–strain

1. Introduction

In China, especially in the cities in the North China region, the acceleration of urbanization has driven the development of underground engineering towards larger scales and deeper excavation. This has given rise to a series of engineering problems related to groundwater, which have become significant challenges, restricting the safe development of urban underground space. In recent years, with the Chinese government's increasing emphasis on groundwater resources, the groundwater levels in many regions of China have witnessed a notable rebound [1–3]. Although the rebound of groundwater levels effectively improves the regional hydrological cycle and promotes ecological restoration, due to variations in the rebound speed, amplitude, and driving factors, it poses potential risks to the safety of buildings [4–6]. For example, the rising groundwater level may increase the saturation of the stratum, thereby softening the soil structure and triggering the building settlement. Simultaneously, it may also enhance the buoyant force, leading to damage to the building structure [7,8]. Consequently, establishing a theory for the coordinated control of groundwater dynamic regulation and the safety of underground building structures has become a crucial scientific issue for ensuring the safe operation and maintenance of underground engineering.

Current research lacks in-depth investigations into how the rebound of groundwater levels specifically impacts buildings, and thus, fails to offer sufficient theoretical support for resolving specific engineering problems. As a result, analyzing the influence of groundwater level rebound on buildings using numerical simulation methods has emerged as a new research focus at present. For instance, Chang [9], Chen [10], Nguyen [11], and Lai [12], have, respectively, employed numerical simulation methods to analyze the impacts of groundwater level rise on tunnel structures, caused by factors such as blocked drainage blind pipes, sea-level rise, and construction activities. Zou has utilized numerical simulation methods to analyze the influence of high-water-head pressure on hydropower station tunnel projects [13]. However, in the current stage of research, static groundwater level boundary conditions are mostly adopted, which only simulate the transient response under a single water level condition. These studies lack dynamic tracking of periodic groundwater level fluctuations and thus fail to explain the cumulative damage effect on buildings caused by repeated deformations that are induced by such periodic fluctuations. Furthermore, the special geological structure of silty soil and sandy clay layers in the North China Plain is inherently different from that in other regions; the direct application of theories developed for other regions will inevitably lead to misjudgments in engineering practice.

For many years, the groundwater levels in Hebei Province have been on a long-term downward trend. During many engineering investigation and design phases, the potential future rebound of groundwater levels was not fully taken into account, leaving latent hazards in engineering structure and protection designs. Particularly under extreme weather conditions, such as the heavy rainfall events in Hebei Province in 2016 and 2021, seepage problems caused by rising groundwater levels have occurred in some areas. These problems include water infiltration in some garages, flooding, and cracking of the floor slab and walls. This paper takes the dynamic recovery of groundwater levels in the North China Plain as the research background, breaks through the limitations of the traditional static water level assumption, focuses on the sensitive stratum medium of the silt layer under long-term dewatering conditions, sets the water level boundary conditions, based on the accumulated groundwater level monitoring data in this area for many years, and uses the FLAC3D three-dimensional explicit finite difference software to construct a multi-field coupling numerical model of groundwater and building structures. It systematically studies the influential law of the entire process of dynamic water level recovery on buildings, and reveals the cumulative damage mechanism of building structures under unsteady water level conditions. It clarifies the evolution characteristics of the stress concentration area of underground structures, providing targeted theoretical basis and engineering practice reference for the safety prevention and control of buildings in the groundwater level recovery area of the North China Plain.

2. Study Area

Shijiazhuang is located on the western edge of the North China Plain. The per capita water resources in this area are approximately 157 m^3, making it an extremely water-scarce region. Previously, the economic model of this area was mainly agricultural production. However, due to the severe shortage of water resources, agricultural irrigation has long relied on the exploitation of groundwater, leading to a continuous decline in the groundwater level. In recent years, with the acceleration of urbanization and the transformation and upgrading of the industrial structure, the demand for agricultural water has decreased year by year. At the same time, the successive commissioning of major water conservancy projects such as the South-to-North Water Diversion Project has effectively reduced the exploitation of groundwater, promoting a gradual rise in the groundwater level. In 2021, the annual precipitation in this area reached its highest level since 1996, with the maximum annual precipitation recorded at the Luquan District rainfall station being 1028 mm. According to the water resources bulletin released by Hebei Province in 2024, the shallow groundwater levels in 129 counties across the province generally showed an upward trend.

2.1. Hydrogeological Conditions

In the study area, groundwater is primarily categorized into two types: Quaternary loose rock pore water and carbonate rock karst fissure water. Pore water is predominantly found within the coarse sand and gravel strata of the Quaternary gravel-bearing pebbles. In contrast, the karst water is primarily located within the Cambrian oolitic limestone and the formation chert-bearing dolomite, as illustrated in Figure 1. Given that the karst aquifer is situated at a depth exceeding 100 m and is subject to regional structural influences, it exhibits minimal direct hydraulic connectivity with the pore water. Consequently, it does not significantly affect the shallow groundwater level variations examined in this study, rendering further discussion unnecessary. Precipitation infiltration serves as the principal source of pore water recharge, contributing approximately 56% of the total recharge, while lateral recharge constitutes about 38%. The groundwater flow in the region progresses from west to east, characterized by a gradient of approximately 6.1‰ and a gentle flow rate.

2.2. Engineering Geological Conditions

Through systematic drilling investigations, this study obtained the spatial distribution characteristics of the strata in the study area. Based on the vertical sequence of the strata from top to bottom, they can be classified into five distinct engineering geological layer units: the fill layer, the loess-like silt layer, the silt layer, the silty clay layer, and the gravel layer. The detailed stratigraphic profile is presented in Figure 2, while the engineering geological properties of each layer are summarized in Table 1.

2.3. Groundwater Level Rise

Based on the dynamic groundwater level data obtained from monitoring points within the study area, it is observed that the groundwater level exhibits a general upward trend. The annual fluctuations in water level are minimal, indicating a relatively stable increase. Notably, in 2021, there was a marked rise in groundwater levels, with an increase exceeding 6 m. Despite this significant change, the burial depth of the water level remained relatively stable at approximately −5 m, as illustrated in Figure 3.

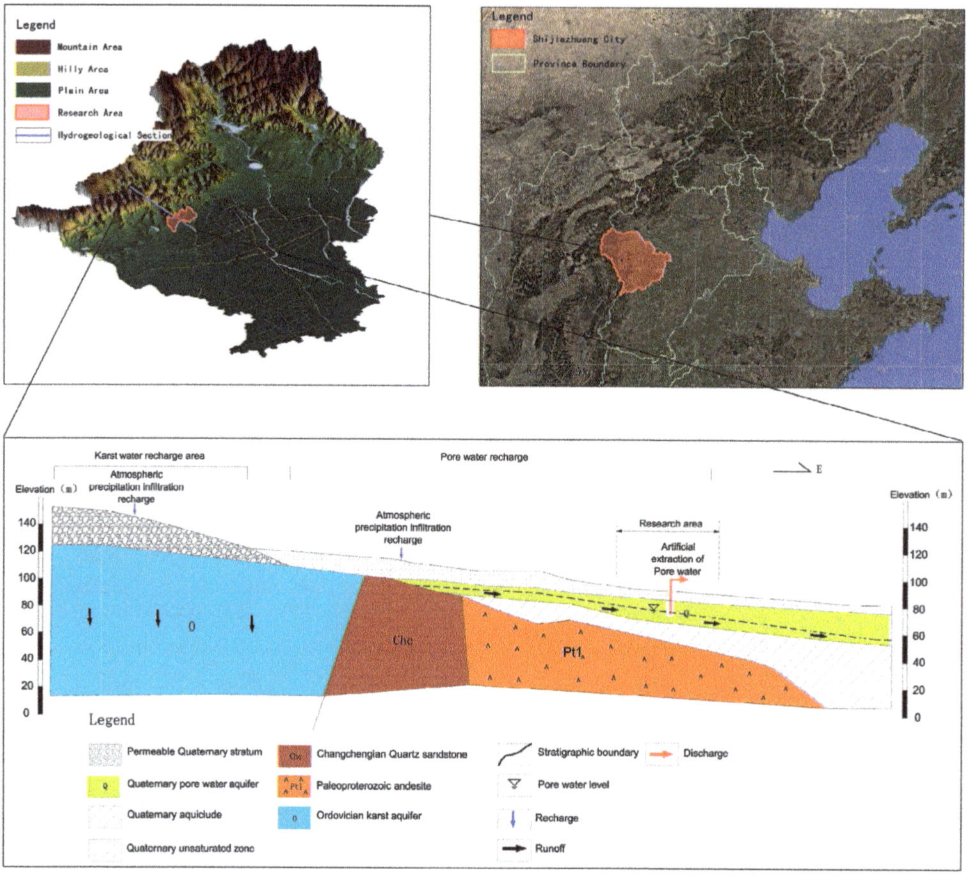

Figure 1. Hydrogeological profile of Shijiazhuang Piedmont area.

Table 1. Engineering geological characteristics of each soil layer.

Serial Number	Layer	Thickness (m)	Characterization
①	Miscellaneous fill	0.60–3.90	The soil is uneven, loose, slightly wet, containing a large amount of bricks and ash, and has poor engineering properties.
②	Loess-like silt	2.10–4.70	Mainly composed of powder particles, containing clay and sand particles, with low strength and toughness.
③	Silt	2.00–5.30	Mainly composed of powder particles, with low strength and toughness, slightly wet, moderately dense.
④	Silty clay	1.50–8.20	Mainly composed of clay and powder particles, locally containing a small amount of sand and gravel, soft plastic, moderate strength and toughness.
⑤	Gravel	-	The original rock composition is mainly composed of limestone and sandstone, accounting for about 50%. The particle size is generally between 2 cm and 6 cm, with a maximum particle size of 8 cm.

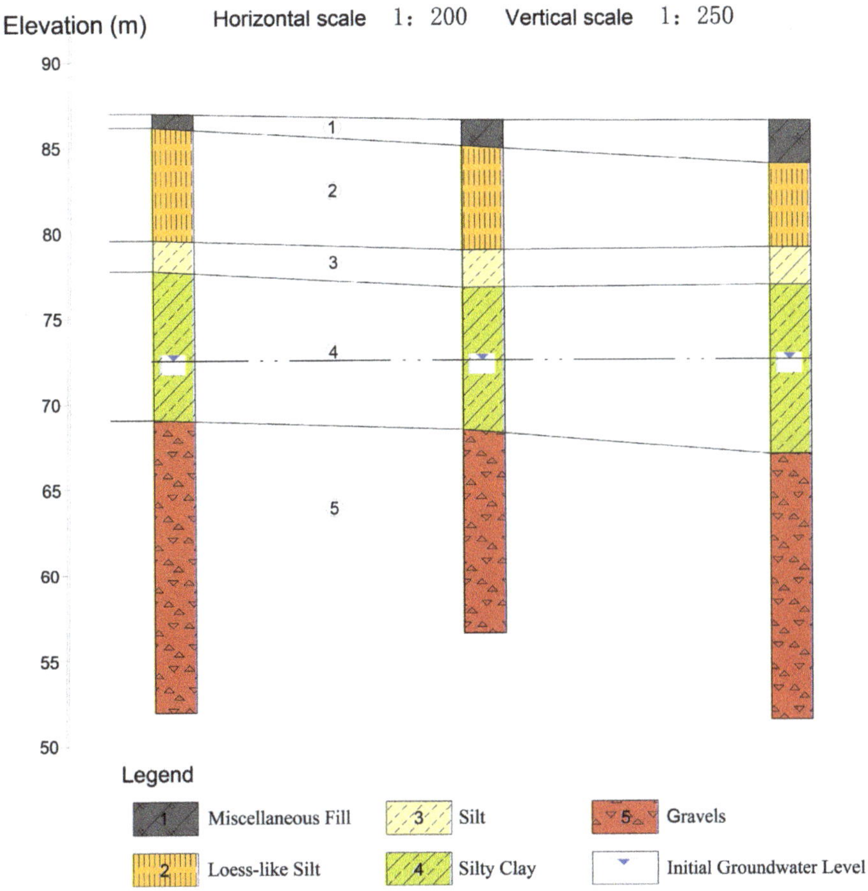

Figure 2. Stratigraphic profile of the study area.

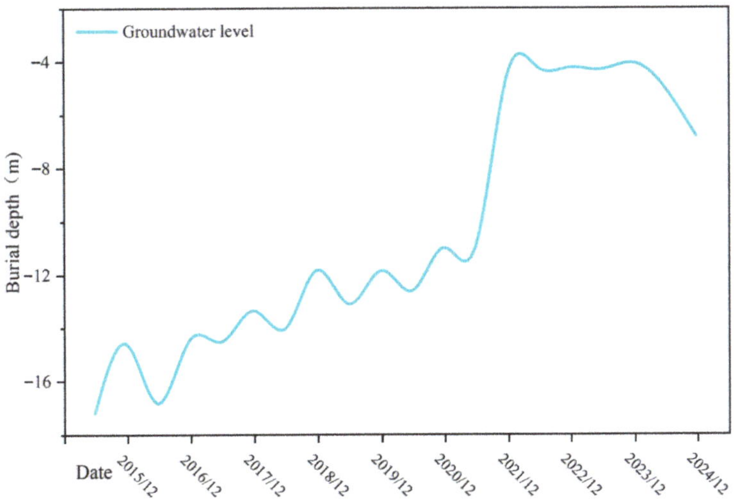

Figure 3. Variation in groundwater level in the study area from 2015 to 2024.

3. Methods

This study primarily addresses the challenge of determining whether the proposed numerical simulation method can accurately capture the effects of long-term dynamic changes in groundwater levels on buildings, thereby ensuring that the simulation results possess both reference value and theoretical significance. Consequently, an in-depth analysis of the numerical simulation method was conducted, taking into account the specific conditions of the research area. This led to the development of a theoretical framework for numerical simulation that is tailored to assess the impact of rising groundwater levels on buildings within the study region.

3.1. Selection of Numerical Simulation Software

This article employs the FLAC3D numerical simulation method to analyze the impact of groundwater level rise on buildings, due to several compelling reasons. Firstly, the FLAC3D method is highly regarded in geotechnical engineering applications, significantly contributing to analyses and research related to foundation deformation, building structure deformation, slope excavation stability, and anchor support stability, among other areas [14–18]. Furthermore, in engineering challenges involving fluid–structure interaction, FLAC3D 6.0 software is based on the explicit finite difference method and can simulate the interaction between fluids and solids by coupling algorithms that combine fluid motion equations with solid mechanics equations, thereby providing valuable insights into issues such as the effects of groundwater seepage on the stability of rock and soil, as well as structural stability [19–24]. Consequently, this study employs the FLAC3D numerical simulation technique to investigate the mechanisms by which rising groundwater levels affect buildings. Relevant research findings are referenced to enhance the credibility of the results, while this study also represents an innovative effort to broaden the application scope of FLAC3D. This approach aims to develop a novel theoretical framework for future analyses and research concerning the impact of groundwater level increases on engineering construction.

3.2. Numerical Simulation Scheme Analysis

The potential impact of rising groundwater levels on buildings can be categorized into three primary aspects [25]. First, an increase in groundwater levels can create a significant hydraulic head difference, resulting in pore water undergoing pressured seepage flow. This dynamic water pressure can exceed the gravitational or cohesive forces of soil particles, leading to piping and subsequent seepage damage. Second, as groundwater levels rise, previously unsaturated soil above the water table becomes saturated, altering its physical and mechanical properties. This transformation results in a reduction in the soil's bearing capacity and strength, potentially causing deformation and structural damage to buildings. Thirdly, when the groundwater level rises above the bottom plate of a building's foundation, the buoyant force acting on the foundation increases. This buoyancy is equivalent to the weight of the foundation displacing the groundwater. Concurrently, the soil surrounding the building's underground structure transitions from an unsaturated to a saturated state. As a result, the pore water pressure progressively increases, and this static water pressure exerts lateral forces on the walls of the building's underground structure.

In the study area, the minimal water gradient and slow runoff velocity, combined with the fact that groundwater level fluctuations are primarily influenced by precipitation, result in a consistent and gradual upward trend. Additionally, typical civil structures in the area have relatively small spans and shallow foundation depths, exerting minimal influence on the seepage field. Consequently, significant differences in hydraulic head or dynamic water pressure around these structures are unlikely. Thus, when evaluating the impact of rising groundwater levels on buildings within the study area, the effects of dynamic water pressure can be disregarded. However, soil saturation emerges as a critical variable in assessing the potential impacts of groundwater level rise on buildings. On a macro scale, the incremental rise in groundwater level affects buildings, while on a micro scale, the soil transitions from an unsaturated to a saturated state. This transition alters the mechanical properties of the soil, including pore water pressure, effective stress, soil structure, and hydrostatic pressure, thereby influencing the structural integrity and stability of buildings [26,27]. Therefore, in this numerical simulation study, the dynamic elevation process of the groundwater level is analogous to modeling the transition of a specific soil layer thickness from an unsaturated to a saturated state. The change in water level height corresponds to the thickness of the soil layer that undergoes this "unsaturated to saturated" transformation.

This study primarily investigates the impact of groundwater level fluctuations on the bearing capacity and strength variations in foundation soils: specifically, silt and silty clay. Additionally, it examines the plastic deformation of subterranean building structures, including concrete engineering, and the surrounding rock and soil. To this end, the Mohr–Coulomb constitutive model has been employed to analyze and assess potential deformation and failure in buildings, due to rising groundwater levels. This analysis is grounded in the Mohr–Coulomb criterion and the maximum tensile stress criterion.

3.3. Construction of 3D Numerical Simulation Model

The subject of this study is an eight-story high-rise residential building, located within the designated research area. Based on the construction specifications, the building employs a raft foundation composed of a uniformly thick reinforced concrete slab, measuring 500 mm in thickness, with foundation dimensions of 42 m × 18 m, a burial depth of 8 m, and a foundation-bearing layer consisting of the ③ silt layer. The underground wall is constructed from cast-in-place concrete, featuring a thickness of 400 mm, an anti-seepage grade of P6, and a permeability coefficient of $\leq 10^{-9}$ cm/s. The utilization of a raft foundation facilitates an even distribution of the building load, which is approximately 200 kPa. Detailed data are presented in Table 2.

Table 2. Construction information of the researched building.

Material	Foundation Thickness (mm)	Wall Thickness (mm)	Foundation Size (m)	Foundation Depth (m)	Concrete Strength (MPa)	Anti Leakage Grade (MPa)	Permeability Coefficient (cm/s)	Building Load (Kpa)
Concrete	500	400	42 × 18	8	30	0.6	10^{-9}	200

The physical and mechanical parameters, including Poisson's ratio, density, porosity, elastic modulus, cohesion, and internal friction angle for each stratum, necessary for numerical simulation methods, were derived from in situ experiments and geotechnical tests. The original in situ soil samples were obtained by drilling, sealed promptly, and sent to the laboratory for the determination of soil-layer physical parameters. During the sampling process, disturbance was strictly controlled to ensure the representativeness of the samples. The initial saturation was determined by the drying weighing method. Poisson's ratio and elastic modulus were calculated based on indoor triaxial compression test data, combined with Hooke's law. Cohesion and the internal friction angle were determined through indoor direct shear tests and according to the Coulomb strength criterion. The permeability coefficient (k) was mainly determined by on-site unstable flow pumping tests, and the permeability coefficients of each soil layer were quantitatively calculated using the linear graphical method. Furthermore, since this study primarily simulates the process of groundwater level rise by equivalently modeling the transition from an unsaturated to a saturated state of the strata, it is essential to acquire additional "initial saturation" parameters for each stratum. The detailed data are presented in Table 3.

Table 3. Physical and mechanical parameters of each soil layer.

Layer	Initial Saturation	Poisson's Ratio	Density	Void Ratio	Elastic Modulus	Cohesion	Internal Friction	Permeability Coefficient
	%	-	g/cm^3	-	MPa	kPa	°	cm/s
② Loess-like silt	67.8	0.31	1.86	0.92	12.2	23.3	25.5	3.8×10^{-4}
③ Silt	50.5	0.30	1.88	0.88	13.3	20.19	24.9	8.8×10^{-5}
④ Silty clay	82.4	0.32	2.04	0.82	18.2	22.5	19.7	3.2×10^{-5}
⑤ Gravels	Saturated	0.35	2.20	0.5	30.0	40	35	1×10^{-3}

Drawing upon comprehensive data on engineering geological conditions, geological physical and mechanical properties, and construction details from the study area, a three-dimensional geological model has been developed. The numerical model of the underground structure of the building is constructed, based on the information of the foundation size and burial depth. It mainly consists of two parts: the wall and the base plate. The wall part adopts a grid unit of 1 m × 0.4 m × 0.5 m, and the base plate part adopts a grid unit of 1 m × 1 m × 0.5 m. The numerical model size of the stratum area is 126 m × 72 m × 40 m, and the overall grid unit is 2 m × 2 m × 2 m. Due to the difference in grid scale between the stratum and the underground structure, the grid of the stratum around the building was locally encrypted, and the geometric shape of the model was optimized. Sharp corners were rounded to avoid stress concentration or stress singularity. The entire numerical model contains 97,875 zones and 104,156 grid points.

This numerical simulation adopts the fluid–solid coupling method and comprehensively considers the seepage effect of groundwater. Therefore, both mechanical boundary conditions and seepage boundary conditions need to be applied simultaneously in the model. The mechanical boundary conditions are achieved through velocity constraints: the bottom boundary of the model has a velocity of 0 in the X, Y, and Z directions; the left and right boundaries have velocity constraints in the horizontal X direction (velocity = 0); the front and back boundaries have velocity constraints in the horizontal Y direction (velocity = 0); and the top boundary is set as a free boundary, allowing free deformation. The seepage boundary conditions are set as follows: the top of the model is a permeable boundary, allowing free fluid exchange between nodes and the outside; the front, back, and bottom boundaries are impermeable boundaries with no fluid exchange; the left and right boundaries are set as constant head inflow boundaries to simulate the continuous water-injection process, thereby causing the groundwater level to rise.

This model is bifurcated into two principal components: the geological strata and the subterranean structural elements of the building. Given the significant influence of groundwater level fluctuations on the foundations and walls of these underground structures, this paper primarily focuses on examining the direct impacts of rising groundwater levels on these components. The potential transmission effects of deformation in the walls and foundations on other structural elements of the buildings have not been explored in this study. Consequently, the building structure has been simplified, as depicted in Figure 4.

3.4. Simulation Method for the Dynamic Rise in Groundwater Level

To investigate the effects of the long-term dynamic rise in groundwater levels on buildings, this study encompasses a period spanning nearly a decade, from June 2015 to December 2024. Given the seasonal variations, with the wet season occurring from June to December and the dry season from December to June of the subsequent year, the trend in water level changes during this timeframe remains relatively consistent. Consequently, a segmented simulation approach is employed, utilizing six-month intervals as the fundamental unit of analysis.

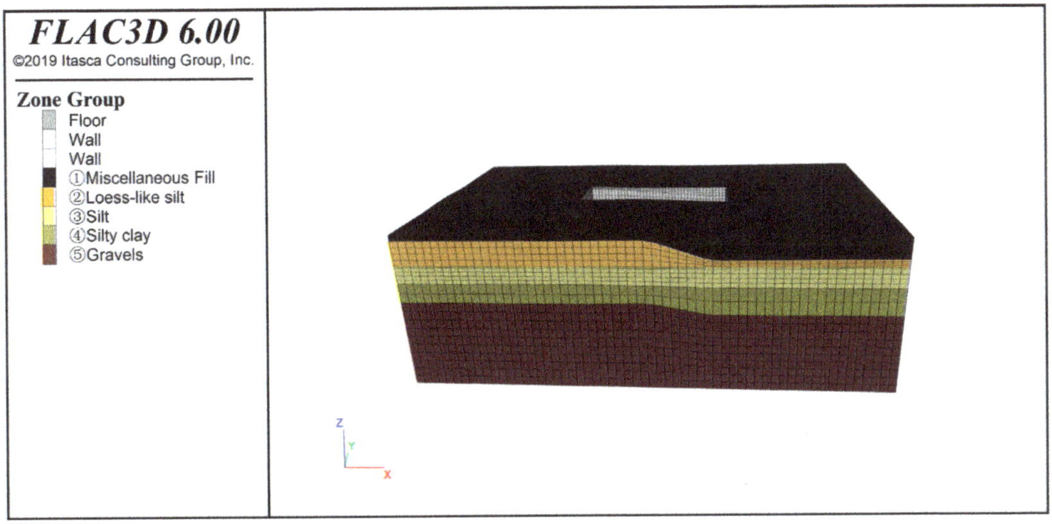

Figure 4. Three-dimensional structure model of numerical simulation.

4. Results and Discussion

Drawing on the dynamic characteristics of the groundwater level rise process within the study area, and integrating findings from numerical simulations, this paper categorizes the entire cycle of groundwater level rise into four distinct stages. The detailed categorization is presented in Table 4, while Figure 5 illustrates the variations in water levels across these stages. Notably, each stage exhibits significant differences in the effects of groundwater recovery speed, amplitude, and distribution area on buildings and the surrounding strata.

Table 4. Stage division of groundwater level dynamic rise process.

Stage	Period	Water Level	Description
First Stage	June 2015 to June 2019	−17.18 m to −13.09 m	The first stage is characterized by gradual fluctuations and a rising trend in groundwater levels. During this stage, there is a general upward trajectory in groundwater levels, with noticeable increases occurring in the wet season and minor declines observed during the dry season. The rebound in groundwater levels is predominantly concentrated within the silty clay layer (Layer ④).
Second Stage	June 2019 to August 2021	−13.09 m to −8.00 m	In the second stage, the groundwater level ascends above the bottom plate of the bearing layer (12.3 m) within the silt layer (③), and continues to rise until it reaches the bottom plate of the building's underground structure, which is buried at a depth of −8 m. At this stage, the groundwater does not exert a buoyant force on the building, as it does not surpass the bottom plate.
Third Stage	August 2021 to June 2022	−8.00 m to −4.24 m	The third stage occurs when the groundwater level exceeds the building floor, thereby generating a buoyant force on the structure. During this phase, the water level progressively rises from the building floor, ultimately reaching a peak that surpasses the floor by approximately 4 m.
Fourth Stage	June 2022 to December 2024	−4.24 m to −6.20 m	The fourth stage is characterized by a stabilization of the water level, which remains relatively constant within the range of −4 m to −6 m and shows a tendency to stabilize.

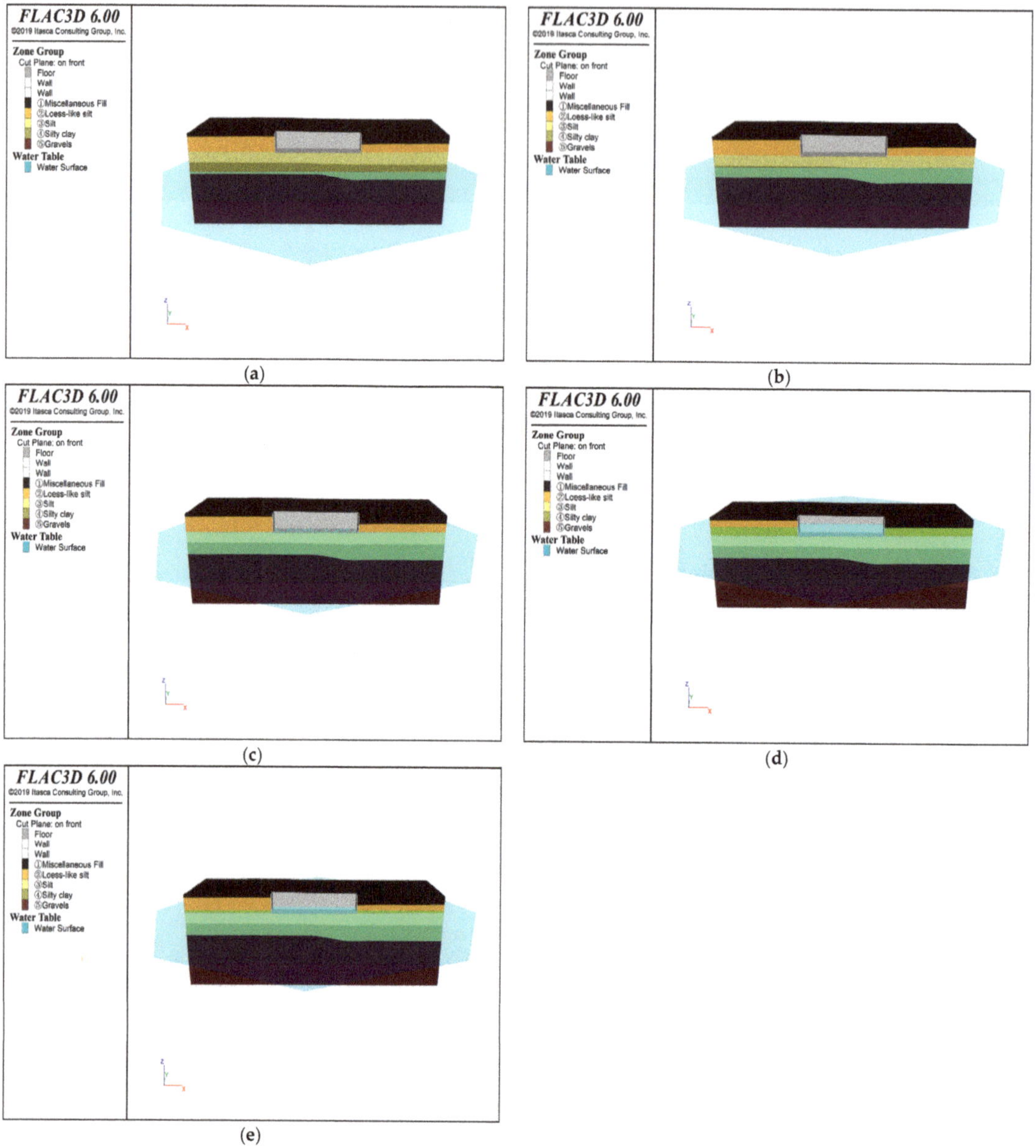

Figure 5. Schematic diagram of groundwater level at different stages: (**a**) initial groundwater level, (**b**) end of first stage groundwater level, (**c**) end of second stage groundwater level, (**d**) end of third stage groundwater level, (**e**) end of fourth stage groundwater level.

4.1. Displacement of Geological Strata

Analyzing the overall trends depicted in Figure 6a–d, it is evident that as the groundwater level incrementally rises, the lower strata of the building undergo settlement deformation, due to the influence of building loads. This results in a gradual expansion of both the magnitude and extent of the settlement. Concurrently, a minor rebound is observed around the building and within the deeper strata, aligning with the findings of Tian [28] regarding ground deformation being induced by rising groundwater levels. The local strata predominantly consist of cohesive soil layers, which exhibit a minimal or delayed response to the rising groundwater level. During the initial phase of groundwater level increase, as illustrated in Figure 6a, the deformation range of the strata is notably limited, primarily occurring in the lower right corner of the building. A comparison with Figure 6a′ reveals that the deformation range closely corresponds with the stratigraphic distribution. The thickness of the silty clay layer in the lower right corner of the building exhibits a gradual increase. Enhanced plasticity of the silty clay upon water absorption results in more pronounced deformation at the lower right corner of the building, although the overall deformation remains relatively minor, with a maximum deformation of only 4 mm. As the groundwater level rises to the second stage, illustrated in Figure 6b,b′, and reaches the foundation bearing layer of silt (layer ③), the settlement range of the stratum expands from a localized area to encompass the entire lower section of the building, leading to a significant increase in stratum deformation. The deformation pattern continues to be influenced by the stratum distribution, with the settlement deformation of the lower left stratum of the building being less than that of the lower right stratum. As the groundwater level rises further, the deformation of the strata increases; however, the range and trend of deformation remain largely unchanged (as depicted in Figure 6c,d).

4.2. Displacement of Underground Structures of Buildings

Influenced by the elevation of the groundwater level, the building demonstrates displacement deformation, as illustrated in Figure 7, indicating an overall settlement deformation. The deformation pattern aligns with that of the geological strata, exhibiting a progressively increasing deformation trend from left to right. The extent of settlement deformation escalates in conjunction with the rise in the groundwater level. Based on the distribution of load-bearing columns within the building, which have an approximate spacing of 8 m, monitoring points have been established on the foundation slab to observe deformation and displacement at various locations, as depicted in Figure 6a. During the initial phase (Figure 6a′), the maximum settlement deformation of the building's underground structure is recorded at 2.2 mm, with both the magnitude and rate of settlement deformation being minimal. During the second stage (Figure 6b,b′), the settlement deformation of the building intensified considerably, reaching a maximum settlement deformation of 30.6 mm and a maximum settlement rate of approximately 0.034 mm/day. The greatest settlement difference was observed between monitoring points five and six, with a differential settlement of 4.13 mm. In the third stage (Figure 6c,c′), the building experienced a maximum settlement of 49.54 mm, with a peak settlement rate of approximately 0.16 mm/day and a maximum differential settlement of 8.05 mm. In the fourth stage (Figure 6d,d′), the settlement rate of the building decreased markedly, with a rate of 0.005 mm/day, while the maximum differential settlement remained within 9 mm. In accordance with the stipulations outlined in the Chinese "Code for Design of Building Foundation" GB50007-2011 [29], the total deformation of raft foundations for high-rise buildings must not exceed 200 mm, and the settlement rate during the stabilization phase should remain between 0.01 and 0.03 mm/d. Furthermore, the differential settlement limit, calculated for a column spacing of 8 m, is set at 16 mm. The total settlement deformation and differential settlement of the subject under investigation, observed during the dynamic rise in the groundwater level, fall within the permissible range specified by the code. However, the settlement rate, which is excessively rapid, poses potential risks.

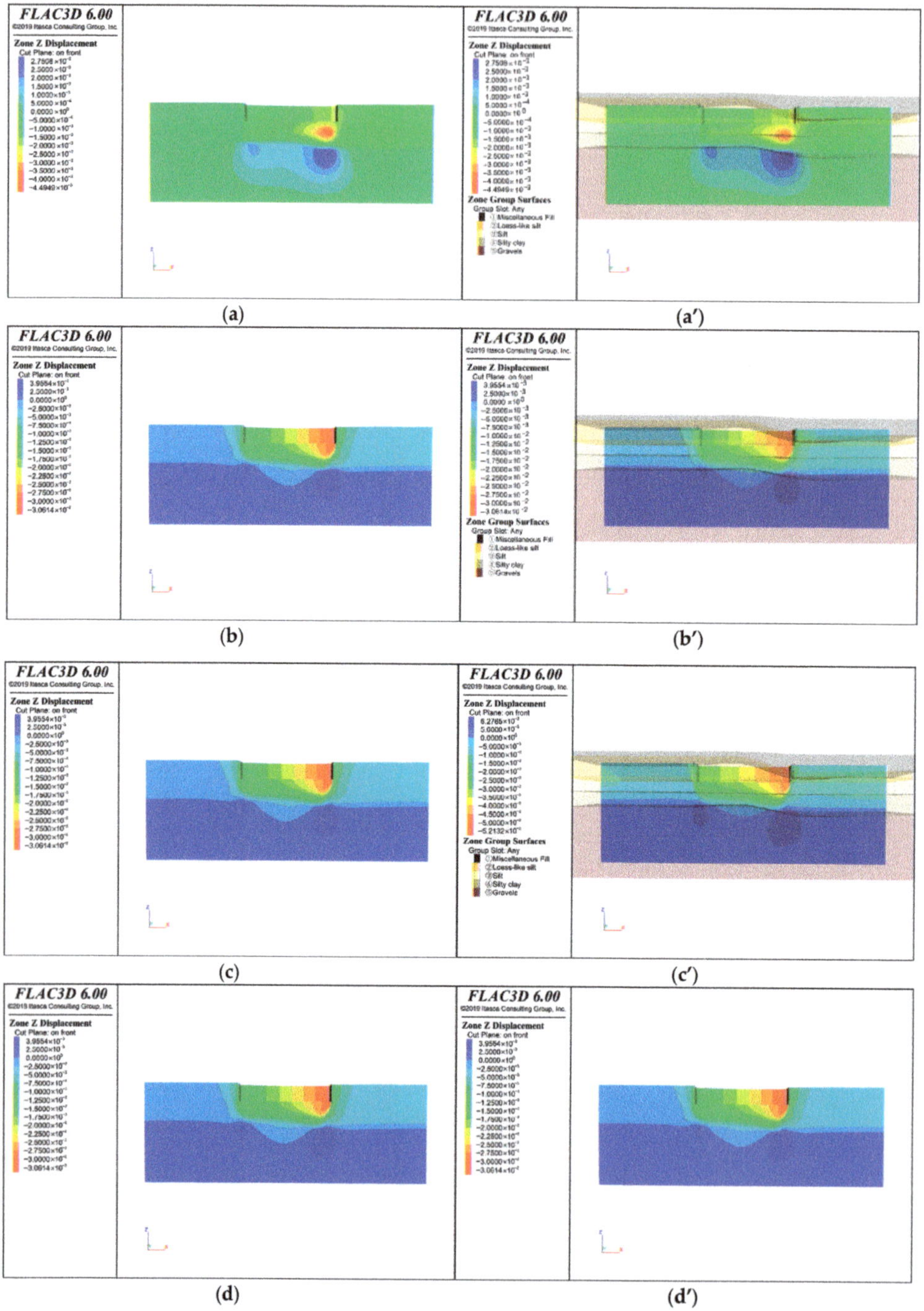

Figure 6. Stratum displacement due to groundwater level rise: (**a**) first stage, (**a′**) first stage superimposed strata perspective, (**b**) second stage, (**b′**) second stage superimposed strata perspective, (**c**) third stage, (**c′**) third stage superimposed strata perspective, (**d**) fourth stage, (**d′**) fourth stage superimposed strata perspective.

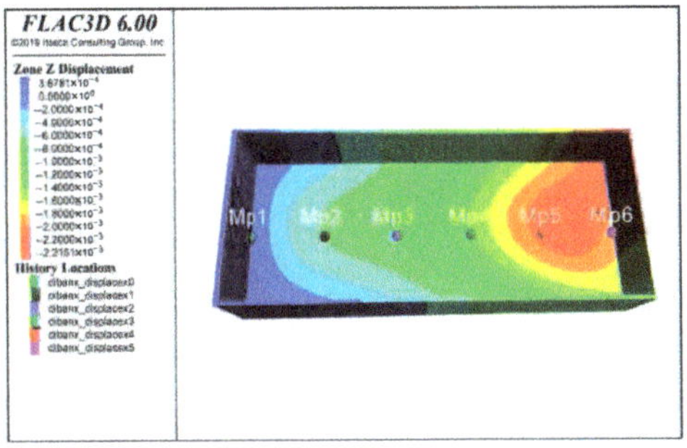

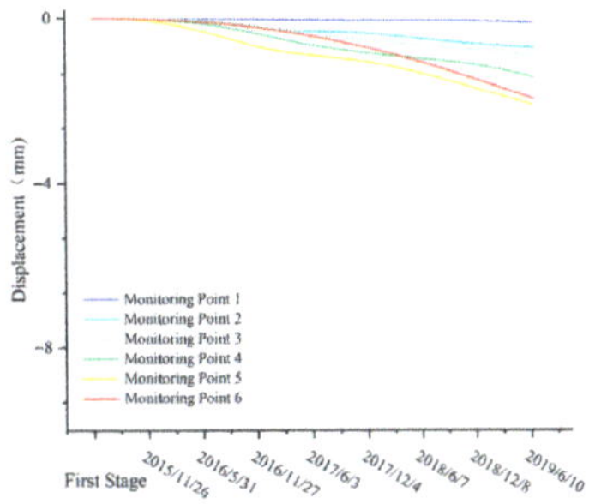

(**a**) (**a′**)

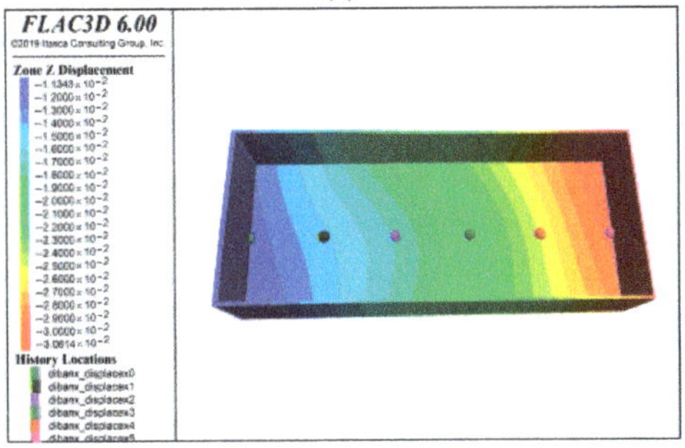

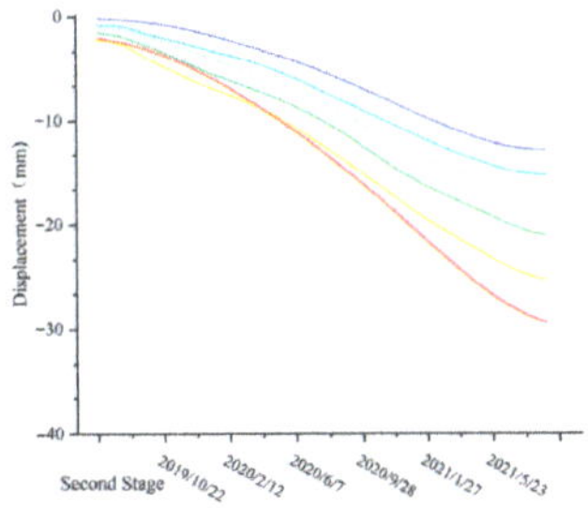

(**b**) (**b′**)

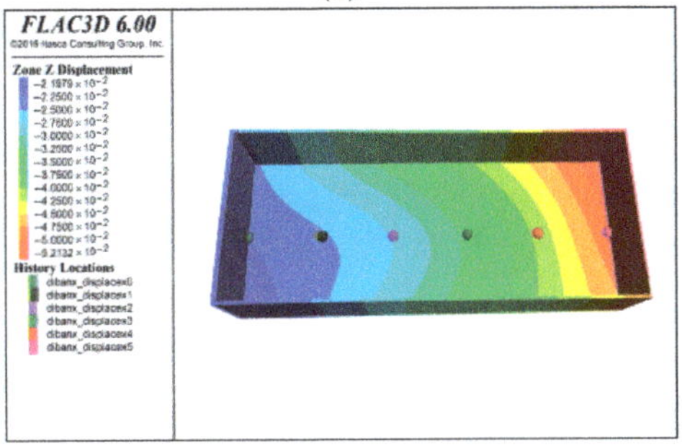

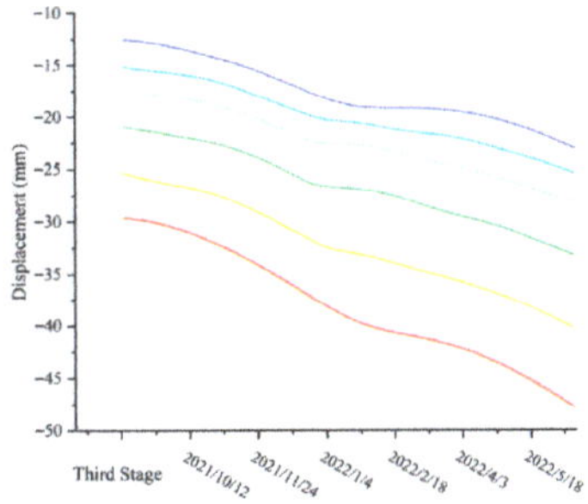

(**c**) (**c′**)

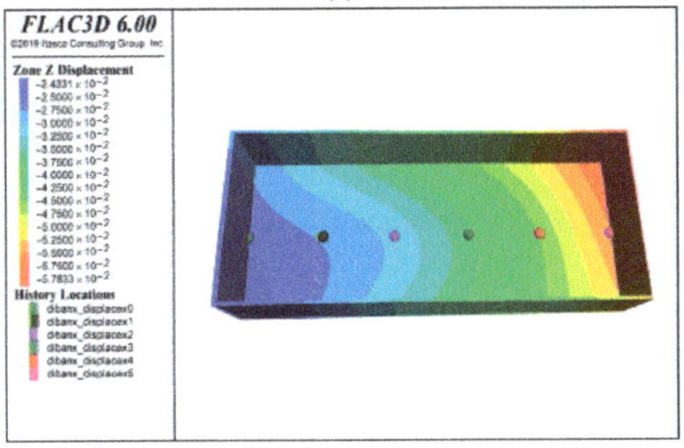

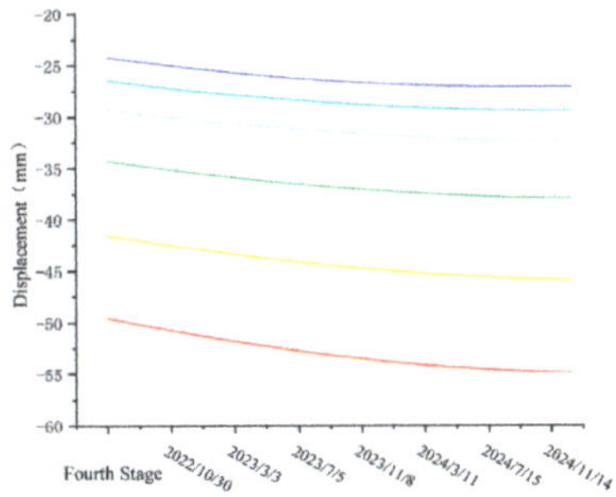

(**d**) (**d′**)

Figure 7. Displacement of the structure, due to groundwater level rise: (**a**) first stage displacement nephogram, (**a′**) first stage monitoring point displacement curve, (**b**) second stage displacement nephogram, (**b′**) second stage monitoring point displacement curve, (**c**) third stage displacement nephogram, (**c′**) third stage monitoring point displacement curve, (**d**) fourth stage displacement nephogram, (**d′**) fourth stage monitoring point displacement curve.

By analyzing the deformation and displacement of buildings in the first and second stages depicted in Figure 8, it is evident that, despite the similar magnitudes of groundwater level rise in both stages (4.09 m and 5.09 m, respectively), there is a notable disparity in the settlement deformation rate and the overall deformation of the buildings. This discrepancy can be attributed primarily to the differing response mechanisms of silty clay and silt in the study area to the rise in groundwater level, which results in the transition of the soil from an unsaturated to a saturated state. In the unsaturated state, silt exhibits high shear strength and low compressibility, due to matrix suction. However, as saturation progresses and the water film thickens, the matrix suction dissipates, leading to a significant reduction in the shear strength of the silt and a potential exponential increase in its compressibility. Over the past few decades, the research area has experienced a sustained decline in groundwater levels, leading to prolonged dewatering of the silty soil layer. This condition may have heightened the sensitivity of the soil's physical properties to variations in moisture content, thereby causing significant displacement and deformation of the silty soil layer as groundwater levels rise. Consequently, buildings in the area are likely to undergo rapid settlement and deformation. In contrast, the silty clay layer, due to its proximity to the groundwater level and high saturation, exhibits relatively less sensitivity to groundwater level fluctuations, resulting in a comparatively smaller degree of displacement and deformation.

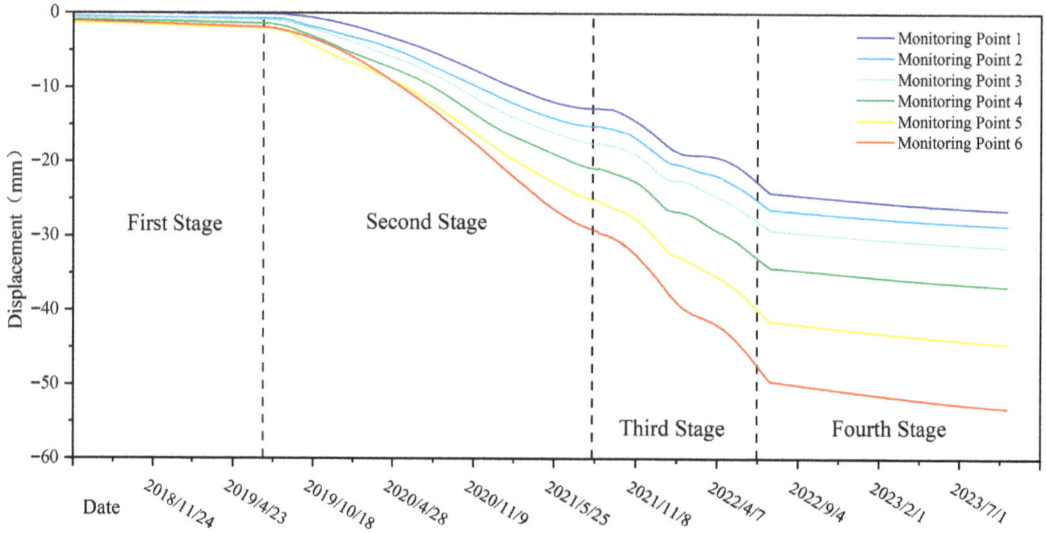

Figure 8. Displacement of the structure during the whole period of dynamic rise in groundwater level.

An analysis of the second and third stages depicted in Figure 8 reveals that, for the building under investigation, the buoyancy force generated when the groundwater level surpasses the building's floor is insufficient to counteract the settlement trend or induce uplift. The building's own load continues to predominantly drive the overall displacement deformation towards settlement.

In the fourth stage of Figure 8, it is evident that as the groundwater level stabilizes, there is a marked reduction in the rate of displacement and deformation, which gradually reaches a state of equilibrium.

4.3. Stress and Strain of Underground Structures of Buildings

An analysis of Figure 9a–d reveals that, as the groundwater level incrementally rises, the building floor experiences a notable stress concentration, indicated by a color transition from light blue to deep blue. Concurrently, the maximum principal stress exhibits a negative increase. The compressive stress on the floor progressively intensifies, reaching a peak value of approximately 0.31 MPa. Throughout the elevation of the groundwater level, the stress variation in the walls remains relatively insignificant; however, there is a discernible concentration of compressive stress at the four corners of the wall. The remaining sections of the wall sustain a relatively uniform distribution of compressive stress: approximately 0.25 MPa. According to the Chinese "Code for Design of Concrete Structures" GB50010-2010 [30], the standard compressive strength of C30 concrete is specified as 20.1 MPa. This standard indicates that the compressive stress exerted on the walls and bottom plates of buildings due to rising groundwater levels is unlikely to significantly affect the integrity of underground structures. However, an analysis of Figure 9a′–d′ reveals a concentration of tensile stress at the junction between the bottom plate and the wall, aligning with the direction of differential settlement. As the groundwater level increases, the area of stress concentration expands, and the tensile stress intensifies, with the maximum tensile stress rising from 1.8 MPa to 2.19 MPa. In accordance with the "Code for Design of Concrete Structures" GB50010-2010, which specifies a standard tensile strength value of 2.01 MPa for C30 concrete, it is observed that as the groundwater level rises, the tensile stress in affected regions progressively increases. Consequently, the concrete structure may transition from elastic to plastic deformation, potentially resulting in the formation of tensile cracks. Such structural changes could contribute to water seepage and leakage issues in buildings.

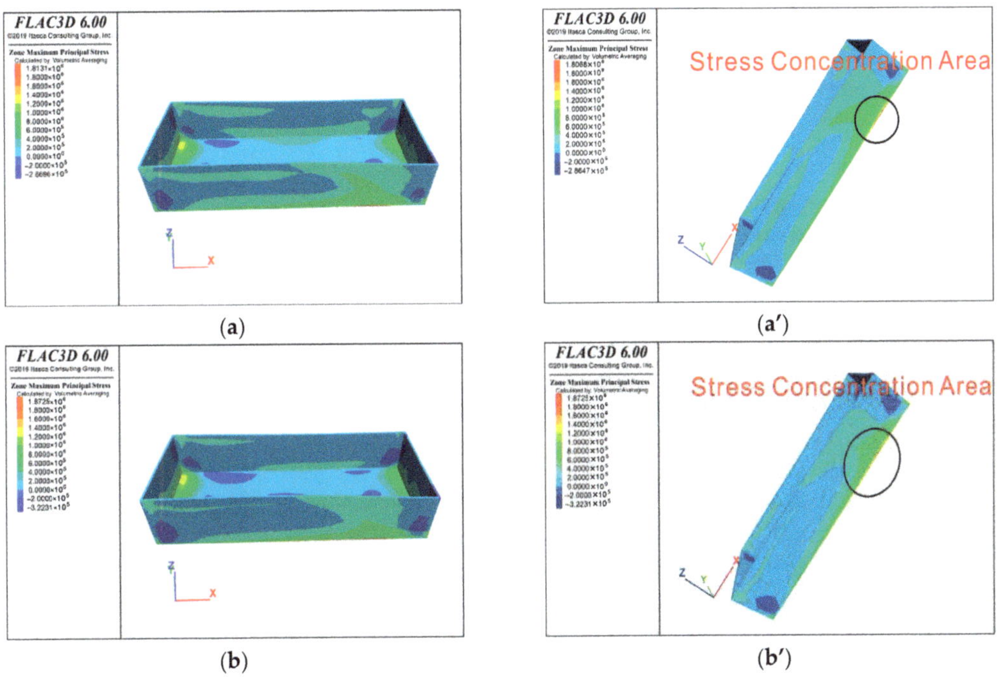

Figure 9. *Cont.*

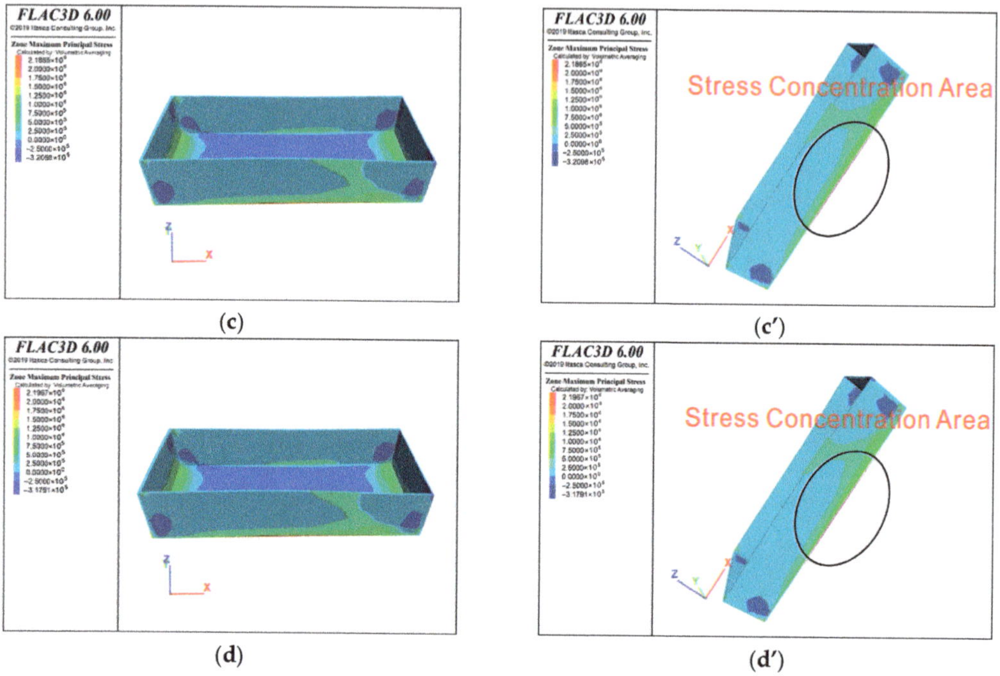

Figure 9. Stress and strain of building structure, due to groundwater level rise: (**a**) first stage maximum principal stress distribution cloud map, (**a′**) first stage stress concentration area, (**b**) second stage maximum principal stress distribution cloud map, (**b′**) second stage stress concentration area, (**c**) third stage maximum principal stress distribution cloud map, (**c′**) third stage stress concentration area, (**d**) fourth stage maximum principal stress distribution cloud map, (**d′**) fourth stage stress concentration area.

5. Conclusions

(1) The impact of the entire process of dynamic groundwater level rebound on buildings exhibits distinct phased characteristics. During the dynamic groundwater level rebound process, the maximum settlement of the building and the maximum differential settlement between columns both occur at the end of the fourth stage, and both remain within the specified range. Nevertheless, the maximum settlement rate of the building is observed in the third stage, reaching 0.16 mm/d, which exceeds the code standard. This situation may potentially pose a certain influence on the structural safety of the building.

(2) There are substantial differences in the response mechanisms of different strata lithologies to the rebound of the groundwater level. Thus, it is necessary to pay particular attention to strata whose physical properties and behaviors are sensitive to changes in the moisture content. In this study, the displacement and deformation induced by the impact of the groundwater level rebound on the silt layer that has been in a drained state for an extended period have exerted a notable impact on the building.

(3) During the groundwater level rebound process, tensile stress concentration zones emerge at the connection between the floor slab and the wall, in the direction consistent with the differential settlement. The maximum tensile stress has exceeded the standard value of the tensile strength of C30 concrete. This might be one of the crucial factors contributing to the leakage problem in the underground structure of the building.

Based on the research conclusions regarding the impact of dynamic groundwater level recovery on buildings, the following core suggestions should be given priority in engineering construction: Firstly, additional dynamic groundwater level monitoring wells should be set up in potential areas of water level recovery, and the stage characteristics and development trends of water level recovery should be accurately identified, based on the measured monitoring data. Secondly, during the investigation stage, the physical and mechanical properties of water-content-sensitive strata (especially the silt layer in a long-term drained state) should be analyzed more intensively, and the foundation design scheme should be optimized accordingly. Thirdly, during the construction and operation periods, the settlement deformation of buildings (with particular attention to the critical stages where the settlement rate exceeds the limit) and the stress state of underground structures (especially the tensile stress concentration at the connection between the bottom plate and the wall, to prevent concrete cracking and water seepage risks) should be monitored in real time. At the same time, it is necessary to avoid applying the experience theories of other regions or misjudging the structural deformation trend, due to the neglect of the buoyancy force effect. Moreover, the engineering countermeasures should be dynamically adjusted according to the different stage characteristics of water level recovery, to ensure the safety of the building structures and the rationality of the engineering construction.

Author Contributions: Conceptualization, H.L. and M.G.; methodology, H.L. and M.Z.; software, H.L. and M.G.; validation, H.L. and B.M.; formal analysis, X.Z.; investigation, J.T. and L.C.; resources, M.G. and L.G.; data curation, M.Z. and H.L.; writing—original draft preparation, H.L.; writing—review and editing, M.G. All authors have read and agreed to the published version of the manuscript.

Funding: This research was funded by the Postdoctoral Research Workstation of Hebei Bureau of Geology and Mineral Resources Exploration, the Hebei Key Laboratory of Geological Resources and Environment Monitoring and Protection and the Hebei Yanzhao Golden Platform Talent Gathering Plan Backbone Talent Project (postdoctoral platform) (b2024003008).

Data Availability Statement: The original contributions presented in the study are included in the article; further inquiries can be directed to the corresponding author.

Conflicts of Interest: The authors declare no conflicts of interest.

References

1. Zhang, M.; Hu, L.; Yao, L.; Yin, W. Numerical studies on the influences of the South-to-North Water Transfer Project on groundwater level changes in the Beijing Plain, China. *Hydrol. Hydrol. Process.* **2018**, *32*, 1858–1873. [CrossRef]
2. Zhao, Y.; Zhu, Y.; Lin, Z.; Wang, J.; He, G.; Li, H.; Li, L.; Wang, H.; Jiang, S.; He, F.; et al. Energy reduction effect of the south-to-north water diversion project in China. *Sci. Rep.* **2017**, *7*, 15956. [CrossRef]
3. Zou, J.; Zhan, C.; Xie, Z.; Qin, P.; Jiang, S. Climatic impacts of the Middle Route of the South-to-North Water Transfer Project over the Haihe River Basin in North China simulated by a regional climate model. *J. Geophys. Res. Atmos.* **2016**, *121*, 8983–8999. [CrossRef]
4. Sukop, M.C.; Rogers, M.; Guannel, G.; Infanti, J.M.; Hagemann, K. High temporal resolution modeling of the impact of rain, tides, and sea level rise on water table flooding in the Arch Creek basin, Miami-Dade County Florida USA. *Sci. Total Environ.* **2018**, *616*, 1668–1688. [CrossRef]
5. Czajkowski, J.; Engel, V.; Martinez, C. Economic impacts of urban flooding in South Florida: Potential consequences of managing groundwater to prevent salt water intrusion. *Sci. Total Environ.* **2018**, *621*, 465–478. [CrossRef] [PubMed]
6. Li, Y.; Wang, R.; Ma, H.; Zhang, J.-M. Rising groundwater table due to restoration projects amplifies earthquake induced liquefaction risk in Beijing. *Nat. Commun.* **2025**, *16*, 1466. [CrossRef] [PubMed]
7. Yang, J.; Cao, G.; Han, D.; Yuan, H.; Hu, Y.; Shi, P.; Chen, Y. Deformation of the aquifer system under groundwater level fluctuations and its implication for land subsidence control in the Tianjin coastal region. *Environ. Monit. Assess.* **2019**, *191*, 162. [CrossRef]
8. Zucca, M.; Crespi, P.G.; Longarini, N. Seismic vulnerability assessment of an Italian historical masonry dry dock. *Case Stud. Struct. Eng.* **2017**, *7*, 1–23. [CrossRef]
9. Chang, K.; Guo, X.; Ma, W.; Peng, Z.; Chi, L. Study on variation laws of external water pressure and lining stress caused by blockage of drainage blind pipe in railway tunnel. *Railw. Stand. Des.* **2025**, *69*, 120–128.

10. Chen, X.; Wang, H. Impact of sea level rise on asphalt pavement responses considering seasonal groundwater and moisture gradient in subgrade. *Transp. Geotech.* **2023**, *40*, 100992. [CrossRef]
11. Nguyen, Q.-P.; Nguyen, V.-M.; Nguyen, K.-C.; Nguyen, N.-H.; Pham, K. Influence of groundwater level rise on the geomechanical changing process in the tunnel surrounding jointed rock mass and in the tunnel support structure. *AIP Conf. Proc.* **2021**, *2420*, 8. [CrossRef]
12. Lai, H.; Tan, Z.; Sun, Y.; Pengzhi, H. Study on water migration law of surrounding rock during construction of water-rich loess tunnel. *China J. Highw. Transp.* **2023**, *36*, 12.
13. Zou, H.; Wu, H.; Yang, X. Key technologies for waterproofing and drainage of tunnel group in CCS hydropower station water diversion and power generation system. *Yellow River* **2020**, *42*, 6.
14. Kim, B.H.; Larson, M.K. Development of a fault-rupture environment in 3D: A numerical tool for examining the mechanical impact of a fault on underground excavations. *Int. J. Min. Sci. Technol.* **2019**, *29*, 105–111. [CrossRef]
15. Zhou, Z.; Zhou, J.; Lai, L.; Xu, M.; Xu, Y. Determination of optimal mining width for coal mining under the slope by of using numerical simulation. *Sci. Rep.* **2024**, *14*, 1124. [CrossRef]
16. Guo, H.; Zhao, X.; Sun, C. Numerical simulation study on the force of overwintering foundation support structure of unsaturated seasonal permafrost under indoor experiments. *R. Soc. Open Sci.* **2024**, *11*, 240992. [CrossRef]
17. Fan, P.; Chen, J.; Chen, J.; Shen, X.; Wang, M. Optimal design of a multistage slope using the multi-verse optimization algorithm. *Sci. Rep.* **2025**, *15*, 6376. [CrossRef] [PubMed]
18. Meng, Z.; Xu, J.; Li, H.; Tao, Z. Improvement of anchor structural unit in FLAC3D and its application to the 110 construction method. *Sci. Rep.* **2025**, *15*, 7690. [CrossRef]
19. Lai, J.X.; Fan, H.B.; Zhou, F. Fluid-Solid Coupling Numerical Simulation for Tunnel in Fracture Zone Based on 2D-FLAC Software. *Adv. Mater. Res.* **2012**, *503*, 167–170. [CrossRef]
20. Wang, W.; Gao, S.; Min, Y.; Liu, L.; Chen, J. Three-dimensional fluid–solid coupling numerical simulation of effects of underlying karst cave on shield tunnel through sand stratum. *Geotech. Geol. Eng.* **2019**, *37*, 4825–4836. [CrossRef]
21. Li, Z.; Luo, Z.; Xu, C.; Tan, J. 3D fluid-solid full coupling numerical simulation of soil deformation induced by shield tunnelling. *Tunn. Undergr. Space Technol.* **2019**, *90*, 174–182. [CrossRef]
22. Lei, M.; Luo, S.; Chang, J.; Zhang, R.; Kuang, X.; Jiang, J. Fluid–Solid Coupling Numerical Analysis of Pore Water Pressure and Settlement in Vacuum-Preloaded Soft Foundation Based on FLAC3D. *Sustainability* **2023**, *15*, 7841. [CrossRef]
23. Cao, Z.; Zhang, S.; Du, F.; Ma, D.; Li, Z.; Huang, C.; Wang, W. Water Inrush Mechanism and Variable Mass Seepage of Karst Collapse Columns Based on a Nonlinear Coupling Mechanical Model. *Mine Water Environ.* **2025**, *44*, 259–274. [CrossRef]
24. Teng, T.; Chen, Y.; Wang, S.; Jia, W.; Wang, Y.; Liu, K.; Li, Z. Water injection softening modeling of hard roof and application in Buertai coal mine. *Environ. Earth Sci.* **2025**, *84*, 54. [CrossRef]
25. Sun, W.; Han, F.; Zhang, Y.; Zhang, W.; Zhang, R.; Su, W. Experimental assessment of structural responses of tunnels under the groundwater level fluctuation. *Tunn. Undergr. Space Technol.* **2023**, *137*, 105138. [CrossRef]
26. Lu, D.; Li, X.; Du, X.; Lin, Q.; Gong, Q. Numerical simulation and analysis on the mechanical responses of the urban existing subway tunnel during the rising groundwater. *Tunn. Undergr. Space Technol.* **2020**, *98*, 103297. [CrossRef]
27. Zhengzheng, C.; Yi, X.; Yi, X.; Feng, D.; Zhenhua, L.; Cunhan, H.; Shuren, W.; Yongqiang, Y.; Wengiang, W.; Minglei, Z.; et al. Diffusion Evolution Rules of Grouting Slurry in Mining-induced Cracks in Overlying Strata. *Rock Mech. Rock Eng.* **2025**, *58*, 6493–6512. [CrossRef]
28. Tian, M.; Zhao, L.; Cui, W.; Gaoxuan, G.; He, L.; Aihua, S.; Xinhui, W.; Hang, C.; Pan, W. Impact of groundwater level rise on land subsidence control under the South-to-North Water Diversion Project: A case study of the Chaobai River groundwater system in Beijing. *Geol. China* **2023**, *50*, 872–886.
29. *GB50007-2011*; Chinese Ministry of Housing and Urban Rural Development, Code for Design of Building Foundation. China Architecture & Building Press: Beijing, China, 2012.
30. *GB50010-2010*; China Academy of Building Research, Code for Design of Concrete Structures. China Architecture & Building Press: Beijing, China, 2016.

water

MDPI

Article

Health Risk Assessment of Groundwater in Cold Regions Based on Kernel Density Estimation–Trapezoidal Fuzzy Number–Monte Carlo Simulation Model: A Case Study of the Black Soil Region in Central Songnen Plain

Jiani Li [1,2], Yu Wang [1,2], Jianmin Bian [1,2,*], Xiaoqing Sun [1,2] and Xingrui Feng [1,2]

1 Key Laboratory of Groundwater Resources and Environment, Ministry of Education, Jilin University, Changchun 130021, China; jnli24@mails.jlu.edu.cn (J.L.); wangyu1224@jlu.edu.cn (Y.W.); sunxq13@jlu.edu.cn (X.S.); fengxr24@mails.jlu.edu.cn (X.F.)
2 Jilin Provincial Key Laboratory of Water Resources and Environment, Jilin University, Changchun 130021, China
* Correspondence: bianjm@jlu.edu.cn

Abstract

The quality of groundwater, a crucial freshwater resource in cold regions, directly affects human health. This study used groundwater quality monitoring data collected in the central Songnen Plain in 2014 and 2022 as a case study. The improved DRASTICL model was used to assess the vulnerability index, while water quality indicators were selected using a random forest algorithm and combined with the entropy-weighted groundwater quality index (E-GQI) approach to realize water quality assessment. Furthermore, self-organizing maps (SOM) were used for pollutant source analysis. Finally, the study identified the synergistic migration mechanism of NH_4^+ and Cl^-, as well as the activation trend of As in reducing environments. The uncertainty inherent to health risk assessment was considered by developing a kernel density estimation–trapezoidal fuzzy number–Monte Carlo simulation (KDE-TFN-MCSS) model that reduced the distribution mis-specification risks and high-risk misjudgment rates associated with conventional assessment methods. The results indicated that: (1) The water chemistry type in the study area was predominantly HCO_3^-–Ca^{2+} with moderately to weakly alkaline water, and the primary and nitrogen pollution indicators were elevated, with the average NH_4^+ concentration significantly increasing from 0.06 mg/L in 2014 to 1.26 mg/L in 2022, exceeding the Class III limit of 1.0 mg/L. (2) The groundwater quality in the central Songnen Plain was poor in 2014, comprising predominantly Classes IV and V; by 2022, it comprised mostly Classes I–IV following a banded distribution, but declined in some central and northern areas. (3) The results of the SOM analysis revealed that the principal hardness component shifted from Ca^{2+} in 2014 to Ca^{2+}–Mg^{2+} synergy in 2022. Local high values of As and NH_4^+ were determined to reflect geogenic origin and diffuse agricultural pollution, whereas the Cl^- distribution reflected the influence of de-icing agents and urbanization. (4) Through drinking water exposure, a deterministic evaluation conducted using the conventional four-step method indicated that the non-carcinogenic risk (*HI*) in the central and eastern areas significantly exceeded the threshold (*HI* > 1) in 2014, with the high-*HI* area expanding westward to the central and western regions in 2022; local areas in the north also exhibited carcinogenic risk (*CR*) values exceeding the threshold (*CR* > 0.0001). The results of a probabilistic evaluation conducted using the proposed simulation model indicated that, except for children's *CR* in 2022, both *HI* and *CR* exceeded acceptable thresholds with 95% probability. Therefore, the proposed assessment method can provide a basis for improved groundwater pollution zoning and control decisions in cold regions.

Keywords: health risk assessment; kernel density estimation–trapezoidal fuzzy number–Monte Carlo simulation; self-organizing map; water quality assessment; groundwater vulnerability

1. Introduction

Groundwater plays a critical role in environmental ecology. In recent years, human activities have increasingly impacted groundwater systems as accelerated economic development worsens environmental degradation. Groundwater, a core component sustaining regional water cycles and ecological equilibrium, is particularly critical in the black soil region of central Songnen Plain. Characterized as a typical cold region, this area experiences distinct freeze–thaw cycles that alter aquifer permeability, accelerate pollutant migration, and trigger seasonal solute release, including the sudden input of de-icing agents in winter and the concentrated leaching of agricultural pollutants during spring. This significantly increases the complexity of groundwater chemical evolution as well as the uncertainty of associated health risk assessments [1]. Groundwater quality evaluation and health risk assessment are foundational aspects of groundwater protection, and assessing contamination susceptibility under hydrogeological and anthropogenic conditions is essential for pollution prevention.

Groundwater quality is a key determinant of a safe drinking water supply and essential to maintaining agricultural productivity. Groundwater quality evaluation techniques have developed considerably in recent years. The entropy-weighted groundwater quality index (E-GQI) provides a reliable tool for quickly assessing the suitability of groundwater for drinking purposes [2]. This method uses weights based on raw data to avoid the subjective biases of traditional methods, making it more objective and intuitive. However, E-GQI evaluation considers numerous water quality parameters that must be selected subjectively. Therefore, this study selected key water quality parameters based on feature importance rankings obtained using a random forest (RF) algorithm and incorporated a vulnerability index to provide a critical dimension of water quality evaluation [3]. This approach can more comprehensively identify potential areas of high-vulnerability, low-quality groundwater risk. Groundwater vulnerability measures the sensitivity of a groundwater system to human activities or natural factors, such as geology, hydrogeology, and pollutant discharge conditions and physicochemical properties. The DRASTIC model is a widely used and mature model for groundwater vulnerability assessment [4]. In China, its application can be divided into several stages spanning from the mid-1990s, when it was based on localized parameter calibration, to the late 1990s, when fuzzy mathematics membership functions were introduced to construct a fuzzy DRASTIC model that addressed the issue of discontinuous representation of geological boundary transition zones [5], to the 21st century, when further improvements integrated land use type indicators and applied ArcGIS for statistical analysis to overcome subjectivity [6,7].

Drinking contaminated groundwater can seriously damage human health, and quantifying pollutant risk levels is a key step in assessing exposure probability and managing potential health risks. The United States Environmental Protection Agency's (USEPA's) classic "four-step method" (FSM) is a commonly used deterministic method for human health risk assessment (HHRA) that effectively distinguishes between carcinogenic and non-carcinogenic risks under various pollution scenarios and exposure pathways. It has been widely adopted owing to its simplicity and comprehensive evaluation capabilities [8]. However, the FSM is subject to limitations when dealing with uncertainty, primarily owing to the randomness of exposure parameters as well as limited sample sizes that fail to capture variability in the risk assessment process, resulting in significant discrepancies between the calculated results and actual risks [9]. Previous stochastic health risk assessments have typically described the uncertainty in pollutant concentrations and exposure parameters using triangular stochastic simulation (TSS). Although the membership function associated with a TSS has the mathematical advantage of fixed endpoints, it struggles to reflect the gradual distribution characteristics of parameters. Currently, the use of improved trapezoidal fuzzy numbers (TFNs) adjusts the distribution boundaries to improve precision when describing the core intervals and extreme value ranges of parameters. However, the determination of the core interval still relies on subjective judgment or empirical assignment, which cannot represent multi-modal, skewed, or heavy-tailed features in actual data [10,11]. Therefore, this study proposed a kernel density estimation–trapezoidal fuzzy number–Monte Carlo simulation (KDE-TFN-MCSS) health risk assessment model. Note that the accurate estimation of the probability distribution function (PDF) for each variable in the model is necessary for the application of a Monte Carlo simulation (MCSS). Critically, the non-parametric kernel density estimation (KDE) method can evaluate data distribution characteristics entirely from samples and estimate the density functions of arbitrary shapes, reducing potential errors caused by the incorrect selection of probability distribution models, known as distribution mis-specification risks. The proposed KDE-TFN-MCS model significantly improves the probability density function estimation accuracy, thereby reducing the misjudgment of high-risk areas and minimizing the impact of parameter uncertainty. As a result, KDE provides the MCSS with more robust risk representation capabilities that are especially well-suited to estimating the probability distributions of complex random variables [12].

The acceleration of urbanization has introduced widespread ammonia nitrogen contamination in the groundwater of the central Songnen Plain owing to intensive agricultural activities, yet groundwater remains an important water resource in the region as an irreplaceable resource for drinking water, irrigation, and industry, Indeed, groundwater is often the primary source of drinking water and a key guarantee for agricultural irrigation, especially in rural areas. However, existing studies on cold-region groundwater quality have primarily evaluated the impact of natural freeze–thaw processes on water chemistry and have not conducted systematic analyses of the coupled effects of anthropogenic pollution (e.g., widespread ammonia nitrogen and rising Cl^- levels) and natural freeze–thaw cycles to determine the specific health risks posed by inherent geogenic contaminants, such as As. Therefore, the groundwater pollution characteristics and health risk evolution mechanisms in the Songnen Plain, a typical cold region, require further investigation. This study used groundwater quality monitoring data collected in the central Songnen Plain in 2014 and 2022 to analyze the water chemistry characteristics and their causes. First, an RF algorithm was used to determine water quality evaluation factors, which were combined with vulnerability indices and entropy-weighted indices to realize groundwater quality assessment. Next, self-organizing maps (SOMs) were employed to identify key risk factors [13], then the results obtained using the FSM and improved KDE-TFN-MCSS models to conduct groundwater health risk assessment were compared. The proposed KDE-TFN-MCSS model was shown to reduce distribution mis-specification risks in the evaluation process and improve the accuracy of the results for complex regional groundwater systems, thereby providing multi-dimensional support for groundwater risk management in cold regions. The research approach applied in this study is detailed in Figure 1.

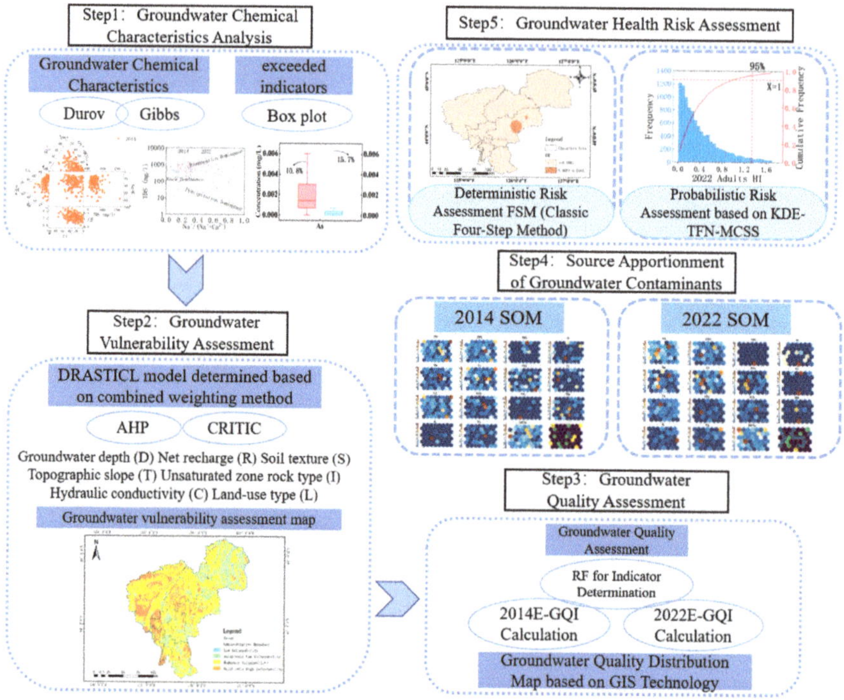

Figure 1. Research approach.

2. Materials and Methods

2.1. Study Area

The central Songnen Plain is an alluvial core area connecting the eastern high plain and western low plain, serving as a key transition zone controlling water/salt migration. Changchun is situated in the central Songnen Plain at the coordinates 43°05′~45°15′ N, 124°18′~127°05′ E and was selected as a representative location in this study. This region covers an area of 24.744 km^2 with terrain that primarily consists of plains and terraces decreasing in elevation from east to west. It has a semi-arid, semi-humid continental climate influenced by the monsoons of the northern temperate zone. The average annual temperature is approximately 4.6 °C and exhibits considerable diurnal variation. Winters are cold and dry, and spring temperatures rise quickly with little precipitation and high evaporation rates. The average annual precipitation of 577.3 mm is concentrated from May to September, and the potential evaporation is 752.8 mm, decreasing from the northwest to southeast.

The primary water system of the target region is a part of the Drinking Horse River basin in the Songhua River basin. The primary water-bearing layer is the Quaternary aquifer, which can be classified into unconfined and confined aquifers based on burial conditions. The thickness of the Quaternary aquifer increases gradually from the southeast to northwest of the region, with lithology primarily consisting of gravel and sand with local deposits of clay and silt. The Quaternary unconfined aquifer is widely distributed with a thickness generally between 15 and 50 m, though it may be thicker in some areas. The aquifer's sediment granularity is coarser in the Yitong and Drinking Horse River basins due to fluvial deposition, which improves permeability. The hydraulic gradient of the unconfined aquifer decreases with the water table depth, which generally ranges from 1.5 to

15 m in most areas. Critically, this type of aquifer is easily affected by surface pollution infiltration. The geographical location of the study area and sampling point distribution within are shown in Figure 2.

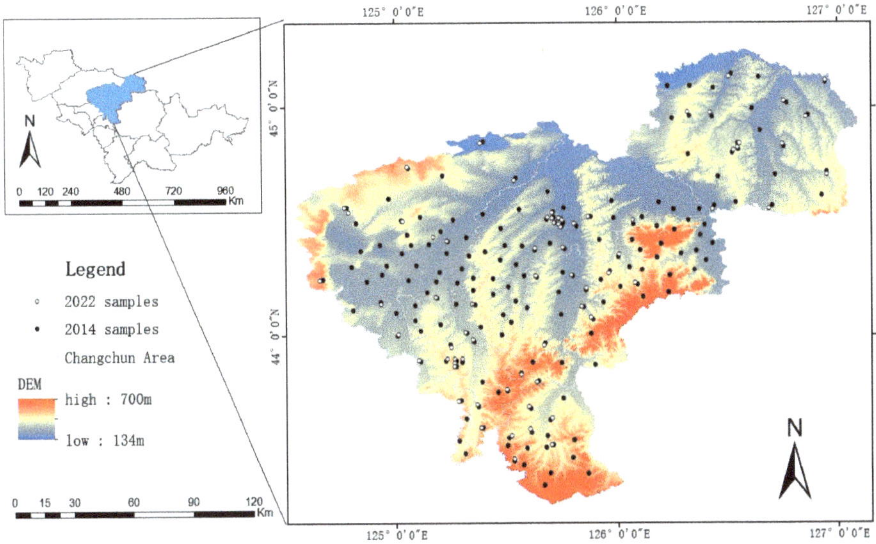

Figure 2. Location of the study area and distribution of sampling points.

2.2. Data Source

This study used groundwater test data collected by our research group during previous work in the study area in 2014 and 2022. All groundwater samples were collected during the wet season (July and August) from similar distribution depths; the sampling sites used in 2022 were a subset of those used in 2014. Data were collected from 209 unconfined sampling points of Quaternary aquifer, and 23 water quality indicators were evaluated comprising routine indicators (pH, total hardness (TH), and total dissolved solids (TDS)), eight major ions (K^+, Ca^{2+}, Na^+, Mg^{2+}, HCO_3^-, CO_3^{2-}, Cl^-, and SO_4^{2-}), three forms of nitrogen (NO_3^-, NO_2^-, and NH_4^+), geogenic indicators (F^-, Fe, Mn, and As), and heavy metals (Hg, Cr^{6+}, Pb, S, and Al). The methods used to obtain these indicators are detailed in Table 1.

Table 1. Water indicators and associated test methods.

Indicator	Test Method	Indicator	Test Method
TH	Ethylenediaminetetraacetic acid titration [14]	HCO_3^-, CO_3^{2-}	Acid–base titration [15]
TDS	Gravimetric method [16]	NH_4^+	Nessler's reagent spectrophotometry [17]
pH	Electrode method [18]	NO_2^-	N-(1-Naphthyl)ethylenediamine dihydrochloride spectrophotometry [19]
Ca^{2+}, Mg^{2+}, K^+, Na^+, Cl^-, SO_4^{2-}, NO_3^-, F^-	Ion chromatography [20,21]	Fe, Mn, As	Inductively coupled plasma optical emission spectrometry [22]
Hg	Atomic fluorescence spectrometry [23]	Pb	Graphite furnace atomic absorption spectrometry [24]
Cr^{6+}	Diphenylcarbazide spectrophotometry [25]	S^{2-}	Methylene blue spectrophotometry [26]
Al	Spectrophotometry (chrome azurol S) [27]		

2.3. Groundwater Evaluation Methods

2.3.1. Improved DRASTICL Model

The DRASTIC model, which is widely used for groundwater vulnerability assessment, is based on seven key evaluation parameters: groundwater depth (D), net recharge (R) directly related to climate and rainfall, aquifer rock type (A), soil texture (S) and topographic slope (T) surface conditions, unsaturated zone rock type (I), and hydraulic conductivity (C). Because the study area lies in a transition zone between mountain and plain regions, the variation in aquifer thickness is not a major factor influencing groundwater vulnerability; therefore, the analysis of aquifer thickness (A) was excluded from this case study. The dynamic effects of freeze–thaw cycles in cold regions on aquifer permeability were considered in this study by adding a land use type (L) parameter to quantify human disturbances such as de-icing agents and agricultural activities, creating a more regionally adaptive DRASTICL model [28,29] defined as follows:

$$DRASTICLindex = D_r D_W + R_r R_w + S_r S_w + T_r T_w + I_r I_w + C_r C_w + L_r L_w \quad (1)$$

where the subscript r represents the rating value of each parameter and the subscript w represents the weight of each parameter.

Thus, the *DRASTICL* vulnerability index can be obtained by calculating and summing the products of the rating values and weights; the higher the index value, the greater the groundwater vulnerability and risk of contamination.

2.3.2. Random Forest Method

The RF algorithm provides an efficient ensemble learning method that comprises dual random sampling of the observations and feature variables in the modeling dataset using the "bootstrap" resampling technique, which generates multiple new sample sets by repeatedly sampling with replacements from the original data [30]. This approach typically creates subsets of the same size as the original dataset, with some observations potentially appearing multiple times and others potentially omitted. This process is generally repeated 1000 to 10,000 times to estimate the uncertainty or distribution of statistical metrics and thereby randomly extract multiple subsets from the original dataset to build a series of decision trees. During the decision tree splitting process, the mean decreased Gini coefficient (*MDG*) is calculated to assess the quality of feature split points as follows:

$$Gini(i) = 1 - \sum_{j=1}^{|y|} \left(X(i,j) / \sum_{j=1}^{n} X(i,j) \right)^2, i \in d \quad (2)$$

where y denotes the total number of distinct categories or classes in the target variable, *X(i,j)* represents the *i*-th index of the *j*-th variable and *d* represents the node number of the regression tree [31].

Each decision tree iterates through splitting until it generates a classification result that matches the tree properties. Finally, the RF algorithm integrates the classification wisdom of all decision trees and uses a voting mechanism to determine the final classification result, improving classification accuracy and reliability.

2.3.3. Entropy-Weighted Groundwater Quality Index (E-GQI)

The E-GQI uses water quality indicators to represent the influence of geological and anthropogenic activities on overall water quality. It assesses water quality based on standard allowable limits and has gained widespread application among scholars owing to its reliability and objectivity. However, the E-GQI considers many different water quality indicators, the selection of which can be subjective. Therefore, this study selected the key indicators based on the importance ranking of features from the RF algorithm and coupled them with a vulnerability index to form a comprehensive evaluation system. The weight of each indicator was set using the entropy weighting method to quantify its dispersion through information entropy. This weighting method relies entirely on the statistical distribution of the water quality monitoring dataset, avoiding the risk of subjective bias inherent to conventional methods. The E-GQI approach applied in this study can be expressed as follows:

$$X_i = \frac{w_i}{\sum_i^k w_i} \tag{3}$$

$$M_i = \frac{m_i \times 100}{A_i} \tag{4}$$

$$GQ_i = X_i \times M_i \tag{5}$$

$$GQI = \sum_{i=1}^{k} GQ_i \tag{6}$$

where w_i denotes the weight of indicator i, X_i denotes its relative value, M_i denotes its evaluated quality level, m_i denotes its measured value, A_i denotes its permissible standard value, GQ_i denotes its quality index, and k denotes the number of evaluated indicators. This study classified GQI values of <50, 50–100, 100–200, 200–300, and >300 as corresponding to water quality categories of "Good," "Fair," "Moderate," "Poor," and "Very Poor," respectively [32]. The water quality indicators and their classification limits adopted in this study are summarized in Table 2.

Table 2. Water indicators and classification limits [33].

Indicator \ Lmits	Class I	Class II	Class III	Indicator \ Lmits	Class I	Class II	Class III
TH	≤150	≤300	≤450	SO_4^{2-}	≤50	≤150	≤250
TDS	≤300	≤500	≤1000	HCO_3^-		-	
pH		6.5 ≤ pH ≤ 8.5		NH_4^+	≤0.02	≤0.1	≤0.5
Ca^{2+}		-		NO_3-	≤2	≤5	≤20
Mg^{2+}		-		NO_2^-	≤0.01	≤0.1	≤1
K^+		-		F^-	≤1	≤1	≤1
Na^+	≤100	≤150	≤200	Fe	≤0.1	≤0.2	≤0.3
Cl^-	≤50	≤150	≤250	Mn	≤0.05	≤0.05	≤0.1
				As	≤0.001	≤0.001	≤0.01

2.3.4. Self-Organizing Map (SOM)

The SOM is an unsupervised learning artificial neural network that learns from data in the input space and maps high-dimensional input data to a low-dimensional discrete space (typically a two-dimensional grid) to provide data clustering and visualization. The SOM network structure consists of input and output layers. Continuous iterative training with a competitive learning algorithm gradually adjusts the weights of the output layer neurons until they accurately reflect the data distribution in the output layer [34]. The structure of a classic SOM network is shown in Figure 3. This study integrated the SOM with the K-means clustering method to overcome the limitations of traditional clustering techniques when identifying water chemistry features. In this application, K-means clustering optimizes the classification boundaries based on the SOM outputs to enhance the interpretability of the water chemistry dataset and provide critical risk factor selection criteria for the subsequent HHRA. Before generating an SOM, three key parameters must be set: (1) the number of neurons, (2) the error quantification parameter for evaluating the topology, and (3) the optimal number of clusters determined by silhouette scoring [35,36].

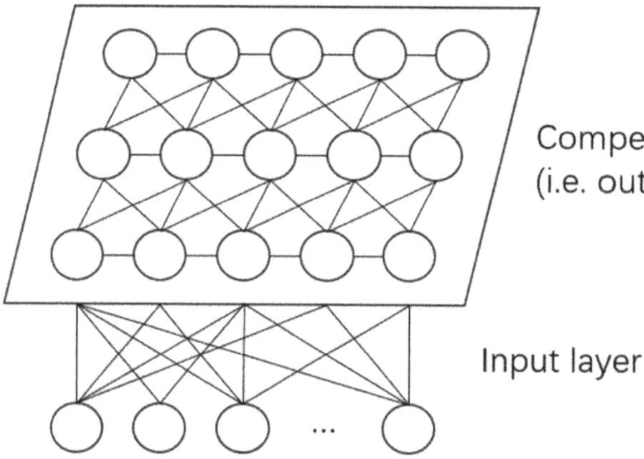

Figure 3. SOM network structure.

2.3.5. Groundwater Health Risk Assessment

The HHRA estimates the health risks of harmful pollutants entering the human body through various exposure pathways. Considering the primary uses of groundwater in the case study region as well as individual indicator variations in the collected data, this study analyzed the drinking water exposure pathway and used the KDE-TFN-MCSS and FSM to, respectively, perform probabilistic and deterministic HHRAs by calculating the non-carcinogenic risk (*HI*) and carcinogenic risk (*CR*) for children and adults [37,38]. The overall health risk from multiple pollutants was considered assuming that the risks from individual pollutants are independent and the total risk follows an additive relationship [39].

The distribution of mis-specification risks in TFN-MCSS was assessed in this study by coupling it with the KDE during groundwater HHRA. For a known sample size n, each sample in $(X_1, X_2, \ldots, X_n)$ was assumed to be independent of the others and follow the same continuous distribution $F(x)$. Under these conditions, the KDE for the unknown kernel density can be defined as follows:

$$\hat{f}_h(x) = \frac{1}{nh}\sum_{i=1}^{n} K\left(\frac{x - X_i}{h}\right) \tag{7}$$

where $\hat{f}_h(x)$ represents the function of the unknown kernel density; the constant h represents the smoothing coefficient of the smoothing curve, also known as the window width; and $K(\cdot)$ represents the kernel density given on the real number domain and is referred to as the kernel function for the unknown probability density.

The results of the KDE depend on the applied kernel function and window width. The specific choice of kernel function does not significantly affect the result; this study used the Gaussian kernel function. When the dataset is close to a normal distribution, the optimal $h = 1.06\sigma n^{-\frac{1}{5}}$ can be used; however, if the dataset is asymmetric or has multimodal features, the optimal h must be adjusted as follows:

$$h = 0.9\min\{\sigma, IQR/1.34\}\, n^{-\frac{1}{5}} \tag{8}$$

where σ represents the standard deviation of the dataset and *IQR* represents its interquartile range.

The PDFs of the relevant indicators were calculated based on the KDE results using the R 4.4.2 software to determine the boundaries of the core and support intervals, which were used to define the parameters a, b, c, and d for the TFN. Next, the obtained TFN data were converted into interval values using the α-TST truncation technique. If $\forall a$ ($a \in [0, 1]$), TFN $\tilde{A} = (a, b, c, d)$ can be converted into interval values [40] as follows:

$$A^a = \{x|, \mu A(x) \geq a, x \in X\} \tag{9}$$

$$A^a = (a_L^a, a_R^a) = [(b - a)a + a, -(d - c)a + d] \tag{10}$$

in which the membership value a used in this calculation was 0.9.

The Oracle Crystal Ball 11.1.2.4 (OCB) software was used to define probability distributions for the input indicators leveraging the derived TFN parameters and their converted interval values. The software was configured to execute MCSS using a predefined number of trial iterations, typically 10,000, to propagate the uncertainties inherent in the model. Furthermore, OCB was instrumental in generating predictive contours that encompassed frequency distribution charts and cumulative probability plots for the obtained *HI* and *CR* outcomes.

The *HI* and *CR* for drinking water exposure were calculated as follows:

$$CDI_i = \frac{C_i \times IR \times EF \times ED}{BW \times AT} \tag{11}$$

$$CR_i = CDI_i \times SF \tag{12}$$

$$HQ_i = \frac{CDI_i}{RfD} \tag{13}$$

$$CR = \sum_{i=1}^{n} CR_i \tag{14}$$

$$HI = \sum_{i=1}^{m} HQ_i \tag{15}$$

where C_i is the concentration of groundwater contaminant i in mg/L, *IR* is the daily drinking water intake in L/d, *EF* is the exposure frequency, *ED* is the number of years of lifetime exposure to a toxic substance (a), *BW* is the body weight in kg, *AT* is the average exposure time in days, and *RfD* is the reference dose in mg/(kg·d). The specific values of the exposure parameters applied for the non-carcinogenic risk assessment are provided in Table 3.

Table 3. Exposure parameter values for non-carcinogenic risk.

Exposure Parameter	Reference Value	
	Adult	Child
IR (L/d)	[1.48, 1.72]	[0.96, 1.04]
BW (kg)	[61, 69]	[19, 23]
ED (years)	24	6
AT (days)	8760	2190
Rfd_{As} [mg/(kg·d)]	3×10^{-4}	3×10^{-4}
Rfd_{Cl^-} [mg/(kg·d)]	0.10	0.10
Rfd_{NH4^+} [mg/(kg·d)]	0.97	0.97

According to the USEPA health risk assessment standards, the reference threshold for *HI* is 1 and that for *CR* is 0.0001.

3. Results and Discussion

3.1. Groundwater Chemical Characteristics

3.1.1. Water Chemistry Indicator Statistics

The statistical variations in the water quality indicators over time were explored using box plots to analyze the routine indicators (TH, TDS, pH), cations (K^+, Ca^{2+}, Na^+, Mg^{2+}), anions (HCO_3^-, Cl^-, SO_4^{2-}), nitrogen compounds (NO_3^-, NO_2^-, NH_4^+), and geogenic indicators (Fe, F^-, Mn, As), as shown in Figure 4. According to the thresholds specified in the Standard for Groundwater Quality (GB/T 14848-2017) (Table 2) [33], the monitoring data indicated that 37% of the case study area had groundwater quality below Class III, with the primary pollutants comprising TH, NO_3^-, NO_2, NH_4^+, Fe, and Mn. In both sampling periods, the indicators exceeding the Class III threshold comprised TH, TDS, pH, Na^+, SO_4^{2-}, NO_3^-, NO_2^-, NH_4^+, F^-, Fe, Mn, and As, but the exceedance rates were lower in 2022 than in 2014, reflecting overall improvement in water quality. Specifically, the contents of the routine indicators, eight major ions, three nitrogen compounds, and geogenic indicators all decreased between 2014 and 2022, with a notable reduction in groundwater salinity primarily owing to increased precipitation in 2022, which accelerated groundwater renewal and dilution. However, the concentrations of geogenic indicators remained above the standard limits, and the exceedance rate of As even increased, indicating a high level of pollution in the area. Furthermore, the monitoring data from 2022 indicated that the thresholds for all three nitrogen compounds were exceeded at 15 groundwater monitoring points. This pollution exhibited point-source diffusion following a strip-like migration pattern, with the average NH_4^+ concentration increasing significantly from 0.06 to 1.26 mg/L to exceed the Class III limit of 1.0 mg/L, indicating an accelerating trend in NH_4^+ pollution in local areas that can be attributed to increased agricultural runoff. Notably, NH_4^+ can, under certain conditions, convert into nitrite in water. The consumption of nitrate over the long term can produce strong carcinogenic substances that pose severe health risks to humans and harm aquatic life.

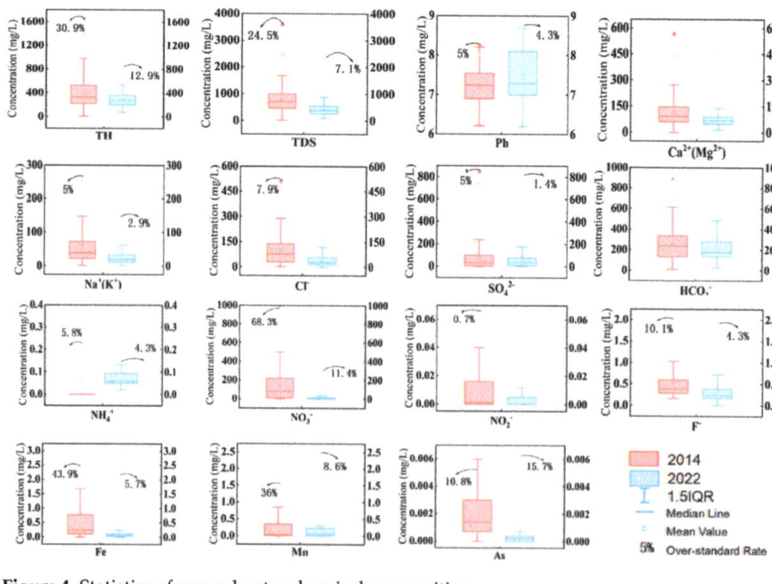

Figure 4. Statistics of groundwater chemical composition.

3.1.2. Water Chemistry Types

The Durov diagrams of the case study area's groundwater in 2014 and 2022, shown in Figure 5, were plotted using the Origin 2022 software to determine the groundwater chemistry types. The results indicated that the water chemistry type in this area in both years was primarily HCO_3^--Ca^{2+} with moderately weak alkaline water. The TH in the study area generally fell within the moderately high range, indicating hard water. Furthermore, the pH range shifted from 6.2–8.2 in 2014 to 6.2–8.8 in 2022, indicating increased alkalinity.

Finally, statistical analyses revealed that the groundwater TDS and salinity in 2022 were lower than those in 2014 owing to increased precipitation in the study area in 2022, which accelerated groundwater renewal and dilution.

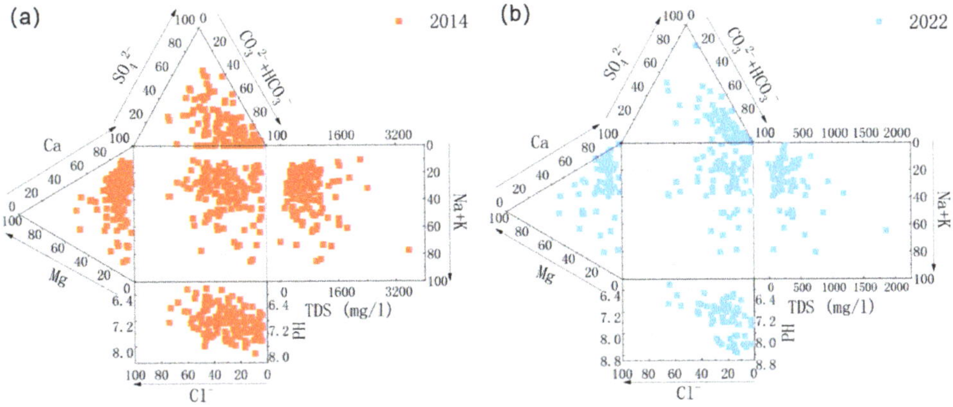

Figure 5. Groundwater Durov diagrams for (**a**) 2014 and (**b**) 2022.

3.1.3. Water Chemistry Origin Analysis

The Gibbs diagram can be used to determine the characteristics of groundwater evolution owing to evaporation crystallization, rock dissolution, and multi-factor influences [41], as shown in Figure 6. Rock dissolution and evaporation crystallization were the two primary factors influencing groundwater chemical evolution in the case study area. Ion ratio analysis indicated that groundwater chemistry was primarily controlled by silicate rock dissolution, with evaporation crystallization being a secondary factor. The observed decrease in groundwater salinity was partly attributed to reduced evaporation crystallization in 2022 compared to 2014.

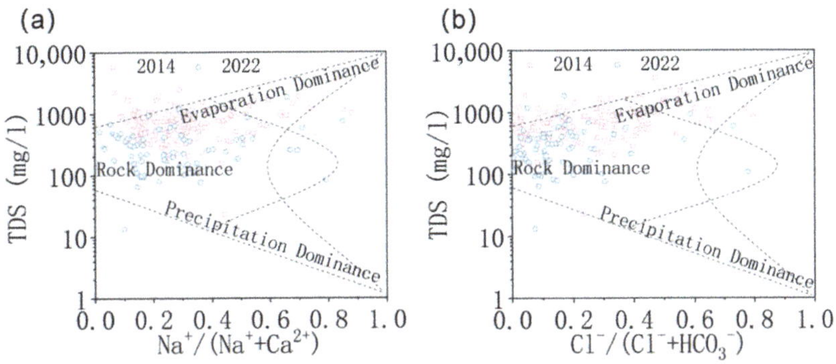

Figure 6. Groundwater Gibbs diagrams for (**a**) 2014 and (**b**) 2022.

3.2. Groundwater Quality Assessment

3.2.1. Groundwater Vulnerability

This study used hydrogeological surveys and field observation data in combination with regional geological and environmental characteristics to classify and assign values to groundwater vulnerability assessment parameters. The ArcGIS 10.8.2 geostatistical analysis module was employed to plot the rating map for each parameter, and a combined weighted sum approach was used to generate the groundwater vulnerability index distribution map using the DRASTICL model (Figure 7) [42]. The groundwater vulnerability in the study area was classified into five levels from high to low. Higher and moderate vulnerability areas were the most common and widely distributed in the central and western regions; lower vulnerability areas were the second most common and primarily concentrated in the northeast; low vulnerability areas were the least common and only sporadically appeared in the northeast. The overall case study area did not exhibit any high-vulnerability regions. The results of this assessment were used as factors in the subsequent water quality evaluation.

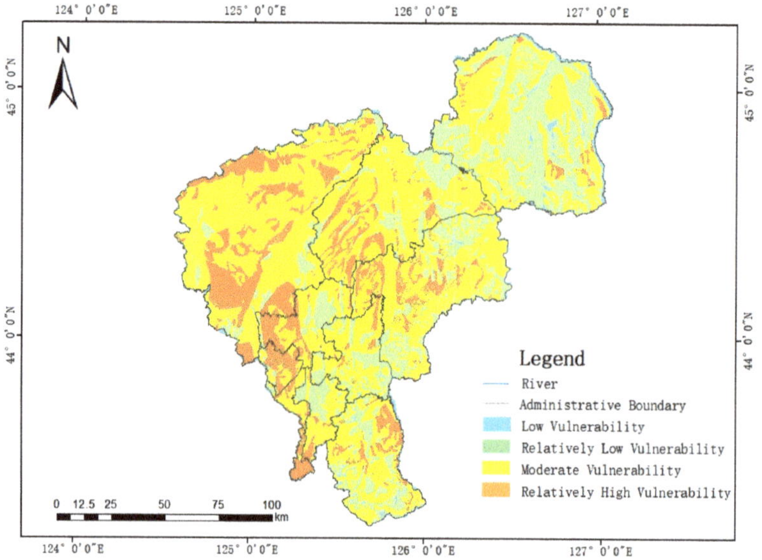

Figure 7. Groundwater vulnerability index distribution.

3.2.2. Selection of Groundwater Quality Indicators

A feature importance analysis of water quality indicators was performed using the RF algorithm in the SPSSPRO software as described in Section 2.3.2, resulting in the ranking shown in Figure 8. The top 16 ranked indicators were Mg, Mn, NH_4^+, As, NO_3^-, TDS, Fe, Cl^-, TH, Ca, SO_4^{2-}, K, F^-, HCO_3^-, Na, and NO_2^-. Among these indicators, the geogenic and nitrogen indicators had relatively high importance rankings. These key indicators and the vulnerability index discussed in Section 3.2.1 were selected as groundwater quality evaluation factors [43].

3.2.3. Groundwater Quality Assessment Results

The top ten groundwater quality indicators ranked by vulnerability index and feature importance were selected as evaluation factors, and their weights were determined using the E-GQI discussed in Section 2.3.3.

The weighted quality indices were aggregated using Equations (3)–(6), and spatial interpolation was performed using Kriging in ArcGIS 10.8 to generate the groundwater quality distribution maps shown in Figure 9 for spatial pattern analysis. In 2014, groundwater quality in the case study area was generally poor and predominantly classified into Classes IV and V, with limited occurrences of Class III in the western and northeastern regions and no instances of Classes I or II. By 2022, groundwater quality had generally improved, predominantly falling into Classes I–IV following a banded distribution from southwest to northeast. However, slight deterioration was observed in some central and northern zones, where Class V persisted.

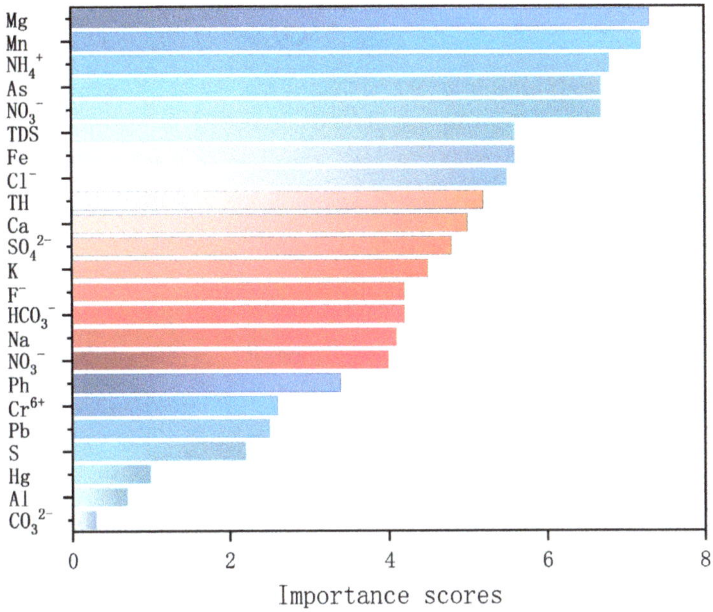

Figure 8. Feature importance ranking of water quality indicators.

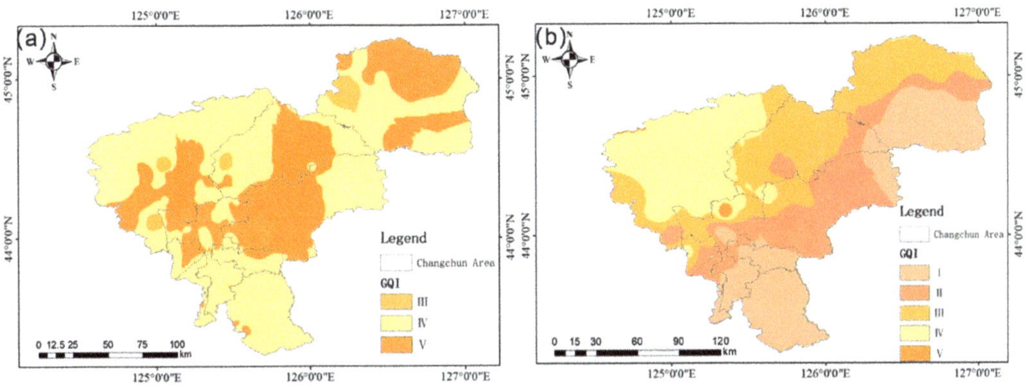

Figure 9. E-GQI water quality distribution maps for (**a**) 2014 and (**b**) 2022.

3.3. *Groundwater Health Risk Assessment*

3.3.1. Source Apportionment of Groundwater Contaminants

An SOM analysis was conducted using the R software with optimal neuron counts empirically determined by $5\sqrt{n}$. Topological structures comprising 7×8 and 6×7 were constructed using a learning rate of 0.7 with 10,000 iterations to generate neuron distribution maps for 2014 and 2022 (Figures 10 and 11).

Figure 10. Neuron SOMs for 2014 groundwater quality indicators. (Note: Red and blue colors represent high and low ion concentrations, respectively, in the groundwater samples mapped to the neurons. The yellow, green, purple, and blue colors in the lower-right corner represent Categories 1, 2, 3, and 4, respectively).

Figure 11. Neuron SOMs for 2022 groundwater quality indicators (Note: Red and blue colors represent high and low ion concentrations, respectively, in the groundwater samples mapped to the neurons. The yellow, green, purple, and blue colors in the lower-right corner represent Categories 1, 2, 3, and 4, respectively).

In 2014, the concentrations of TH, TDS, Ca^{2+}, Na^+, and Cl^- exhibited similar spatial distributions, suggesting a co-clustering mechanism driven by natural groundwater dissolution processes as well as anthropogenic chloride inputs from de-icing salts and chlorinated industrial effluents [44–46]. Notably, the study area lies on the southeastern margin of the Songliao Basin, where Cretaceous sandstone formations contribute Ca^{2+}

through limestone weathering, thereby increasing TH. As Ca^{2+} and Cl^- tend to migrate synergistically by forming soluble $CaCl_2$ complexes, enhancing their spatial correlation, their high-concentration zones were predominantly observed along urban industrial belts and major traffic corridors, where chlorinated effluents and de-icing salts are present in winter. The primary sources of Na^+ include domestic sewage (e.g., sodium from table salt) and industrial discharges (e.g., detergents). The HCO_3^-, Mg^{2+}, and SO_4^{2-} indicators exhibited similar distributions. The HCO_3^- primarily originated from carbonate dissolution or CO_2–rock interactions, the SO_4^{2-} from gypsum ($CaSO_4$) dissolution or sulfide oxidation, and the Mg^{2+} from dolomite or silicate weathering, collectively reflecting the dominant influence of geochemical weathering. The As, NH_4^+, NO_3^-, NO_2^-, and K^+ indicators exhibited distinct localized hotspots within different regions of the SOM topology, with low overall concentrations but high point values. These anomalies were driven by various mechanisms: the high As^{3+} concentrations were primarily of geogenic origin; the presence of NO_3^- and NH_4^+ reflected nitrogen leaching from fertilizers and domestic wastewater; the NO_2^- anomalies were attributed to incomplete nitrification of fertilizers or sewage [47,48], but could also result from dry and wet deposition following coal combustion [49,50]; the high K^+ levels were attributed to enrichment from suburban agricultural potassium fertilizer application or sewage leakage; and finally, elevated levels of Fe and Mn were also geogenic and predominantly controlled by the reducing conditions within the primary aquifer system.

In 2022, the TH, TDS, Ca^{2+}, Mg^{2+}, and Cl^- indicators exhibited co-located high-concentration zones, particularly around the urban periphery and near industrial corridors. These patterns can be attributed to long-term overextraction of groundwater, which promotes preferential dissolution of dolomite ($CaMg(CO_3)_2$) from Cretaceous aquifers, replacing earlier calcite-dominated hardness. Additionally, persistent application of chloride-based de-icing agents and leakage from expanded sewage pipelines contributed to elevated Cl^- levels, indicating a shift in the mineral weathering sequence of the aquifer. Furthermore, the correlation between TDS and SO_4^{2-} concentrations was significantly stronger in 2022 and exhibited enhanced spatial consistency, primarily owing to the continuous dissolution of the sulfates, such as $CaSO_4$, in Quaternary alluvial sediments, as well as increased replacement of traditional organic fertilizers with sulfur-based fertilizers, which promoted SO_4^{2-} transport via soil infiltration. Other localized high-value indicators, such as As, Fe, and Mn, although largely geogenic, may also share common input sources with NH_4^+ or be subject to coupled geochemical processes, such as co-release under reducing aquifer conditions.

The results of SOM-based source tracing indicated a transition in groundwater hardness composition from Ca^{2+}-dominated to Ca^{2+}–Mg^{2+}-synergistic control between 2014 and 2022, driven by Cl^- input from winter de-icing agents and preferential dissolution of dolomite in aquifers under cold-region conditions [51]. Moreover, the intensification of SO_4^{2-} pollution increased its contribution to TDS levels, whereas concentrations of Fe, Mn, As, and NH_4^+ remained spatially localized. In particular, the "polarized" migration of point-source As and NH_4^+ pollution was restricted, exacerbating spatial heterogeneity in regional health risks.

3.3.2. Deterministic Health Risk Assessment

Based on the water quality screening and SOM-based source apportionment results, NH_4^+, As, and Cl^- were selected as representative contaminants for a deterministic assessment of the cumulative health risks associated with groundwater ingestion. The evaluation results are shown in Figures 12 and 13. This assessment was conducted using the USEPA's FSM to calculate the *HI* and *CR* levels. The spatial distributions of these risks for children and adults were interpolated using ArcGIS 10.8.2. The results indicated that, under the ingestion exposure pathway, both *HI* and *CR* exhibited similar spatial trends for adults and children: overall, most of the study area presented acceptable risk levels for both populations. In 2014, elevated health risks were observed in the central and eastern regions, where *HI* was higher for children than for adults but *CR* was higher for adults than for children. In 2022, *HI* exceeded the threshold in the central-western region, and *CR* surpassed the limit in the northern region. The cumulative *HI* remained obviously higher in children and *CR* exhibited a slight predominance in children, both with comparable magnitudes. The observed difference between the *HI* levels for adults and children may be attributed to children's lower body weight and underdeveloped immune systems.

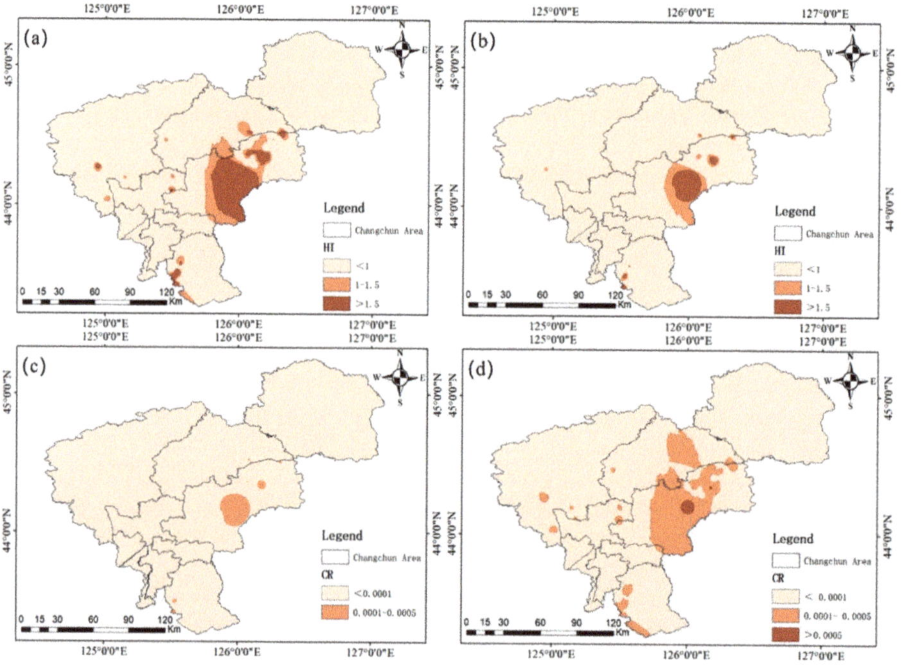

Figure 12. FSM-based deterministic health risk assessment map for 2014: (**a**) *HI* assessment map for children, (**b**) *HI* assessment map for adults, (**c**) *CR* assessment map for children, (**d**) *CR* assessment map for adults.

3.3.3. Probabilistic Health Risk Assessment

(1) TFN and Interval Value Construction Based on KDE

This study employed KDE with a Gaussian kernel function to estimate the probabilistic distributions of NH_4^+, As, and Cl^- concentrations in 2014 and 2022, resulting in the PDFs illustrated in Figure 14. The points with the maximum positive and negative slopes on the rising and falling edges of the KDE curve were designated as "b" and "c," forming the core of the TFN; points where the KDE slope dropped below 10% of the maximum on each side were defined as "a" and "d," forming the support interval. The resulting TFNs are shown in Figure 15. The fuzziness of contaminant concentration decreases as the value of "a" increases and is reflected by the narrowing of the interval between the upper and lower bounds. Given the variability in the groundwater data from the different years, most TFNs were trapezoidal in shape. However, when the most probable interval value (TMPIV) was narrow, some TFNs approximated triangles with extremely short upper bases.

Using α-TST, each TFN was converted into an interval value, with the results presented in Table 4. These interval values were subsequently used as input for the MCSS to generate probabilistic risk assessment outcomes [10].

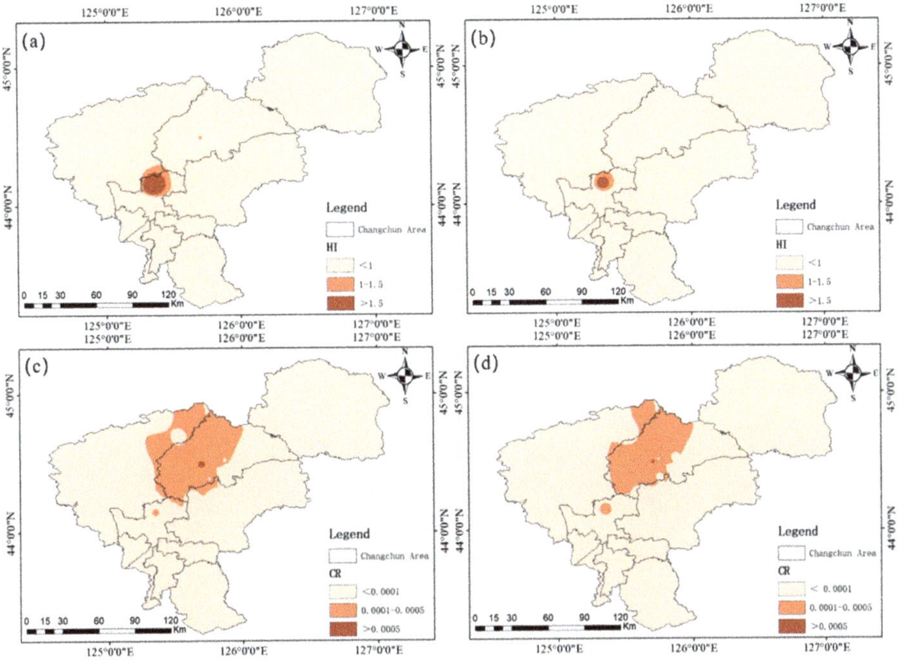

Figure 13. FSM-based deterministic health risk assessment map for 2022: (a) *HI* assessment map for children, (b) *HI* assessment map for adults, (c) *CR* assessment map for children, (d) *CR* assessment map for adults.

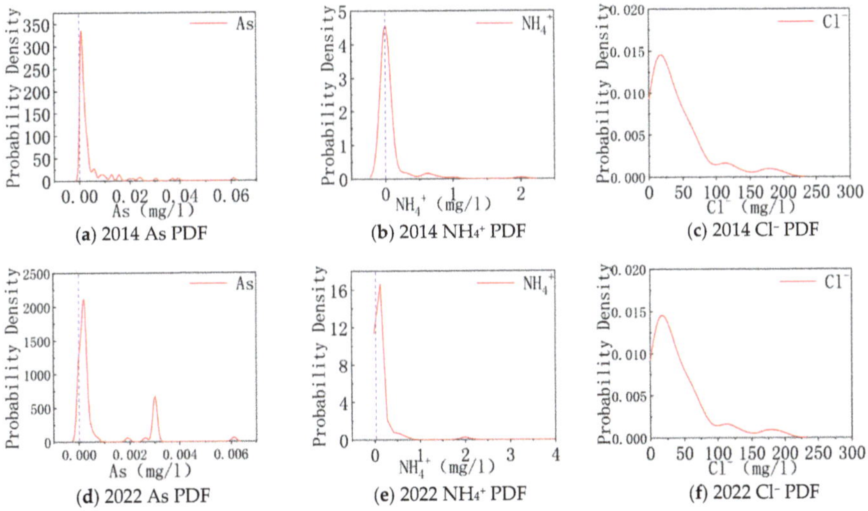

Figure 14. KDE PDF plot.

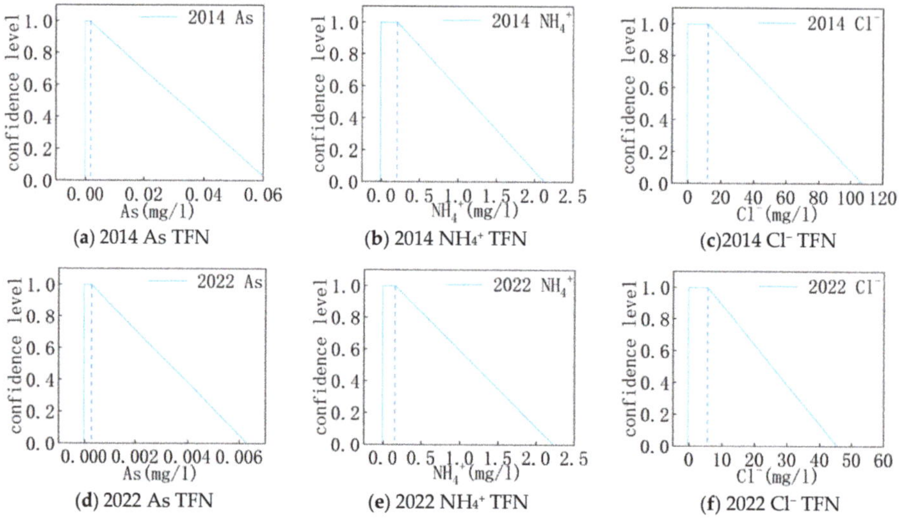

(**a**) 2014 As TFN (**b**) 2014 NH_4^+ TFN (**c**) 2014 Cl^- TFN
(**d**) 2022 As TFN (**e**) 2022 NH_4^+ TFN (**f**) 2022 Cl^- TFN

Figure 15. TFN plot.

Table 4. Results of TFN conversion to interval values.

Year	2014		2022	
Interval Value (mg/L)	$a^a{}_L$	$a^a{}_R$	$a^a{}_L$	$a^a{}_R$
As	0	0.007873775	0	0.000914422
NH_4^+	0	0.41469583	0	0.378673057
Cl^-	0	22.1351929	0	9.775605

(2) Health Risk Assessment Results Based on KDE-TFN-MCSS

The probabilistic health risks for children and adults in 2014 and 2022 were estimated using the proposed KDE-TFN-MCSS model; the results are shown in Figure 16. The 95th percentile of the simulated output was adopted as the high-risk threshold. The probability distributions of the total *HI* and *CR* for both age groups were left-skewed, indicating higher probabilities of low to moderate risk levels. Except for the 2022 *CR* in children, the 95th percentile values for all scenarios exceeded the acceptable risk threshold. A comparison between years revealed that overall health risks in 2022 were lower than those in 2014, with a slight increase observed only in the adult *CR*. Within the same year, children were exposed to higher *HI* levels than adults, whereas adults were exposed to significantly higher *CR* levels.

Both the FSM and KDE-TFN-MCSS models revealed spatial heterogeneity and uncertainty in groundwater health risks across the central Songnen Plain. The deterministic FSM assessment indicated that the *HI* levels in central-eastern regions exceeded the safety threshold ($HI > 1$) in 2014. By 2022, the high-risk zones expanded westward to the central-western region, and the *CR* levels in parts of the north surpassed the threshold ($CR > 0.0001$), which corresponded closely with the increase in NH_4^+ concentrations from 0.06 to 1.26 mg/L and geogenic As enrichment. However, as the FSM is based on fixed-point estimates, it fails to quantify parameter variability, potentially underestimating extreme risks in localized high-concentration areas. By contrast, the proposed probabilistic assessment model used KDE to derive the distributions of NH_4^+ and As, incorporated TFNs to account for parameter fuzziness, and used an MCSS to reveal that, except for the *CR* for children in 2022, the 95th percentile values for *HI* and *CR* exceeded the acceptable thresholds. Furthermore, the left-skewed risk distributions suggested a higher probability of low to moderate risks, reflecting the spatial heterogeneity of NH_4^+ pollution driven by combined agricultural and domestic sources, along with localized enrichment of As and Cl^-.

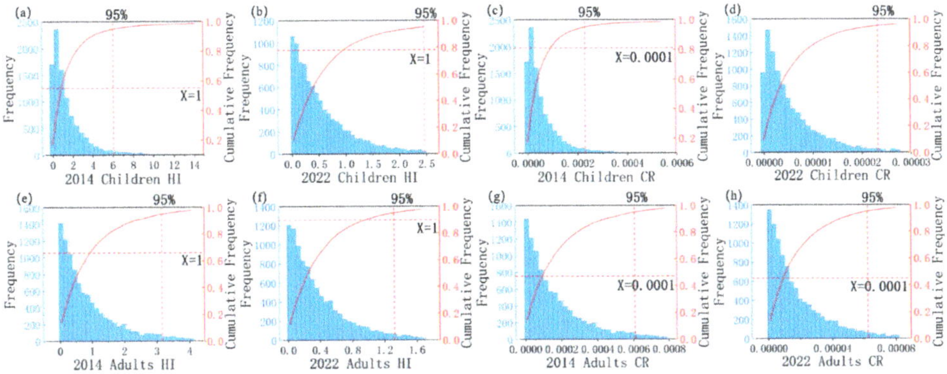

Figure 16. Probabilistic risk assessment plot.

Additionally, both models consistently exhibited higher *HI* levels in children and higher *CR* levels in adults. However, the probabilistic model quantified risk probabilities more precisely. For example, the 95th percentile of the *CR* for adults in 2022 was 0.00012, slightly exceeding the safety threshold and indicating that adults are more susceptible to long-term carcinogenic accumulation, whereas children's lower body weight and immature immune systems render them more vulnerable to non-carcinogenic hazards. The divergence between the results obtained by the two evaluated methods highlights the limitations of deterministic models in cold-region aquifer systems, where freeze–thaw cycles alter aquifer permeability, enhancing pollutant migration and increasing parameter variability. For example, the water quality deterioration observed in 2022 in the central region can be attributed to de-icing salt infiltration and preferential thaw-induced pore development in the vadose zone. In this complex environmental context, the use of a probabilistic assessment method, which integrates both natural and anthropogenic uncertainties, proved more suitable for risk characterization.

3.4. Comprehensive Analysis

In 2022, areas of groundwater quality deterioration and elevated *HI* levels in the central Songnen Plain spatially overlapped with high-vulnerability zones identified by the DRASTICL model, indicating that regions with greater vulnerability were more susceptible to anthropogenic influence. Indeed, shallow aquifers, highly permeable vadose zones, and intensive agricultural land use facilitated the vertical migration of NH_4^+ and Cl^-, directly linking water quality degradation to increased health risks. Furthermore, SOM-based source identification revealed that NH_4^+ anomalies coincided with leakage from decentralized rural sanitation systems, whereas high Cl^-–Mg^{2+} co-concentrations in industrial zones promoted the formation of soluble $CaCl_2$ complexes, enhancing pollutant diffusivity. Further analysis indicated that the mineralization of organic matter and the hydrolysis of urea in agricultural return flow significantly contributed to the NH_4^+ load in the aquifer, especially during the post-fertilization and irrigation seasons. This led to NH_4^+ accumulation and elevated *HI* levels in highly vulnerable areas of the central-western region. Simultaneously, SOM analysis revealed that the spatial distribution of As-related CR hotspots was closely associated with specific hydrogeochemical conditions, notably the prevalence of reducing environments and the presence of aquifers with high natural arsenic background levels. Although mean concentrations of As declined in 2022, its high carcinogenic slope factor (SF = 1.5) continued to drive *CR* exceedances in localized areas, particularly among adults with prolonged exposure. Sulfate reduction under reducing conditions as well as Cl^- input were found to promote As mobilization, confirming that freeze–thaw cycles enhance hydraulic connectivity in aquifers, thereby facilitating As activation. By contrast, low-vulnerability zones in the eastern region—characterized by deep-seated, compact aquifers—maintained Class I or II water quality with the lowest risk levels, underscoring the natural buffering role of geological barriers [36].

Freeze–thaw cycles in cold plains regions seasonally alter permeability, accelerating Cl^- and NH_4^+ infiltration and enhancing As release under reducing conditions, producing a synergistic amplification of natural and anthropogenic risks. This spatially coupled mechanism of "high vulnerability–water quality degradation–pollutant input" reflects a systemic transformation of groundwater risk in the central Songnen Plain from natural weathering dominance to strong anthropogenic interference [52].

4. Conclusions

This study used groundwater quality monitoring data collected from the case study area in 2014 and 2022 to derive vulnerability indices and evaluate aquifer vulnerability using a modified DRASTICL model considering regional anthropogenic activities and land use types. The relative importances of key water quality indicators were determined via the RF algorithm to conduct an E-GQI evaluation, and an SOM analysis was undertaken to characterize groundwater pollution patterns. The refined KDE-TFN-MCSS model was proposed to reduce the uncertainty and limitations associated with the TFN-MCSS model for health risk assessment and enhance the accuracy of probabilistic evaluation. Finally, the results of the conventional FSM deterministic evaluation were compared with those of the proposed probabilistic KDE-TFN-MCSS model when assessing the groundwater environment of the central Songnen Plain, providing theoretical support for winter-period pollution control in the northeastern industrial regions of China. The key conclusions of this study are as follows:

(1) The dominant hydrochemical type in the study area was determined to be HCO_3^-–Ca^{2+}, reflecting moderately to weakly alkaline groundwater. No significant shifts in water chemistry were observed between 2014 and 2022, and water hardness remained high. Rock weathering and evaporative crystallization were identified as the primary factors driving hydrochemical evolution. The indicators exceeding their thresholds comprised TH, TDS, pH, Na^+, SO_4^{2-}, NO_3^-, NO_2^-, NH_4^+, F^-, Fe, Mn, and As. Although exceedance rates declined and overall water quality improved from 2014 to 2022, geogenic and nitrogenous pollutants remained significant. Notably, NH_4^+ concentrations rose markedly from 0.06 mg/L in 2014 to 1.26 mg/L in 2022, exceeding the Class III limit (1.0 mg/L).

(2) In 2014, groundwater quality was poor overall, dominated by Class IV and V water with minor distribution of Class III water in the west and northeast and no Class I–II water. In 2022, Class I–IV water predominated, exhibiting a southwest–northeast banded pattern. Although general water quality improved, localized deterioration was observed in central and northern zones.

(3) The study area exhibited a systemic shift from natural weathering dominance to increasing anthropogenic disturbance. The results of the SOM analysis revealed localized high concentrations of As and NH_4^+ indicative of legacy industrial and agricultural pollution, whereas the Cl^- distribution reflected urbanization and de-icing agent inputs.

(4) The NH_4^+, As, and Cl^- indicators were selected for health risk assessments using the conventional FSM and proposed KDE-TFN-MCSS models. Under the ingestion

pathway, most areas in the central Songnen Plain exhibited acceptable risk levels for both children and adults, with a higher probability of low to moderate risks. The *HI* levels were consistently higher in children, whereas the *CR* levels were higher in adults. Although overall groundwater quality improved in 2022, the unique coupling of freeze–thaw cycles and urbanization in the case study cold region led to persistent NH_4^+ and As accumulation in localized areas. Therefore, zonal management of de-icing zones and legacy industrial belts in the case study region is recommended to balance groundwater development with ecological safety.

The findings of this study informed the following general recommendations for groundwater management in cold regions: (1) implement zonal management strategies targeting high-vulnerability areas to restrict intensive agricultural activities and reduce NH_4^+ and Cl^- inputs from de-icing agents and fertilizers; (2) establish dynamic monitoring networks focusing on seasonal freeze–thaw cycles to track pollutant migration and As mobilization in reducing environments; and (3) prioritize remediation in zones where carcinogenic and non-carcinogenic risks exceed thresholds, especially for children and adults with long-term exposure.

The results of this study confirmed that the proposed KDE-TFN-MCSS model provides a robust tool for quantifying uncertainty in health risks, enabling policymakers to design adaptive policies that address both anthropogenic and geogenic pollution. Future research should integrate climate projections to evaluate how warming temperatures and extreme precipitation events alter groundwater quality dynamics in cold regions.

Author Contributions: Conceptualization, J.L. and Y.W.; methodology, J.L. and Y.W.; software, J.L.; validation, J.L., Y.W. and X.F.; formal analysis, data curation, X.S.; writing—original draft preparation, J.L.; writing—review and editing, J.L. and Y.W.; visualization, J.L. and X.F.; supervision, Y.W.; project administration, J.B.; funding acquisition, J.B. All authors have read and agreed to the published version of the manuscript.

Funding: This study was funded by the National Key Research and Development Program of China (Grant No. 2022YFD1500500) and the Projects of Jilin Provincial Department of Science and Technology (Grant No. 20240101059JC). The authors thank all the experts who helped with this article.

Data Availability Statement: The original contributions presented in this study are included in the article. Further inquiries can be directed to the corresponding author.

Conflicts of Interest: The authors declare no conflicts of interest. The funders had no role in the design of the study; in the collection, analyses, or interpretation of data; in the writing of the manuscript; or in the decision to publish the results.

References

1. Dharmani, B.A.; Verma, M.; Rani, S.; Narang, A.; Singh, M.R.; Saya, L.; Hooda, S. Unravelling groundwater contamination and health-related implications in semi-arid and cold regions of India. *J. Contam. Hydrol.* **2024**, *261*, 104303.
2. Sanjupriya, S.; Poonkothai, M.; Karunanidhi, D.; Rao, N.S.; Subramani, T.; Marghade, D. Evaluating nitrate contamination in groundwater and its health threats from a semi-arid province of southern India using GIS techniques with a special focus on entropy water quality index. *Environ. Geochem. Heal.* **2025**, *47*, 346. [CrossRef]
3. Ahn, S.H.; Jeong, D.H.; Kim, M.; Lee, T.K.; Kim, H.K. Prediction of groundwater quality index to assess suitability for drinking purpose using averaged neural network and geospatial analysis. *Ecotoxicol. Environ. Saf.* **2023**, *265*, 115485. [CrossRef]
4. Jia, X.Q.; Luo, M.M. Groundwater vulnerability assessment in the typical karst watershed of Xiangxi River based on an improved DRASTIC model. *Geol. Sci. Technol. Inf.* **2019**, *38*, 255–261.
5. Chen, S.Y.; Fu, G.T.; Zhou, H.C.; Wang, G.-L. Fuzzy analytical evaluation model and method for aquifer vulnerability. *J. Hydraul. Eng.* **2002**, 23–30.
6. Yu, L.H.; Tao, Z.B.; Hu, S.T.; Gao, S.; Yang, N.; Yu, X.J. Application of the DRASTIC model in groundwater vulnerability assessment. *Yellow River* **2020**, *42*, 45–46+50.

7. Xu, C.; Zhou, J.Y.; He, X.J.; Yang, Z.-T.; Zhao, K.-Z.; Jiang, Y.-N. Groundwater vulnerability assessment in Shaanxi Province based on an improved DRASTIC model. *China Rural. Water Hydropower* **2020**, *3*, 44–51.
8. Liao, X.U.; Li, Y.; Miranda-Avilés, R.; Puy-Alquiza, M.J.; Bian, J.; Anguiano, J.H.H.; Muñoz, A.H.S.; Datta, S.; Zha, X.; Liu, J.; et al. Assessments of Pollution Status and Human Health Risk of Potentially Toxic Elements in Primary Crops and Agricultural Soils in Guanajuato, Mexico. *Water Air Soil Pollut.* **2023**, *234*, 670. [CrossRef]
9. Tenebe, I.T.; Emenike, P.C.; Babatunde, E.O.; Neris, J.B.; Fred-Ahmadu, O.H.; Dede-Bamfo, N.; Etu, E.-E.; Ogarekpe, N.M.; Emakhu, J.; Benson, N.U. Assessing the state of rainwater for consumption in a community in dire need of clean water: Human and health risk using HERisk. *Water Pract. Technol.* **2022**, *17*, 2005–2022. [CrossRef]
10. Li, T.; Bian, J.; Ruan, D.; Xu, L.; Zhang, S. Groundwater health risk assessment and its temporal and spatial evolution based on trapezoidal fuzzy number-Monte Carlo stochastic simulation: A case study in western Jilin province. *Ecotoxicol. Environ. Saf.* **2024**, *282*, 116736. [CrossRef]
11. Ruan, D.; Bian, J.; Wang, Y.; Wu, J.; Gu, Z. Identification of groundwater pollution sources and health risk assessment in the Songnen Plain based on PCA-APCS-MLR and trapezoidal fuzzy number-Monte Carlo stochastic simulation model. *J. Hydrol.* **2024**, *632*, 130897. [CrossRef]
12. Yan, J.; Chen, J.; Zhang, W. A new probabilistic assessment process for human health risk (HHR) in groundwater with extensive fluoride and nitrate optimized by non parametric estimation method. *Water Res.* **2023**, *243*, 120379. [CrossRef]
13. Mohammadpour, A.; Keshtkar, M.; Samaei, M.R.; Isazadeh, S.; Khaneghah, A.M. Assessing water quality index and health risk using deterministic and probabilistic approaches in Darab County, Iran; A machine learning for fluoride prediction. *Chemosphere* **2024**, *352*, 141284. [CrossRef] [PubMed]
14. *GB/T 7477-1987*; Water Quality—Determination of Calcium and Magnesium—Atomic Absorption Spectrometric Method. Standardization Administration of the People's Republic of China: Beijing, China, 1987.
15. *DZ/T 0064.49-1993*; Methods for Analysis of Groundwater Quality—Part 49: Determination of Carbonate, Bicarbonate and Hydroxide Ions—Titrimetric Method. Ministry of Geology and Mineral Resources of the People's Republic of China: Beijing, China, 1993.
16. *GB/T 5750.4-2006*; Standard Examination Methods for Drinking Water—Organoleptic and Physical Parameters. Standardization Administration of the People's Republic of China: Beijing, China, 2006.
17. *HJ 535-2009*; Water Quality—Determination of Ammonia Nitrogen—Nessler's Reagent Spectrophotometry. Ministry of Environmental Protection of the People's Republic of China: Beijing, China, 2009.
18. *HJ 1147-2020*; Water Quality—Determination of pH Value—Electrode Method. Ministry of Ecology and Environment of the People's Republic of China: Beijing, China, 2020.
19. *GB/T 7493-1987*; Water Quality—Determination of Nitrogen (Nitrite)—Spectrophotometric Method. Standardization Administration of the People's Republic of China: Beijing, China, 1987.
20. *HJ 812-2016*; Water Quality—Determination of Water-Soluble Anions (F^-, Cl^-, $NO_2{}^-$, Br^-, $NO_3{}^-$, $PO_4{}^{3-}$, $SO_3{}^{2-}$, $SO_4{}^{2-}$)—Ion Chromatography Method. Ministry of Ecology and Environment of the People's Republic of China: Beijing, China, 2016.
21. *HJ 84-2016*; Water Quality—Determination of Inorganic Anions (F^-, Cl^-, $NO_2{}^-$, Br^-, $NO_3{}^-$, $PO_4{}^{3-}$, $SO_3{}^{2-}$, $SO_4{}^{2-}$)—Ion Chroma-tography Method. Ministry of Ecology and Environment of the People's Republic of China: Beijing, China, 2016.
22. *HJ 776-2015*; Water Quality—Determination of 32 Elements—Inductively Coupled Plasma Optical Emission Spectrometry. Ministry of Environmental Protection of the People's Republic of China: Beijing, China, 2015.
23. *HJ 694-2014*; Water Quality—Determination of Mercury, Arsenic, Selenium, Bismuth and Antimony—Atomic Fluorescence Spectrometry. Ministry of Ecology and Environment of the People's Republic of China: Beijing, China, 2014.
24. *GB/T 5750.6-2006*; Standard Examination Methods for Drinking Water—Metal Parameters. Standardization Administration of the People's Republic of China: Beijing, China, 2006.
25. *GB/T 7467-1987*; Water Quality—Determination of Chromium (VI)—1,5-Diphenylcarbohydrazide Spectrophotometric Method. Standardization Administration of the People's Republic of China: Beijing, China, 1987.
26. *GB/T 16489-1996*; Water Quality—Determination of Sulfide—Methylene Blue Spectrophotometric Method. Standardization Administration of the People's Republic of China: Beijing, China, 1996.
27. *HJ 586-2010*; Water Quality—Determination of Cyanide Easily Released—Spectrophotometric Method. Ministry of Environmental Protection of the People's Republic of China: Beijing, China, 2010.
28. Saravanan, S.; Pitchaikani, S.; Thambiraja, M.; Sathiyamurthi, S.; Sivakumar, V.; Velusamy, S.; Shanmugamoorthy, M. Comparative assessment of groundwater vulnerability using GIS-based DRASTIC and DRAS-TIC-AHP for Thoothukudi District, Tamil Nadu India. *Environ. Monit. Assess.* **2022**, *195*, 57. [CrossRef] [PubMed]
29. Sarkar, M.; Pal, S.C. Application of DRASTIC and Modified DRASTIC Models for Modeling Groundwater Vulnerability of Malda District in West Bengal. *J. Indian Soc. Remote. Sens.* **2021**, *49*, 1201–1219. [CrossRef]

30. Wang, Z.; Wang, Y.; Shi, M.; Ji, W.; Li, R.; Wang, X. Coordinated analysis of groundwater spatiotemporal chemical characteristics, water quality, and potential human health risks with sustainable development in semi-arid regions. *Environ. Geochem. Heal.* **2024**, *46*, 370. [CrossRef]
31. Gu, Z.Q.; Bian, J.M.; Wang, Y.; Sun, X.; Ruan, D. Groundwater quality assessment and index optimization of water quality monitoring in the water source area of Changbai Mountain. *China Environ. Sci.* **2023**, *43*, 5257–5264.
32. Alasiri, A.; Kumar, P.M. Investigating the influence of solar distillation on improving the groundwater quality index in the southern region of Riyadh, Saudi Arabia. *Groundw. Sustain. Dev.* **2024**, *25*, 101155. [CrossRef]
33. *GB/T 14848-2017*; Standard for Groundwater Quality. Standardization Administration of the People's Republic of China: Beijing, China, 2017.
34. Yuan, W.; Wang, Z.; Zhang, T.; Liu, Z.; Ma, Y.; Xiong, Y.; An, F. Assessment and Prediction of Groundwater Vulnerability Based on Land Use Change—A Case Study of the Central Urban Area of Zhengzhou. *Water* **2024**, *16*, 3716. [CrossRef]
35. Balamurali, M.; Silversides, K.L.; Melkumyan, A. A comparison of t-SNE, SOM and SPADE for identifying material type domains in geological data. *Comput. Geosci.* **2019**, *125*, 78–89. [CrossRef]
36. Yan, Y.; Zhang, Y.; Yang, S.; Wei, D.; Zhang, J.I.; Li, Q.; Yao, R.; Wu, X.; Wang, Y. Optimized groundwater quality evaluation using unsupervised machine learning, game theory and Monte-Carlo simulation. *J. Environ. Manag.* **2024**, *371*, 122902. [CrossRef] [PubMed]
37. Zhang, Y.Y.; Zhou, Y.Z.; Zhou, J.L.; Zeng, Y.; Ding, Q.; Jiang, F. Hydrochemical characteristics and health risk assessment of groundwater in the Barkol Lake Basin Plain, Xinjiang. *J. Irrig. Drain.* **2024**, *43*, 95–104.
38. Liang, Z.Q.; Yang, J. Health risk and ecological risk assessment of ammonia nitrogen in surface water of Tianjin. *Environ. Sci. Surv.* **2024**, *43*, 68–73.
39. Xu, L.W.; Bian, J.M.; Sun, X.Q.; Lou, Y.Q.; Sun, G.J. Impact of irrigation return flow on regional groundwater quality and health risk assessment. *China Environ. Sci.* **2023**, *43*, 1688–1695.
40. Debnath, K.; Debnath, P.; Choudhury, S.; Saha, A.K.; Majumdar, A. A framework of trapezoidal fuzzy best-worst method in location selection for surface water treatment plant. *Pollution* **2024**, *10*, 1–15.
41. Chen, W.; Yu, S.W.; Liao, J.; Wang, S.C.; Liu, G.N. Source apportionment and health risk assessment of heavy metals in groundwater in northeastern Nanchang. *Environ. Sci.* **2025**, *46*, 854–863.
42. Azlaoui, M.; Karef, S.; Foufou, A.; Haied, N.; Zeddouri, A.; Bengusmia, D. Integration of land use/land cover factors with machine learning in groundwater vulnerability assessment models for semi-arid regions Algeria. *Desalin. Water Treat.* **2025**, *323*, 101256. [CrossRef]
43. Xiong, Y.; Zhang, T.; Sun, X.I.; Yuan, W.; Gao, M.; Wu, J.; Han, Z. Groundwater Quality Assessment Based on the Random Forest Water Quality Index—Taking Karamay City as an Example. *Sustainability* **2023**, *15*, 14477. [CrossRef]
44. Jiang, Q.; Liu, Q.; Liu, Y.U.; Zhu, J.; Chai, H.; Chen, K. Chemical composition of groundwater and its controlling factors in the Liuzhuang coal mine, Northern Anhui Province, China. *Water Supply* **2023**, *23*, 4937–4956. [CrossRef]
45. Rudolph, D.L.; Melchin, J.; Stone, M.; Sarwar, G.; Hodgins, E. Efficacy of urban road salt reduction strategies on public supply well quality. *Sci. Total. Environ.* **2023**, *900*, 166466. [CrossRef] [PubMed]
46. Ming, R.T. Research on Migration and Accumulation of Deicing Agents in Ski Resorts Based on SWAT Model. Master's Thesis, Beijing University of Civil Engineering and Architecture, Beijing, China, 2021.
47. Li, Y. Research on River Health Assessment in Changchun City Based on PSR and FAHP-CRITIC Models. Master's Thesis, Dalian University of Technology, Dalian, China, 2022.
48. Liu, X.Y. Study on Distribution and Transport Patterns of Nitrate-Nitrogen Pollutants in Groundwater of Typical Watersheds in Changchun. Master's Thesis, Jilin University, Changchun, China, 2023.
49. Yang, Z.P.; Lu, W.X.; Liu, X.R.; Xin, X. Source apportionment of heavy metals in topsoil of Changchun urban area. *Urban Environment Urban Ecol.* **2009**, *5*.
50. Wang, X.; Xiao, C.; Liang, X.; Li, M. Groundwater Quality Assessment in the Northern Part of Changchun City, Northeast China, Using PIG and Two Improved PIG Methods. *Int. J. Environ. Res. Public Health* **2022**, *19*, 9603. [CrossRef]
51. Baraza, T.; Cassidy, K.J.; Hasenmueller, E.A. Road salt applications mobilize trace elements from roadside soil to shallow groundwater. *Sci. Total. Environ.* **2024**, *942*, 173435. [CrossRef] [PubMed]
52. Xie, X.; Ji, H.; Yang, Q.; Bao, S.; Han, C.; Qu, W.U. A hybrid approach for identifying the seasonal variation of groundwater quality, source apportionment and health risk in a coastal area driven by natural and anthropogenic factors. *J. Hazard. Mater.* **2025**, *496*, 139283. [CrossRef] [PubMed]

water

MDPI

Article

Surface Water–Groundwater Interactions in a Sahelian Catchment: Exploring Hydrochemistry and Isotopes and Implications for Water Quality Management

Issoufou Ouedraogo [1,2], Marnik Vanclooster [3], Frederic Huneau [4,*], Yuliya Vystavna [5], Seifu Kebede [6] and Youssouf Koussoubé [2]

1 Université Yembila Abdoulaye Toguyeni, Mining Engineering Department, BP 54, Fada N'Gourma P.O. Box 54, Burkina Faso; ouedraogo.issoufou03@gmail.com
2 Université Joseph Ki-Zerbo, Laboratoire Géoscience et Environnement, Ouagadougou P.O. Box 7021, Burkina Faso; youssouf.koussoube@gmail.com
3 Université Catholique de Louvain, Earth and Life Institute, Croix du Sud 2, Box 2, 1348 Louvain-la-Neuve, Belgium; marnik.vanclooster@uclouvain.be
4 Université de Corse Pascal Paoli, CNRS UMR 6134 SPE, Equipe Hydrogéologie, Campus Grimaldi, BP52, 20250 Corte, France
5 International Atomic Energy Agency, Isotope Hydrology Section, Division of Physical and Chemical Sciences, Department of Nuclear Sciences and Applications, Wagramer Strasse 5, P.O. Box 100, 1400 Vienna, Austria; y.vystavna@iaea.org
6 University of KwaZulu Natal, Centre for Water Resources Research, Private Bag X01, Scottsville, Pietemartizburg 3209, South Africa; kebedegurmessas@ukzn.ac.za
* Correspondence: huneau_f@univ-corse.fr

Abstract

The Sahel Transboundary Taoudéni Basin, covering about 20% of Burkina Faso, hosts vital aquifers critical for water security and development. Effective groundwater monitoring is essential for sustainable resource management. In the Kou sub-basin, groundwater quality assessment is increasingly important. This study integrates hydrochemistry, water stable isotopes ($\delta^{18}O$, $\delta^{2}H$), GIS, and multivariate statistics to understand subsurface geochemical processes. A total of 48 samples—43 groundwater and 5 surface water—were analyzed for 19 hydrochemical parameters and isotopes. In surface water, $\delta^{18}O$ ranged from −5.96‰ to −5.09‰, and $\delta^{2}H$ from −37.65‰ to −29.15‰. In groundwater, $\delta^{18}O$ ranged from −5.93‰ to −4.39‰, and $\delta^{2}H$ from −34.62‰ to −25.05‰. The spatial distribution of $\delta^{18}O$ and $\delta^{2}H$ was mapped using inverse distance weighted (IDW) interpolation in ArcGIS 10.8. A $\delta^{2}H$ vs. $\delta^{18}O$ plot showed groundwater values clustered near the Global Meteoric Water Line, indicating minimal evaporation during recharge. Groundwater chemistry was dominated by $Ca^{2+} > Na^{+} > Mg^{2+} > K^{+}$ and $HCO_3^{-} > NO_3^{-} > Cl^{-} > SO_4^{2-}$. Key hydrogeochemical processes include water–rock interaction (leaching, weathering, ion exchange) and anthropogenic pollution. Isotopic signatures reveal heterogeneous recharge sources and aquifer connectivity. These findings enhance the understanding of water sources and geochemical processes in the Kou basin, supporting informed groundwater resource management.

Keywords: surface water-groundwater; hydrochemistry; stable isotopes; Sahel region; Kou basin; Burkina Faso

1. Introduction

Groundwater is a major component of available freshwater, and its use is increasing due to climate change, reduced surface runoff, and growing water shortages in river and lake catchments [1]. According to BGR and UNESCO [2], the interconnections between surface waters and groundwater and human interactions with them call for an integrated management approach. Surface Water (SW) and GroundWater (GW) should therefore be managed coherently and jointly to ensure water supply quality [3,4].

Recent studies have highlighted the role of lacustrine groundwater discharge (LGD) in regulating nutrient dynamics in lakes, with spatial variability controlled by water depth and connectivity to aquifers [5]. Such findings underscore that groundwater discharge can significantly influence water quality and ecological functioning, particularly in floodplain lakes, emphasizing the need to consider similar processes in river and groundwater management.

In the Sudano-Sahelian zone of West Africa, water demand has steadily increased since the 1970s, while rainfall has declined markedly over the same period [6,7]. This reduction in precipitation, largely attributed to climate change, has decreased surface runoff, potentially limiting the sustainability of current and future water resources [8]. Concurrently, demographic growth and associated land use changes have intensified pressure on natural resources, altering hydrological responses across the region [9,10]. The Upper Mouhoun–Sourou (Ms-S) complex, spanning Burkina Faso and Mali, exemplifies these dynamics: previous studies have quantified climate-driven trends in rainfall and runoff [8] and assessed the impacts of land use change on river flows and soil water retention [9,10]. Despite these advances, critical knowledge gaps remain, particularly regarding the integrated dynamics of surface water, groundwater, and socio-economic water demands in the Ms-S complex. Addressing these gaps is essential for sustainable water resource management under evolving climatic and environmental pressures.

Ref. [8] investigated the impacts of climate change on water resources in the Kou basin, predicting decreased rainfall, increased runoff, and reduced groundwater recharge, which may lead to soil leaching and reservoir sedimentation. At the same time, the Mouhoun-Sourou complex plays a crucial role for irrigation, pastoralism, industry, and mining, where surface water and groundwater are jointly exploited to meet diverse water needs. Despite these insights, environmental and climatic pressures continue to reduce spring flows and threaten local ecosystems, highlighting persistent gaps in understanding the interactions between water resources and socio-economic activities in this region.

Several studies, especially in hydrogeology, have aimed to characterize the aquifer dynamics and potential within the Ms-S complex, which roughly corresponds to the southeastern margin of the Taoudeni sedimentary basin [11–16]. According to Derouane and IAEA [17], this southeastern border can be described as a "unique multilayered aquifer," with ancient groundwater and minimal modern recharge. The Taoudeni basin itself is a transboundary system, also known as the Upper Mouhoun basin (previously called the Black Volta Basin), extending across Burkina Faso into Ghana.

Hydrochemical and isotopic data play a valuable role in both the quantitative and qualitative analysis of water resources, providing essential insights for developing sustainable water management strategies. Additionally, hydrogeochemical investigations reveal that trace element concentrations in rivers and groundwater are strongly influenced by rock–water interactions, geothermal inputs, and other natural enrichment mechanisms [18]. These studies highlight the need to disentangle anthropogenic versus natural sources of contaminants when assessing groundwater vulnerability and water quality in semi-arid and endorheic basins. Recently, numerous researchers have utilized geochemical and isotopic methods to investigate water mineral quality, aquifer interactions, as well as the processes of recharge and evaporation affecting water bodies [11,14,19–25]. Kumari et al. [26] highlight that among the physical tracers available, the stable isotopes of oxygen ($\delta^{18}O$) and hydrogen ($\delta^{2}H$) in water molecules are widely used to analyze hydrological and hydrogeological processes. Coplen et al. [27] note that the ratio of heavy to light stable isotopes ($^{18}O/^{16}O$ and $^{2}H/^{1}H$) provides a powerful tool for addressing practical challenges in environmental isotope geochemistry, with applications spanning hydrology, climatology, and biogeochemistry. As water moves through different stages of the hydrological cycle, isotope fractionation occurs, leading to distinct isotopic signatures in each phase [26]. Research by [28–31] has shown that these isotopes behave conservatively during mixing, making them valuable for studying groundwater–surface water interactions, the relative contributions to surface flow, recharge patterns, and estimating water residence times within catchments. When combined with additional parameters such as chloride concentration, electrical conductivity, and temperature, stable isotopes ($\delta^{18}O$ and $\delta^{2}H$) serve as important indicators of hydrogeochemical processes in a region [28,31–33]. This study focuses on the Kou basin, located at the southwestern boundary of the transboundary Taoudéni sedimentary basin in Burkina Faso, to enhance understanding of the region. The sandstone formations within the Kou basin contain important aquifers that supply drinking water to Bobo-Dioulasso and surrounding communities [34]. In recent decades, these groundwater resources have faced growing anthropogenic pressures, including increased extraction rates and higher levels of pollutant inputs into aquatic systems. To aid resource managers in overseeing and managing these challenges, multivariate hydrochemical analyses combined with stable isotope data ($\delta^{18}O$ and $\delta^{2}H$) integrated within a GIS framework have been utilized to map the spatial patterns of pollution and water use.

Groundwater in the Kou basin is increasingly threatened by anthropogenic pressures, climate variability, and land use changes. However, limited research has examined how surface water–groundwater interactions and aquifer connectivity shape water quality at different depths. Among these processes, lacustrine groundwater discharge (LGD) plays a critical role, not only regulating nutrient fluxes but also mediating the transport and enrichment of potentially toxic elements (PTEs), with implications for human health and microbial dynamics in aquatic systems [35]. Understanding these interactions is therefore essential for assessing both the ecological and socio-economic risks associated with groundwater use, particularly in regions subject to intensive agriculture, industrial expansion, or mining activities.

To address these gaps, this study aims to identify and characterize the interactions between surface water and groundwater, assess the hydrochemical and isotopic signatures of water sources, and evaluate how environmental changes affect groundwater quality. By linking these processes, the study seeks to provide critical insights into the vulnerability of groundwater resources and support the development of sustainable, integrated water management strategies in the basin.

2. Materials and Methods

2.1. Study Area

2.1.1. General Information

The study area is the Kou River basin, part of the National Mouhoun basin in Burkina Faso (Figure 1). The surface area of this small watershed of 1800 km^2 represents 0.66% of the country and holds the second largest city of Burkina Faso (Bobo-Dioulasso), a former public irrigated rice perimeter and several expanding informal agricultural zones. The basin is located between Longitudes 4°08′ W and 4°36′ W and Latitudes 10°55′ N and 11°32′ N. The Kou watershed is situated in the Sudanian climatic zone. The climate is characterized by the alternation of two seasons, namely a rainy season (4 to 5 months) that extends from June to October, and a dry season (7 to 8 months) that extends from October to May. The alternation of these seasons is conditioned by the annual movement of the Intertropical Convergence Zone (ITCZ). The alternation of wet and dry seasons is driven by the annual migration of the Intertropical Convergence Zone (ITCZ). Average yearly rainfall in the region ranges from 900 to 1100 mm, while potential evapotranspiration averages around 2000 mm per year [36]. Specifically, Tirogo et al. 16] reported that between 1961 and 2014, the Bobo Dioulasso station recorded an average annual rainfall of approximately 1025 mm. Rainfall decreases from south to north, with isohyets generally spanning between 900 mm and 1100 mm. Monthly temperatures typically vary from 25 °C to 31 °C. Annual rainfall variations are influenced by prevailing climatic conditions [14]. During years with higher precipitation, the isotopic composition of local rainfall tends to show more negative values, which is attributed to cooler temperatures and increased humidity that reduce evaporation [37]. Moreover, the isotopic characteristics of rainfall are closely connected to its atmospheric source. In the Sahel region, rainfall primarily originates from either the Guinean monsoon or easterly disturbances associated with the African Easterly Jet and Tropical Easterly Jet [38]. The dominance of these weather systems during the rainy season directly impacts both the amount of rainfall and its isotopic signature [39].

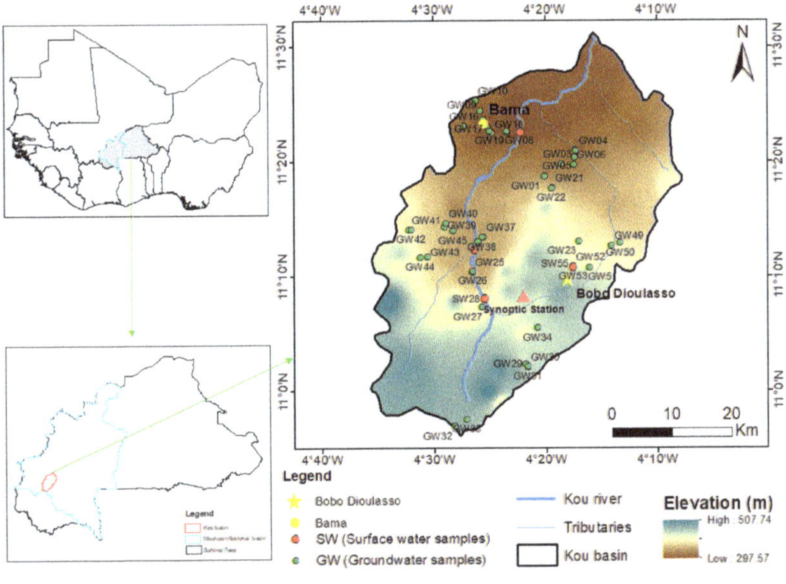

Figure 1. A topographic map showing the distribution of the water sampling. The pink triangle indicates the location of the regional weather station.

Precipitation measurements for the Kou Basin in this research were obtained from the Bobo-Dioulasso synoptic station, spanning the period 1987–2016 (see Figure 2). The soils within the Kou catchment are predominantly ferruginous and ferralitic, with depths ranging from 0.1 to 1.2 m [40]. These soil types are chiefly found in the alluvial plains of the Kou watershed, where they support most of the current hydro-agricultural infrastructures [41].

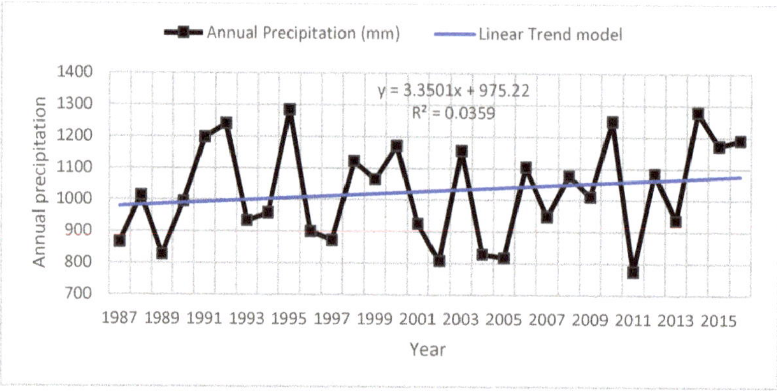

Figure 2. Annual precipitation from 1981–2016.

2.1.2. Geological and Hydrogeological Setting

The geology of the study area primarily consists of a combination of sandstones, shales, and carbonate rocks. Determining the thickness of these lithological formations in the Kou region is challenging, as highlighted by [34]. Ouedraogo [42] identified five distinct geological formations: the Kawara-Sindou sandstone (GKS) with thicknesses between 90 and 350 m; glauconitic fine sandstone (GFG) ranging from 100 to 500 m; quartz-grained sandstone (GGQ) with thicknesses of 300 to 600 m; a formation of siltstones, argillites, and carbonates (SAC1) approximately 300 m thick; and pink fine sandstone (GFR) with a thickness near 100 m (see Figure 3). According to Bronner et al. [43], sediment thickness north of Bobo-Dioulasso may reach up to 2000 m. The basin is also traversed by extensive post-tectonic dolerite dykes, which appear on the surface as chains of hills or isolated hills, oriented from south–southwest to north–northeast. These geological units are highly fractured and segmented by approximately parallel faults with vertical displacements, trending southeast to northwest.

These five geological formations represent the main aquifer systems in the region. Their properties and hydrogeological behavior have been examined in several investigations [11,17,44–46].

Lorenzini [46] describes the Kou basin as being composed of alternating permeable and less permeable geological deposits, which support the formation of multiple aquifer layers. The abundance of faults in the area enhances groundwater flow and establishes important hydraulic links between these aquifers [45]. Additionally, the aquifer system is locally confined due to the presence of clay layers interspersed between the primary aquifer units [16]. In more than two-thirds of the area, the sedimentary aquifer is, nonetheless, unconfined [47]. According to this author, this multilayer aquifer system shelters a substantial water reserve given its sandstone structure and fractures. Tirogo et al. [16] affirm that the thickness of the unsaturated zone varies between 0 and 20 m, but in a few places, it can reach 60 m. In the downstream section of the Kou catchment, within the alluvial plain, the water table lies very close to the surface, typically less than 3 m deep. Within the Kou basin, all piezometers are drilled into the upper aquifer layers, at depths shallower than 200 m [16]. Earlier research often focused on treating the aquifer as a single layer within a multilayer system [11,48], a perspective supported by isotopic analysis conducted in the area [14]. Faults are recognized as playing a significant role in groundwater recharge and in the emergence of water at springs [14,45]. In summary, the study area lies on the southeastern margin of the Taoudeni sedimentary basin, featuring a monoclinal geological structure with a gentle dip of approximately 2° toward the northwest (see Figure 4).

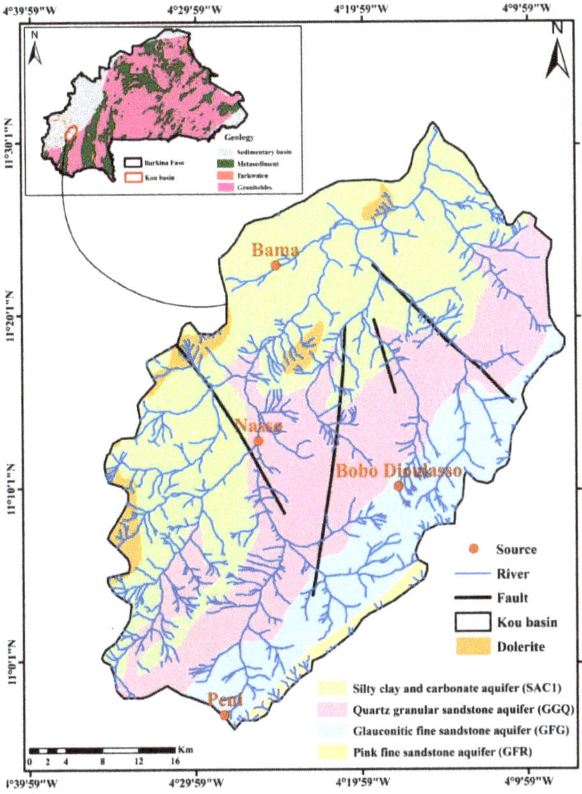

Figure 3. Geologic map in Kou basin.

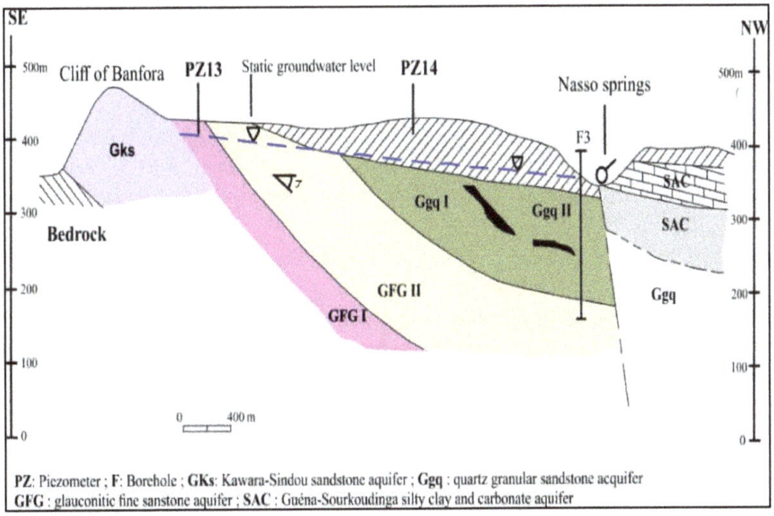

Figure 4. Synthetic geological section of the Bobo-Dioulasso area (Source). Hydrogeological units are referred to by their abbreviated names (see text in Section 2.1.3).

Table 1 provides a summary of the main hydrogeological characteristics of different aquifer levels. The hydrogeological setting in the regional basin of sedimentary deposits of the STBA (Southeastern Taoudeni Basin Aquifer) was extensively documented by [47].

Table 1. Main hydrogeological characteristics of the SBTA (after [47] cited by [14]).

Unit	[a] Number of Borehole	[b] Depth (m)	Mean Yield (m^3/h)	Mean Specific Yield (m^3/h/m)	T (10^{-4} m^2/s)	K (10^{-6} m/s)
GI	21	75	5.1	0.6	0.5	0.2
GKS	69	48	5.1	0.4	2.7	1.2
GFG	166	80	9.1	0.5	2.8	0.5
GGQ	271	62	13.0	1.0	8.5	3.0
SAC1	179	66	13.4	1.4	4.9	2.2
GFR	58	62	6.8	1.3	1.7	1.7

Notes: [a] The number of boreholes refers to the listing made Gombert [47] over the territory of Burkina Faso only. [b] Average depth of the boreholes.

2.1.3. Hydrology and Anthropogenic Pressure

The Kou basin is rich in water resources, including numerous springs, a perennial river, and an accessible shallow aquifer [49]. These reserves, among the most significant in West Africa [16], hold substantial socio-economic importance for Burkina Faso, particularly for Bobo-Dioulasso and surrounding areas. The basin provides drinking water, irrigation for cultivated areas, industrial supply, and ecological functions [36]. Major perennial sources, such as Guinguette (6000 m^3/h), ONEA1 (850 m^3/h), and ONEA2 (550 m^3/h), are critical for meeting the water demand of Bobo-Dioulasso, which now hosts over 900,000 inhabitants [50,51].

Despite this abundance, growing irrigation needs driven by population growth and inefficient practices are depleting both surface and groundwater [52]. Five major irrigated zones lie along the Kou River, where surface water supplies nearly 65% of irrigation demand, exceeding groundwater use [53]. Agricultural expansion and demographic pressure have contributed to a reduction in river low flows, from 2.3 m^3/s in 1960 to 1.6 m^3/s by 2008 [44,46,54]. Although perennial flow is sustained by groundwater recharge from springs near Nasso [55], overexploitation threatens long-term water availability.

These pressures have also generated conflicts over resource allocation. The basin contains nearly 2000 hectares of hydro-agricultural infrastructure, including private gardens and a 1200-hectare rice irrigation scheme. As the water demand of Bobo-Dioulasso intensifies, concerns over water quality have become more pressing. Dione [56] showed that industrialization, combined with inadequate sanitation, leads to untreated discharges of domestic and industrial wastewater. In parallel, the intensive use of fertilizers and pesticides further contributes to water degradation.

Subsequent studies reinforced these findings. Bieupoudé [57] identified vulnerable areas such as the alluvial plains upstream of the Nasso source and downstream of Bobo-Dioulasso, while Ko [58] highlighted industrial effluents as a key source of pollution containing heavy metals, organic matter, and toxic chemicals. Together with agricultural runoff, pesticides, and municipal waste, these discharges constitute major threats to both surface and groundwater in the basin. Overall, human activities have a clear and cumulative impact on water quality, underscoring the urgent need for improved resource management and pollution control measures.

2.2. *Sampling Method and Isotope Analysis*

2.2.1. Sampling Procedures

Water samples were collected from surface water (n = 5) and groundwater (n = 43) in September 2022 during the wet season. Sampling coordinates (latitude, longitude, and altitude) were recorded using a GPS (WGS-84 datum). Physicochemical parameters—including temperature, pH, electrical conductivity, turbidity, dissolved oxygen, and total dissolved solids—were measured in situ after purging the boreholes using a Multi 3430-WTWTM device. Electrodes were rinsed with distilled water prior to each measurement. These parameters help distinguish between different aquifers and identify zones of preferential water supply or potential contamination (Dakouré, 2003; Ben, 2011 [11,59]). Alkalinity was determined in the field by volumetric titration, using a digital titrator HACH (Hach Company, Loveland, CO, USA).

Samples for hydrochemical analysis were collected following IAEA guidelines (IAEA, 2007) [60]. Water was transferred into 125 mL polyethylene bottles, filtered in the field through 0.45 μm syringe filters for cation and anion analysis, and acidified with a few drops of concentrated ultra-pure nitric acid (HNO_3) to preserve cations. All bottles were carefully labeled, stored in coolers during transport, and kept refrigerated until delivery via DHL to the laboratory in France. These procedures ensure sample integrity and reproducibility, consistent with standard protocols for stable isotope and hydrochemical investigations.

2.2.2. Analytical Techniques

Hydrochemical and stable isotope analyses of water samples were performed at the Hydrogeology Department (CNRS UMR 6134 SPE), University of Corsica, France. The concentration of dissolved major ions was determined using a Dionex ICS 1100 chromatograph. Ion balances ranged from −11% to +13%, within the 5% charge balance threshold recommended for quality assurance (Freeze & Cherry, 1979) [61]. These values are consistent with geological variability, as noted by Favreau (2000) [62], who highlighted the influence of bedrock types and weathering on acceptable ion balance limits in West African basins.

For δ^2H and $\delta^{18}O$, samples were analyzed using a LGR IWA-45EP isotope analyzer, with values reported relative to VSMOW (Vienna Standard Mean Ocean Water) and precision better than 1‰ for δ^2H and 0.1‰ for $\delta^{18}O$ (Penna et al., 2010) [63]. Isotopic data were interpreted against the Local Meteoric Water Line (LMWL) derived from Bobo-Dioulasso precipitation data and the Global Meteoric Water Line ($\delta^2H = 8\delta^{18}O + 10$; [64]. Deuterium excess (d-excess = $\delta^2H - 8 \times \delta^{18}O$) was used to assess evaporation and secondary processes (Kumari et al., 2021; Dansgaard, 1964) [26,65].

Hydrochemical facies were represented using a Piper diagram generated with DIAGRAMMES software (Roland Simler, version 31-10-2014, Université d'Avignon) [66], enabling classification of the main water types within the study area.

2.2.3. Spatial Analysis

Spatial distributions of δ^2H, $\delta^{18}O$, and d-excess in groundwater were mapped using the Inverse Distance Weighted (IDW) interpolation method in ArcGIS 10.8. IDW estimates values at unsampled locations based on neighboring measurements, with closer points exerting greater influence. This method allowed the identification of spatial patterns and trends in isotopic composition across the Kou Basin.

3. Results and Discussion

3.1. *Groundwater Depth and Direction of Flows*

The groundwater flow direction has been determined based on depth to groundwater observations collected from different boreholes and wells collected in the Kou basin area. Golden Surfer version 28 was used to translate the XYZ data into a clear surface and contour map, supporting a better understanding of the overall flow direction across the study area. Groundwater in the Kou basin varies ranges from 20 to 200 m (see Figure 5). The map shows that major groundwater flow direction is from south to north of the basin.

Tirogo et al. [16] affirm that the Kou catchment is located upstream of a large sedimentary basin, and the groundwater flow is from southwest to northeast (SW–NE) between 499 m and 286 m.a.s.l., with a gradient of 3‰.

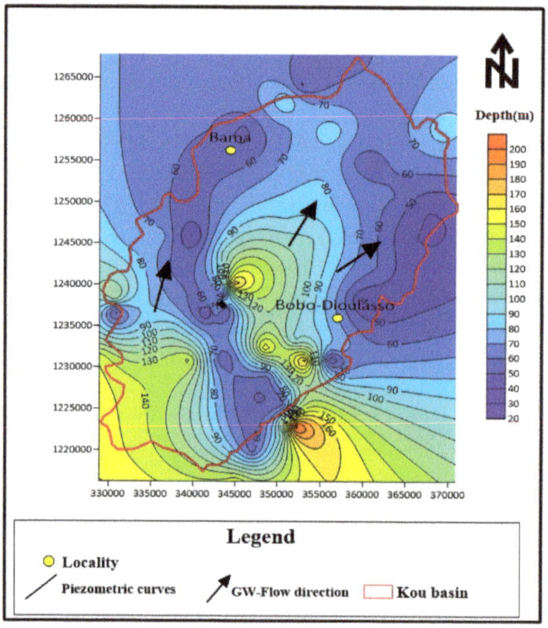

Figure 5. Contour map of the groundwater depth in the study area.

3.2. *Hydrochemical Characteristics*

3.2.1. Descriptive Statistics

This study focuses on freshwater resources, specifically groundwater (boreholes and wells) and surface water within the Kou basin. Descriptive statistics of the measured parameters are summarized in Table 1. Groundwater samples (GW) exhibit pH values ranging from 3.2 to 7.5, while surface water samples (SW) range from 5.4 to 7.1. Electrical conductivity (EC) in GW varies between 8 and 1080 μS/cm, compared to a narrower range of 55.2–83.6 μS/cm in SW. Total dissolved solids (TDS) in GW range from 8 to 1080 mg/L, with an average of 178.74 mg/L, while in SW values extend from 55 to 1616 mg/L. In situ turbidity (Turb) and dissolved oxygen (DO) in GW vary from 0.01 to 126.1 NTU and 3.24 to 6.6 mg/L, with averages of 6.7 NTU and 4.8 mg/L, respectively. For SW, Turb ranges from 2.1 to 843.9 NTU (average 217.4 NTU), while DO values range from 2.6 to 6.88 mg/L (average 5.3 mg/L).

Furthermore, by analyzing results illustrated in Table 2, the relationship between the average concentrations of GW anions was sorted as $HCO_3^- > NO_3^- > Cl^- > SO_4^{2-}$. The concentration of HCO_3^- range from 0.0 mg/L to 309.4 mg/L, with an average concentration of 66.8 mg/L. We observe that the NO_3^- level ranges between 0.1 mg/L to 222.9 mg/L, with an average concentration of 19.1 mg/L. The concentration of Cl^- ranges from 0.1 mg/L to 113.0 mg/L, with an average concentration of 6.8 mg/L. The concentration of SO_4^{2-} ranges from 0.1 mg/L to 11.0 mg/L, with an average concentration of 1.2 mg/L. For the cations, the relationship of the average concentrations in groundwater was $Ca^{2+} > Na^+ > Mg^{2+} > K^+$. Ca^{2+} and Na^+ are dominant, and their concentrations range from 0.1 mg/L to 65.5 mg/L and 0.8 mg/L to 87.7 mg/L, respectively, with average values of 12.8 mg/L and 8.5 mg/L. For other chemical elements, such as F^-, Li^+, Br^-, and NO_2^-, the values obtained during the analysis are below the permissible limits set by the World Health Organization [67] drinking water standards.

Figure 6 displays boxplots illustrating four physical–chemical characteristics (TDS, pH, T and NO_3^-) concerning various lithologies of aquifers in the Kou basin. High concentrations of NO_3^- are observed in GGQ aquifer lithology in the basin. Huneau et al. [14] affirm that in the GFG, nitrate levels can reach concentrations up to 25 mg/L. These concentrations typically indicate poor conditions in the vicinity of the borehole rather than general contamination at the aquifer level. Additionally, the pH, TDS, and temperature in this particular aquifer (GGQ) are lower compared to the other two formations (GFQ and SAC1).

According to Huneau et al. [14], in their study, most samples originating from carbonated lithologies exhibited circumneutral pH values (SAC1, GFG), whereas waters from the most siliceous levels showed a clear tendency toward more acidic pH (GGQ).

Huneau et al. [14] reported that groundwater in the GGQ region generally exhibits the lowest temperatures, typically between 28 and 30 °C, a range that likely indicates its deep source.

In the study area covered by the RAF/7/011 project from IAEA in [68], they observed that pH values generally fell within the normal range of 5 to 7.5 for all formations. However, certain points exhibited acidic pH levels (<4.5), particularly in the sandstone and lemon facies formations. Following Huneau et al. [14], the pH values of most samples from carbonates were close to neutral. Conversely, the pH of water at the most siliceous levels tended to be more acidic.

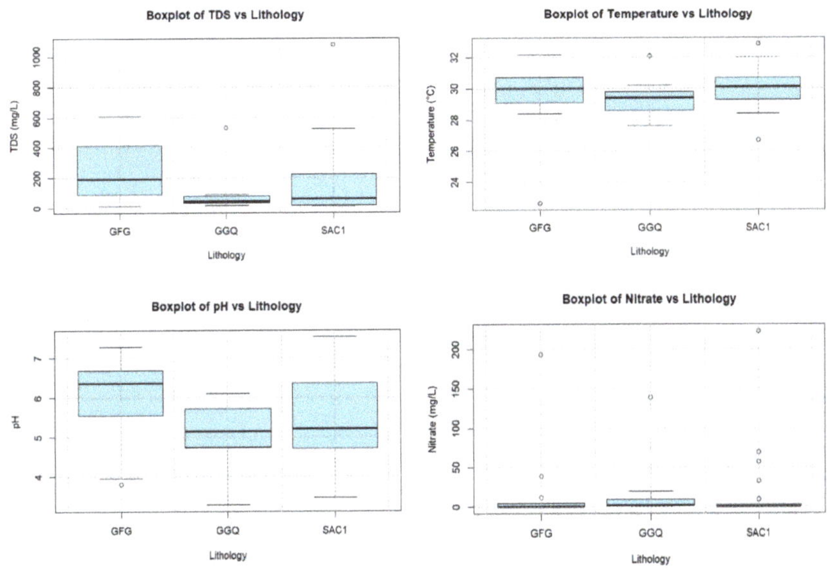

Figure 6. Box plots displaying the four physical–chemical parameters of groundwater from Kou basin according to main lithostratigraphy contexts (GFG: Fine glauconitic sandstones; GGQ: Sandstones with quartz granules; SAC1: Siltstones–claystones–carbonates).

Table 2. Descriptive statistics of hydrochemical parameters and isotope composition in water.

	Temp (°C)	pH (-)	EC (μS/cm)	Turb (NTU)	DO	TDS	HCO_3	F^-	Cl^-	NO_2^-	Br^-	NO_3^-	SO_4^{2-}	Li^+	Na^+	NH_4^+	K^+	Mg^{2+}	Ca^{2+}	$\delta^{18}O$ (‰)	δ^2H (‰)	d-Excess (‰)
Groundwater ($n = 43$)																						
Mean	29.8	5.6	178.3	6.7	4.8	178.3	66.8	0.1	6.8	0.1	0.3	19.1	1.2	0.0	8.5	0.5	3.9	6.0	12.8	−29.7	−5.0	10.4
CV × 100	5.7	20.6	130.5	299.6	19.1	130.5	147.1	75.5	283.5	16.2	17.4	258.3	208.4	69.4	192.2	309.0	112.5	148.2	140.3	−8.0	−7.2	9.0
SD	1.7	1.1	232.8	19.9	0.9	232.8	98.3	0.1	19.3	0.0	0.0	49.3	2.4	0.0	16.4	1.4	4.4	8.9	17.9	2.4	0.4	0.9
Minimum	22.6	3.28	8.0	0.01	3.24	8.0	0.0	0.1	0.1	0.1	0.3	0.1	0.1	0.0	0.8	0.1	0.1	0.0	0.1	−34.6	−5.9	8.3
Maximum	32.9	7.53	1080.0	126.1	6.6	1080.0	309.4	0.3	113.0	0.1	0.5	222.9	11.0	0.0	81.7	6.7	18.3	33.4	65.5	−25.1	−4.4	12.9
Surface water ($n = 5$)																						
Mean	30.2	6.4	447.8	217.4	5.3	447.8	83.3	0.1	37.2	0.1	0.3	15.5	16.0	0.0	35.2	0.2	9.2	3.8	14.9	−31.4	−5.4	11.8
CV × 100	10.9	10.6	149.9	164.7	38.9	149.9	77.1	39.5	158.7	0.0	0.0	165.3	173.5	0.0	175.6	107.8	86.0	48.5	108.2	−11.7	−7.5	11.2
SD	3.3	0.7	671.2	358.1	2.0	671.2	64.2	0.0	59.1	0.0	0.0	25.7	27.8	0.0	61.8	0.2	7.9	1.9	16.1	3.7	0.4	1.3
Minimum	28	5.5	55.2	2.1	2.6	55.0	28.4	0.1	0.6	0.1	0.3	0.0	0.1	0.0	1.6	0.1	2.6	1.6	5.2	−37.7	−6.0	10.1
Maximum	36	7.1	1616.0	843.9	6.9	1616.0	162.1	0.1	137.6	0.1	0.3	59.2	64.7	0.0	144.1	0.5	18.9	5.8	43.1	−29.2	−5.1	13.7
WHO Standards	-	6.5–8.5	1000	5	5–7	1000	350	0.7–1.7	250	0.2	0.1	50	200	1	200	4	12	150	200	-	-	-

3.2.2. Correlation Matrix Analysis

Pearson's correlation analysis carried out in the Kou basin highlights the relationships between various water quality parameters. Table 3 presents the correlation matrix for all hydrochemical variables measured in groundwater samples. The results reveal both positive and negative correlations. Given the sample size of $n = 43$, correlation coefficients (R) are considered statistically significant at the 1% level when they exceed 0.39. Several important hydrochemical correlations (R > 0.39), indicated in Table 3, stand out.

A moderate positive correlation is observed between calcium (Ca^{2+}) and sulfate ($SO_4{}^{2-}$) ions (R = 0.54). Kouanda [8] previously emphasized the predominance of Ca^{2+} over $SO_4{}^{2-}$ in the area, attributing it primarily to the dissolution of evaporite minerals such as gypsum and anhydrite as a secondary process. Additionally, Kouanda pointed out that other geochemical reactions contribute significantly to the Ca^{2+} concentrations in the region's aquifers. The main source of calcium, according to this author, is the dissolution of carbonate minerals including dolomite, calcite, and aragonite. These findings are consistent with our observations and align with the dominance of calcic-magnesium bicarbonate (Ca-Mg-HCO_3) facies, particularly in the Infra-Cambrian basin portion of Burkina Faso. Similar conclusions have been drawn from studies such as [33], who investigated coastal aquifers in India and linked positive Ca^{2+}–$SO_4{}^{2-}$ correlations to anhydrite and gypsum dissolution.

A strong correlation between chloride (Cl^-) and nitrate ($NO_3{}^-$) ions (R = 0.86) points to possible influences from fertilizer leaching and irrigation return flows, which may pose long-term risks for groundwater quality. Serge et al. [69] and Kouanda [8] have both noted that such strong Cl^-–$NO_3{}^-$ correlations often indicate an anthropogenic source of chlorides. This pattern also applies to potassium (K^+), which exhibits a positive correlation with nitrate (R = 0.61). Given the agricultural nature of the study area particularly in rural zones such as the commune of Bama, where heavy use of fertilizers like urea and NPK is common [70,71] these correlations help explain the development of chloride facies observed in the groundwater (see Figure 7).

Other elements, such as Ca^{2+} and Mg^{2+} (R = 0.71), Mg^{2+} and $HCO_3{}^-$ (R = 0.89), and Mg^{2+} and $HCO_3{}^-$ (R = 0.75), exhibit a good and acceptable positive correlation. According to Kouanda [8], these ions share a common origin with minerals Ca^{2+}, Mg^{2+}, and $HCO_3{}^-$, resulting from the dissolution of carbonate minerals, primarily dolomite ($CaMg\,(CO_3)_2$).

The correlation indicates an acceptable relationship between $SO_4{}^{2-}$ and Cl^- (R = 0.58) and between $SO_4{}^{2-}$ and Na^+ (R = 0.58). The presence of chloride and sulfate in groundwater samples in Sahelian Region indicated the existence of an advanced stage of water-rock interaction through the dissolution of sulfate minerals (gypsum and anhydrite) [72].

Table 3. Correlation matrix between physico-chemical parameters in groundwater.

	Temp	pH	EC	Turb	DO	TDS	HCO_3^-	F^-	Cl^-	NO_2^-	Br^-	NO_3^-	SO_4^{2-}	Li^+	Na^+	NH_4^+	K^+	Mg^{2+}	Ca^{2+}
Temp	1																		
pH	−0.16	1.00																	
EC	0.06	0.35	1.00																
Turb	0.09	0.06	−0.08	1.00															
DO	0.10	−0.39	−0.54	0.17	1.00														
TDS	0.06	0.35	1.00	−0.08	−0.54	1.00													
HCO_3^-	0.04	0.70	0.57	−0.06	−0.37	0.57	1.00												
F^-	0.24	0.05	0.28	−0.04	−0.09	0.28	0.26	1.00											
Cl^-	−0.01	−0.01	0.75	−0.05	−0.38	0.75	−0.05	−0.02	1.00										
NO_2^-	0.29	−0.12	−0.12	−0.03	0.14	−0.12	−0.11	−0.07	−0.06	1									
Br^-	−0.11	−0.07	0.10	−0.03	−0.15	0.10	−0.10	−0.07	0.24	−0.03	1.00								
NO_3^-	0.09	−0.20	0.71	−0.04	−0.33	0.71	−0.15	0.25	0.86	−0.06	0.16	1.00							
SO_4^{2-}	−0.16	0.25	0.59	−0.12	−0.46	0.59	0.29	−0.06	0.58	−0.08	0.64	0.38	1.00						
Li^+	0.20	0.10	0.20	0.01	0.12	0.20	0.29	0.71	−0.05	−0.04	−0.04	0.08	−0.09	1.00					
Na^+	0.11	0.19	0.70	−0.08	−0.36	0.70	0.27	0.12	0.69	−0.08	0.11	0.59	0.58	−0.05	1.00				
NH_4^+	−0.12	−0.01	0.51	−0.05	−0.31	0.51	−0.09	−0.07	0.78	−0.05	0.68	0.56	0.70	−0.02	0.38	1.00			
K^+	−0.15	0.14	0.53	−0.10	−0.34	0.53	0.03	0.23	0.53	−0.13	0.26	0.61	0.31	0.12	0.25	0.57	1.00		
Mg^{2+}	0.06	0.57	0.62	−0.04	−0.36	0.62	0.89	0.39	0.01	−0.10	−0.05	0.04	0.21	0.43	0.07	0.00	0.21	1.00	
Ca^{2+}	0.08	0.47	0.92	−0.08	−0.54	0.92	0.75	0.23	0.54	−0.12	−0.04	0.47	0.54	0.18	0.57	0.30	0.31	0.74	1

Note: underline in the table footer: correlations ($R > 0.39$).

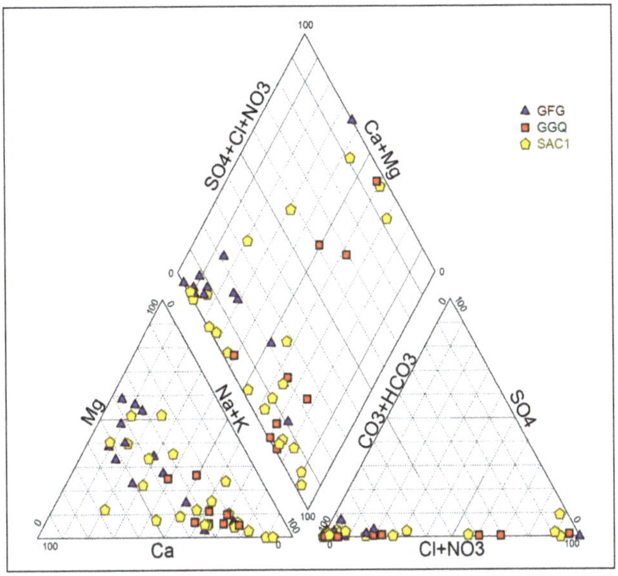

Figure 7. Piper diagram illustrates the groundwater types in relation to main lithology of aquifers.

3.2.3. Groundwater Types Classification Based on the Lithology

Figure 7's Piper diagram displays the diversity of groundwater types corresponding to different litho-stratigraphic units within the study area. Two dominant water types emerge from the plot: Ca-HCO_3 and Na-HCO_3. A previous regional investigation by [14] also identified these water types, noting a transition from Ca–Mg–HCO_3 to Na–K–HCO_3 compositions. According to Huneau and Travi [73], such water signatures likely reflect intensified interactions between groundwater and clay minerals, driven by processes such as isomorphic substitution, cation exchange, and silicate weathering. Moreover, Huneau et al. [14] suggested that these transformations relate to groundwater residence time within the aquifer, serving as an indirect measure of the extent of water–rock interactions.

In the present study, the appearance of a secondary water type characterized by HCO_3–Cl–NO_3–SO_4 composition points to anthropogenic impacts on groundwater quality, as also noted by [14]. They further explained that elevated levels of nitrate, sulfate, and chloride contribute to the progression from the original Ca–Mg–HCO_3 facies toward a Ca–Mg–SO_4–Cl–NO_3 water type. Additionally, our results are consistent with those obtained by [8] regarding the sedimentary aquifer of the Taoudeni Basin in Burkina Faso.

3.3. Isotopic Characterization of Surface Water and Groundwater

3.3.1. Isotope Signature

Table 4 presents the isotopic composition data for groundwater (GW) and surface water (SW) in the Kou basin. Analysis of stable isotope values in groundwater shows $\delta^{18}O$ ranging from −5.93‰ to −4.39‰, with an average value of −5.00‰, while $\delta^{2}H$ varies between −34.62‰ and −25.05‰, averaging −29.65‰. The average d-excess in groundwater is 10.35‰. For surface water, $\delta^{18}O$ values range from −5.96‰ to −5.09‰, with a mean of −5.93‰, and $\delta^{2}H$ spans from −37.65‰ to −29.15‰, averaging −31.4‰. The mean d-excess in surface water is 11.5‰. Additional statistical summaries for $\delta^{2}H$ and $\delta^{18}O$ across different water sample types are provided in Table 1. Variations in isotopic composition indicate different hydrological processes influencing the water sources [74].

Table 4. Stable isotope compositions of GW-SW in the study area.

Sample Name	δ^2H (‰)	δ^{18}O (‰)	d-Excess (‰)
SW20	−37.65	−5.96	10.06
GW13	−34.62	−5.93	12.88
GW25	−34.04	−5.77	12.15
SW28	−31.75	−5.68	13.74
GW21	−32.78	−5.60	12.07
GW38	−34.37	−5.57	10.24
GW22	−32.24	−5.48	11.61
GW45	−33.34	−5.43	10.14
GW19	−32.63	−5.42	10.74
GW04	−32.62	−5.36	10.26
GW42	−32.19	−5.36	10.69
GW35	−31.09	−5.23	10.82
GW36	−32.39	−5.23	9.51
GW16	−30.46	−5.20	11.16
SW54	−29.20	−5.14	11.96
SW24	−29.27	−5.10	11.55
GW37	−30.94	−5.09	9.82
SW55	−29.15	−5.09	11.57
GW32	−30.50	−5.07	10.11
GW23	−29.03	−5.06	11.52
GW26	−29.90	−5.05	10.57
GW51	−29.54	−5.05	10.90
GW33	−29.59	−5.05	10.84
GW44	−30.48	−5.04	9.84
GW50	−28.58	−4.98	11.30
GW01	−29.79	−4.97	9.96
GW49	−29.39	−4.96	10.32
GW43	−29.05	−4.91	10.27
GW34	−29.18	−4.91	10.10
GW10	−28.15	−4.90	11.08
GW31	−28.25	−4.89	10.86
GW41	−29.40	−4.87	9.61
GW06	−28.26	−4.86	10.66
GW52	−28.07	−4.85	10.76
GW17	−29.60	−4.83	9.06
GW29	−28.34	−4.83	10.31
GW05	−27.15	−4.72	10.60
GW08	−28.95	−4.71	8.77
GW02	−27.70	−4.70	9.91
GW30	−27.44	−4.69	10.08
GW07	−28.05	−4.66	9.23
GW09	−27.23	−4.64	9.88
GW27	−27.58	−4.63	9.50
GW39	−28.28	−4.56	8.25
GW40	−27.54	−4.53	8.70
GW03	−25.05	−4.50	10.97
GW18	−25.92	−4.41	9.37
GW53	−25.36	−4.39	9.78

As noted by Serge et al. [69], shallow groundwater chemistry is often largely influenced by anthropogenic factors [75,76]. The aquifers' characteristics allow surface contaminants to infiltrate through permeable zones, with one key process involving residence time that facilitates chemical interactions between groundwater and surrounding rock formations [75,77,78]. Another process appears linked to biological activity, where plant metabolism and organic matter decomposition release carbon dioxide. When dissolved in water, this CO_2 generates bicarbonate ions [69]. Dakoure [11] similarly emphasized this mechanism within the sedimentary aquifers of Western Burkina Faso.

Carbon-14 dating conducted by [14] in the Taoudeni sedimentary basin points to ongoing recharge of the system over long-time scales, despite climatic fluctuations in the Sahel region that have impacted infiltration and recharge dynamics. Additionally, Trabelsi et al. [72] suggest that the depleted stable isotope signatures found in geochemically evolved groundwater of the Sahel are indicative of fossil waters recharged during cooler and more humid periods predating the Last Glacial Maximum (LGM).

In a study from South Africa, Mahlangu et al. [79] identified processes influencing isotopic variations including rainfall recharge, evaporation from shallow groundwater and surface waters, and mixing between these water bodies. The IAEA [68] report further indicates that current recharge is minimal or absent across most of the aquifer, with recent replenishment detected only at limited, localized sites.

The isotopic values obtained here are consistent with other regional research. For instance, Trabelsi et al. [72] recorded $\delta^{18}O$ values from −5.2‰ to −2.2‰ and δ^2H values from −37.7‰ to −18.8‰ throughout the aquifers [79]. Similarly, Huneau et al. [14] documented $\delta^{18}O$ between −7.2‰ and −3.1‰ and δ^2H ranging from −48.8‰ to −20.3‰ in the same area.

For interpreting stable isotope data, we derived the local meteoric water line (LMWL) for the Bobo-Dioulasso station using precipitation records from 1987 to 2016 provided by the DEIE (Direction d'Études et d'Information sur l'Eau) in Burkina Faso, as expressed in Equation (1):

$$\delta^2H = 1.84 + 6.28\delta^{18}O \quad (1)$$

Figure 8 depicts the correlation between the $\delta^{18}O$ and δ^2H stable isotope values of the water samples, alongside the Global Meteoric Water Line (GMWL) established by [64], and the linear equations derived from three earlier studies carried out in this region:

(i) Bobo-Dioulasso station: $\delta^2H = 8\delta^{18}O + 10.2$ [14];
(ii) Bamako station: $\delta^2H = 8.1\delta^{18}O + 11.9$ [37];
(iii) Barogo station: $\delta^2H = 7.7\delta^{18}O + 7.8$ [80].

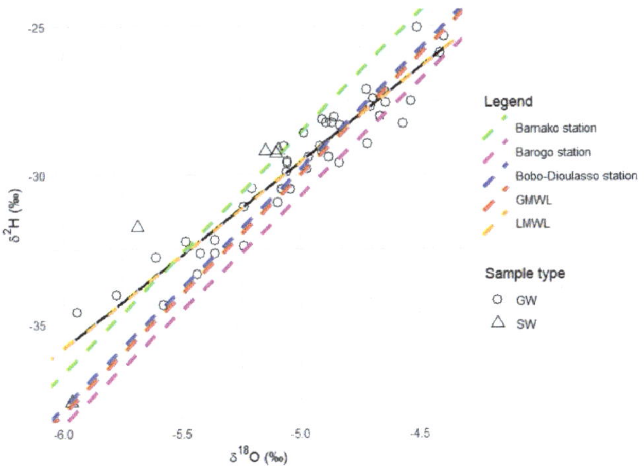

Figure 8. δ^2H versus $\delta^{18}O$ for water samples from the Kou Basin. The Global Meteoric Water Line (GMWL [64]), Local Meteoric Water Line (LMWL, this study), and reference lines from Barogo [80], Bobo-Dioulasso [14], and Bamako [37] are shown for comparison.

The bivariate plot in Figure 8 shows that the majority of water samples align closely with the regression line (*black line in Figure 8*) established by [14] for this region. The computed coefficient of determination is $R^2 = 0.89$, indicating a strong linear relationship. The slope of the regression line, 6.28, is lower than that of both the Global Meteoric Water Line (GMWL) proposed by [64] and the Local Meteoric Water Line (LMWL) defined for Bobo-Dioulasso by [14]. The lower slope observed in Equation (iii) may reflect the effect of evaporation on raindrops prior to their infiltration into the aquifer system in the study area. In their investigation of the Taoudeni aquifer system in the Sahel, Trabelsi et al. [72] reported a local meteoric water line (LMWL) with a slope of 6.68, noting that values below 8 are indicative of isotopic fractionation and enrichment in heavier isotopes due to evaporation processes. Similarly, Taupin et al. [81] highlighted that in the Sahelian climate of West Africa, factors such as limited rainfall, elevated temperatures, and low humidity particularly at the beginning and end of the rainy season foster isotopic enrichment through evaporative effects.

In this regional context, Huneau et al. [14] observed that groundwater isotope signatures typically fall within the range of both the regional and global meteoric water lines, suggesting minimal evaporative alteration during recharge. Supporting this idea, Song et al. [82] propose that limited isotopic variability in groundwater, compared to that of precipitation, may be explained by the selective infiltration of rainwater a phenomenon also documented in other semi-arid settings, such as the Lake Chad Basin [83]. The deuterium excess (d-excess) in our study area varies from 8.3‰ to 12.9‰, with an average of 10.4‰. Figure 9 presents the relationship between d-excess and $\delta^{18}O$ values, with regression lines illustrating the linear trends. As noted by Mamand and Mawlood [84], typical global atmospheric water vapor exhibits a d-excess value of around 10‰ under relative humidity conditions near 85%. In the context of this study, d-excess values falling below this threshold suggest that evaporative processes may have influenced groundwater recharge. This interpretation is consistent with findings by [82] and is further supported by several studies conducted across the Sahel region [83,85,86].

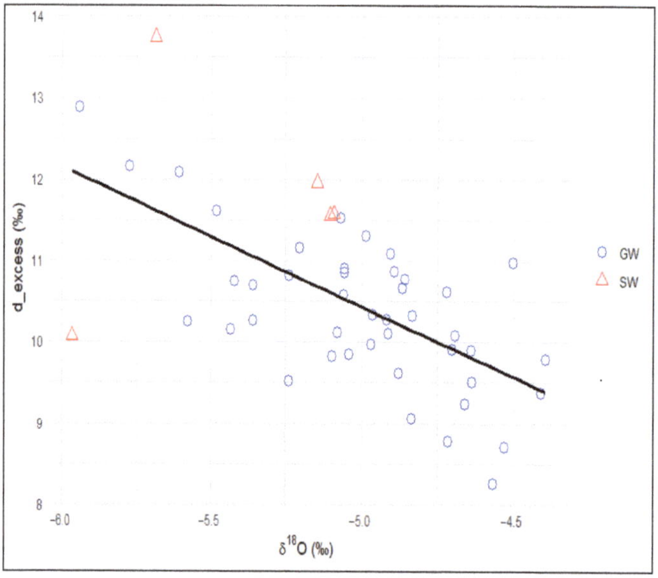

Figure 9. d-excess and $\delta^{18}O$ isotope bivariate plot.

In their 2017 study in Sahelian regions, the IAEA's RAF/7/011 project estimated that 70% of water samples from crystalline formations exhibit a d-excess value exceeding 8‰, suggesting a mild evaporation process. Additionally, in the regional Sahelian conditions reported by RAF/7/011, groundwater from the ICP (folded Infracambrian) and the CTQ (Terminal Quaternary Continental) in Burkina Faso appears to be less affected by evaporation compared to Mali, possibly due to deeper wells.

In our study, 90% of groundwater samples (40 out of 43) display a d-excess greater than 8‰, with 28 out of 43 samples showing a d-excess higher than 10‰. This indicates that other factors or processes may influence the isotopic composition of groundwater samples.

It is important to emphasize, as noted by [14], that the Sahel region has undergone several climatic fluctuations, which have significantly influenced groundwater infiltration and recharge dynamics [87].

3.3.2. Spatial Pattern of δ^2H and $\delta^{18}O$ in Groundwater

The isotopic values of δ^2H, $\delta^{18}O$, and d-excess in groundwater were used as geospatial inputs to generate distribution maps across the Kou basin. As illustrated in Figure 10A, the spatial distribution of $\delta^{18}O$ shows higher concentrations ranging from −5.01‰ to −4.45‰, predominantly in the southeastern and northern parts of the basin. Notably, steep isotopic gradients are observed over short distances between sampling points such as GW27 and GW25; GW52/GW53 and GW23; GW42 and GW45; as well as GW21/GW22 and GW02. These abrupt variations suggest the possible presence of impermeable geological barriers or structural discontinuities influencing groundwater flow. Sampling points GW25 and GW27 are associated with the GFG aquifer at a depth of 80 m and the SAC1 aquifer at 66 m, respectively, as presented in Table 1. A notable difference in isotopic composition over a short distance between these points may reflect a hydrogeological barrier. A comparable situation was reported by [33] in their study in India, where combined hydrochemical and isotopic analyses ($\delta^{18}O$ and δ^2H) revealed sharp gradients between nearby sampling locations. This was interpreted as evidence of a low-permeability boundary that limits horizontal groundwater flow. Furthermore, isotopically depleted groundwater samples were considered indicative of recharge directly from rainfall. While conducting field measurements, we observed that GW52 is situated in a family courtyard, surrounded by numerous septic tanks close to the borehole. Additionally, for GW53, we noted the presence of septic tanks in the courtyard, approximately 15 m from the well. GW42 is another well. During the campaign, an observation revealed the presence of animal waste and human droppings in the surrounding environment around the well. In these wells, the $\delta^{18}O$ values are elevated, as indicated by the red color in Figure 10A. According to IAEA [68], the wells in the studied area are estimated to be shallow, ranging between 25 and 40 m. However, the study reveals significant variability in well depths within the two formations of GFG and SAC1, measuring 80 and 90 m, respectively.

Similar patterns were identified in the spatial distribution of deuterium (δ^2H) values across the Kou basin, as illustrated in Figure 10B. Mamand and Mawlood [84] point out that variations in groundwater temperature, linked to different aquifer systems, can influence the distribution of stable isotopes. Elevated groundwater temperatures are often associated with subsurface processes such as water–rock interactions, which may alter the isotopic composition.

As shown in Figure 10C, for groundwater at the Kou basin, the deuterium excess ranged between 8.28‰ and 12.87‰.

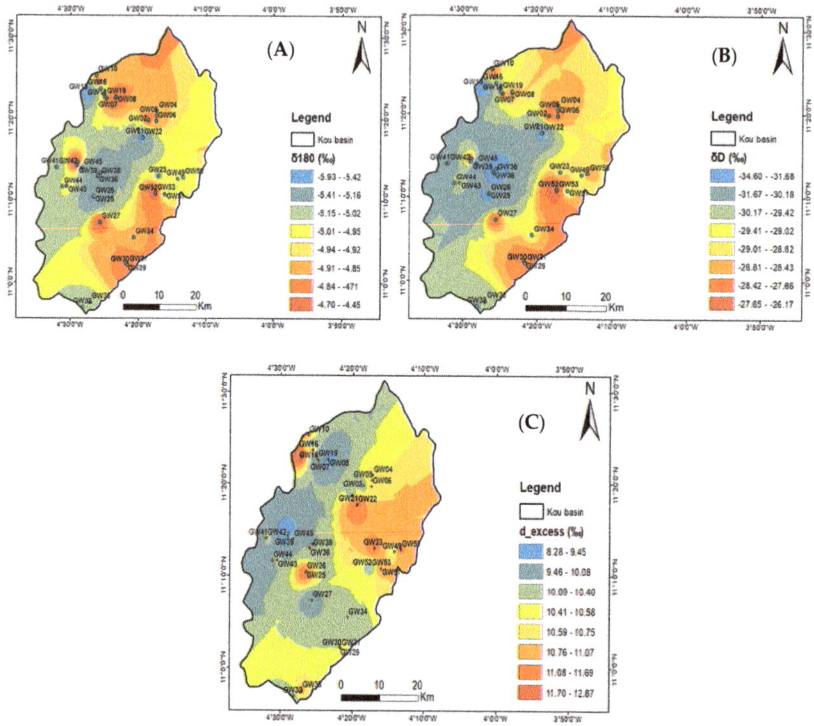

Figure 10. Spatial variation (**A**) $\delta^{18}O$, (**B**) $\delta^{2}H$ and (**C**) d-excess in the groundwater in Kou basin.

3.3.3. Identification of Groundwater Mineralisation

Figure 11 presents the relationship between d-excess and electrical conductivity (EC), with regression lines illustrating the linear trends. According to Liu et al. [88], evaporation in groundwater or surface water typically leads to a reduction in d-excess values while EC increases. In this study, a slight upward trend and a weak positive correlation ($R = 0.06$) were observed between these two parameters. Based on the interpretation of [88], the observed relationship suggests that evaporation plays a key role in influencing the isotopic and chemical characteristics of both groundwater and surface water. Kouanda [8] further emphasizes that, while natural processes primarily govern the mineralization within the Upper Mouhoun-Sourou Complex, anthropogenic pollution is also present. Isotopic analyses using tritium and stable isotopes point to significant recent recharge, alongside the coexistence of older groundwater and mixtures of modern and ancient waters.

Figure 12 shows the variation of $\delta^{18}O$ with depth to groundwater, with regression lines indicating trends and revealing the mixing of groundwater from different aquifer levels (shallow to deep).Although earlier hydrogeochemical and isotopic studies support the interpretation of the STBA as a largely unified and homogeneous multilayered aquifer system, Huneau et al. [14] observed that variations in lithostratigraphic levels correspond to differing hydrodynamic properties, such as permeability, which suggest significant hydraulic connectivity between these layers. Therefore, based on the analysis of groundwater lithology in relation to depth, we conclude that the aquifer formations within the Kou basin are hydraulically interconnected.

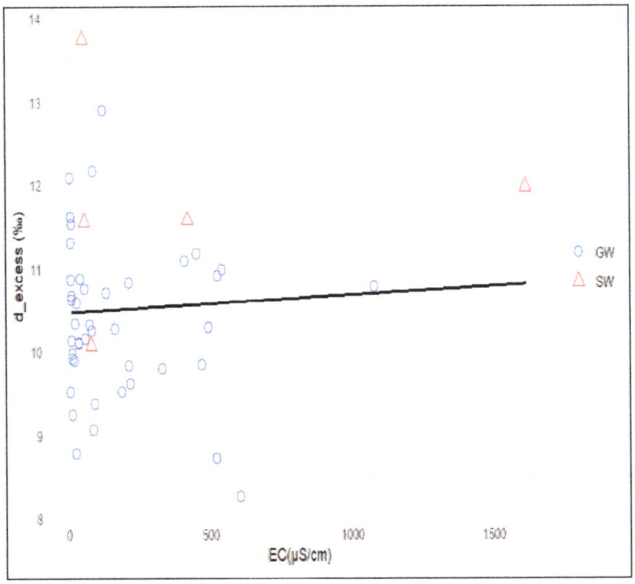

Figure 11. Relation between d-excess and EC values for groundwater and surface water.

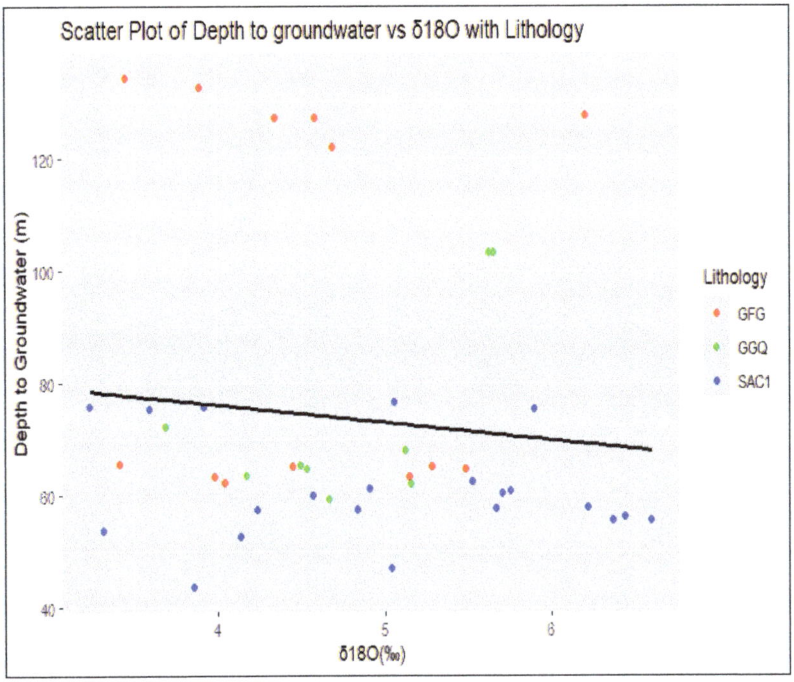

Figure 12. Comparison of δ18O variation with depth to groundwater in relation to the main lithological formations.

According to the IAEA report published [68], there are several shallow wells ranging between 25 and 40 m in the area. The well depths exhibit significant variability in the two formations of GFG and SAC1, measuring 80 m and 90 m, respectively.

3.4. Implication of SW-GW Interactions and Prospects of Water Management

Groundwater resources in the Kou basin, which serve domestic, industrial, and agricultural purposes, are vulnerable to contamination by nitrates that can accumulate over time. The impact of human activities on both surface water and groundwater is a crucial factor in managing water quality. In this study, nitrate concentration was selected as an indicator of anthropogenic pollution because agricultural practices and urban development are the primary sources of nitrate contamination in groundwater. Descriptive statistics reveal that nitrate levels often exceed the WHO's recommended limit for drinking water. Under natural conditions, nitrate concentrations are usually below 5 mg/L, as nitrate acts as a relatively stable tracer in the environment [89]. Elevated nitrate concentrations in some samples suggest significant influences from farming and industrial activities, particularly in urban and agricultural zones (Rawat et al., 2022; Lawniczak et al., 2016 [90,91]). High nitrate intake poses serious health risks, including methemoglobinemia a condition characterized by impaired oxygen transport in the blood and symptoms such as rapid heartbeat, weakness, dizziness, and fatigue [92]. The spatial distribution map of nitrate concentrations, overlaid with land use data (Figure 13), shows that 11.6% of groundwater samples surpass the WHO's permissible limit of 50 mg/L for NO_3^-. The highest nitrate levels were detected at well sites GW51 (138.89 mg/L), GW52 (222.89 mg/L), and GW53 (69.06 mg/L), while among boreholes, GW39 exhibited the highest concentration at 38.79 mg/L. These results are consistent with findings by [93], who highlighted multiple anthropogenic sources of groundwater contamination near water extraction points such as wells and drillings. Contributing factors include waste from wildlife and humans, leaking latrines, solid waste disposal, and wastewater discharge. Rosillon et al. [93] further reported extremely elevated nitrate concentrations in the region, sometimes exceeding 500 mg/L, with a maximum value reaching 860 mg/L.

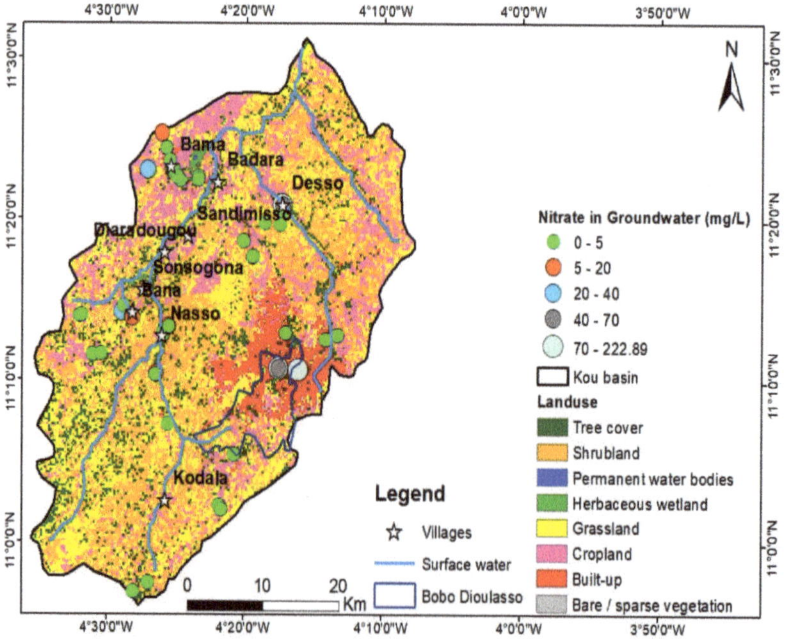

Figure 13. Land-use type and spatial distribution of NO_3^- concentration in the study.

The distribution variation in the boxplot of nitrate over land use and groundwater sample typology is depicted in Figures 14a and 14b, respectively. NO_3^- concentrations are notably higher in well samples compared to boreholes, as illustrated in Figure 14b.

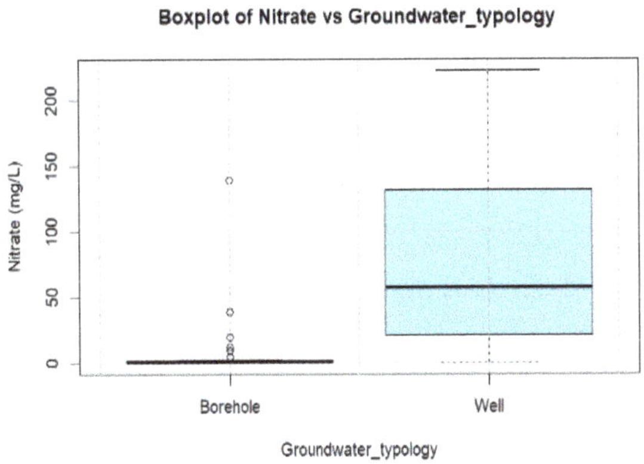

(a)

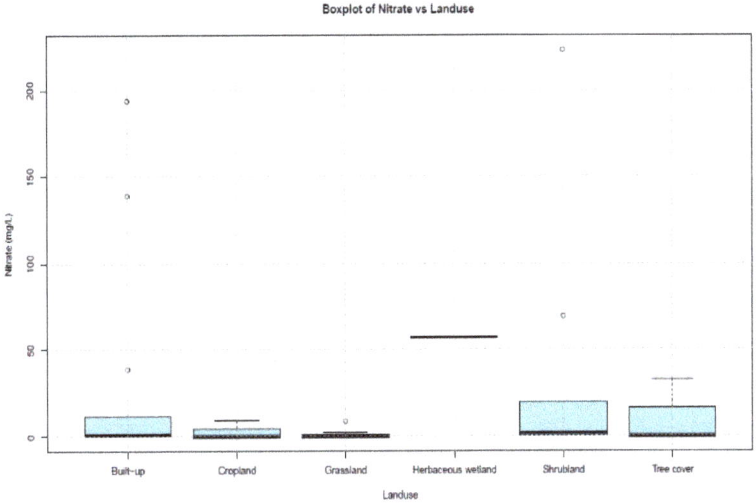

(b)

Figure 14. (**a**) Box plots displaying nitrate parameters according to the main Land contexts and groundwater typology. (**b**) Box plots displaying nitrate parameters according to the main Land contexts and groundwater typology.

In a previous investigation of groundwater in the southeastern part of the Taoudeni sedimentary basin, Huneau et al. [14] observed that human activities significantly impact groundwater characteristics under Sahelian climatic conditions. Contamination tends to be localized near wells or boreholes, often due to insufficient protection measures around these water extraction points. It is common to find livestock watering areas and latrines situated close to boreholes, which contributes to elevated levels of nitrate and sulfate sometimes reaching concentrations as high as 160 mg/L and 300 mg/L, respectively. These elevated values are frequently linked to inadequate well casing and the infiltration of surface water contaminated by manure and wastewater. Since many boreholes sampled are village sources primarily used for drinking water, they are particularly vulnerable to such contamination. Huneau et al. [14] emphasized that pollution is generally confined near pumping sites where protective zones are lacking. Furthermore, the IAEA report [67] highlighted strong correlations among chloride, nitrate, sodium, and potassium concentrations, pointing to anthropogenic origins. Nitrate concentrations in the study area show significant variability, with some locations like SAC1 exhibiting maxima up to 860 mg/L, where nitrates dominate as the primary anion. This contrasts with Huneau et al. [14], whose data reported maximum nitrate levels of 160 mg/L in their sampled boreholes.

4. Conclusions

This study set out to address three fundamental questions: (i) how surface water and groundwater interact within the Kou basin, (ii) which hydrochemical processes govern water quality, and (iii) how anthropogenic activities influence these dynamics.

The Kou basin is a vital source of both groundwater and surface water for Bobo-Dioulasso, the second-largest city in Burkina Faso. Its favorable geological characteristics make the Kou aquifer the primary resource for human development and agriculture, aligning with Sustainable Development Goal 6 (SDG6). This study provides a detailed baseline of groundwater physico-chemical properties and aquifer dynamics, offering concrete data to support informed water management and planning in the basin.

Hydrochemical analysis identified three main water types, Ca-HCO_3, Na-K-HCO_3, and Ca-Mg-HCO_3, reflecting the geological diversity of the region. Mineralization processes are driven primarily by the dissolution of carbonate and evaporite minerals, while elevated concentrations of nitrates, potassium, and chlorides highlight localized anthropogenic pollution. Stable isotope analysis (δ^2H, δ^{18}O, and d-excess) revealed consistent spatial structures, isotopic offsets between groundwater samples, and clear evidence of aquifer interconnectivity. Shallow groundwater was additionally influenced by evaporation, highlighting heterogeneity in aquifer processes.

These results demonstrate pronounced interactions between surface water and groundwater and confirm the significant impact of human activities on water quality, particularly through nitrate contamination. While interpretation of isotopic data is limited by available rainfall measurements, the findings provide a robust framework for understanding aquifer connectivity, recharge processes, and contamination pathways.

In conclusion, this study confirms hydraulic connectivity between aquifer layers and contributions from surface water to groundwater recharge, while also identifying areas affected by water quality degradation. These findings directly address the study's research questions and provide practical guidance for integrated water management. Continuous monitoring of water quality, isotopic signatures, and hydrological variables at key locations is recommended to further reduce uncertainties and ensure the sustainable use of freshwater resources in the Kou basin.

Author Contributions: Conceptualization, I.O., M.V. and S.K.; Data curation, I.O.; Formal analysis, I.O., F.H. and S.K.; Investigation, I.O. and F.H.; Methodology, I.O., M.V., F.H., Y.V. and Y.K.; Resources, I.O. and Y.V.; Software, Y.V.; Supervision, M.V.; Validation, M.V., F.H., Y.V. and Y.K.; Writing—original draft, I.O.; Writing—review and editing, M.V., F.H., Y.V., S.K. and Y.K. All authors have read and agreed to the published version of the manuscript.

Funding: This study received support from the Coordinated Research Project (CRP) of the International Atomic Energy Agency (IAEA), under grant number EVT2404676. The isotopic analyses were funded by the IAEA in cooperation with the Laboratory of Environmental Sciences (UMR 6134 SPE) at the University of Corsica, France, for the analysis of δ^{18}O and δ^2H isotopes as well as chemical parameters.

Data Availability Statement: The original contributions presented in this study are included in the article. Further inquiries can be directed to the corresponding author.

Acknowledgments: This study was conducted as part of the IAEA Coordinated Research Project (CRP) F32010 entitled "Improving Understanding of Nitrate Sources in Connected River and Groundwater Systems Through Linking Nitrate Isotopes and Contaminants of Emerging Concern." The project involved several countries participating in the CRP F32010. In Burkina Faso, the project was led by the University of Fada N'Gourma (Université Yembila Abdoulaye Toguyeni). We extend our sincere thanks to the General Directorate of Water Resources of Burkina Faso (Direction Générale des Ressources en Eau) through Jamel Nebie and Cheick Ouattara for their technical support during data collection.

Conflicts of Interest: The authors declare no conflicts of interest.

References

1. IPCC (Intergovernmental Panel on Climate Change). Climate Change 2022: Impacts, Adaptation and Vulnerability. In *Contribution of Working Group II to the Sixth Assessment Report of the Intergovernmental Panel on Climate Change*; Pörtner, H.-O., Roberts, D.C., Tignor, M., Poloczanska, E.S., Mintenbeck, K., Alegría, A., Craig, M., Langsdorf, S., Löschke, S., Möller, V., et al., Eds.; Cambridge University Press: Cambridge, UK, 2022; p. 3056. [CrossRef]
2. Federal Institute for Geosciences and Natural Resources (BGR); United Nations Educational, Scientific and Cultural Organization—International Hydrological Programme (UNESCO-IHP). River and Groundwater Basins of the World. 2012. Available online: www.whymap.org (accessed on 28 January 2024).
3. Guo, X.; Zuo, R.; Wang, J.; Meng, L.; Teng, Y.; Shi, R.; Gao, X.; Ding, F. Hydrogeochemical Evolution of Interaction between Surface Water and Groundwater Affected by Exploitation. *Groundwater* **2018**, *57*, 430–442. [CrossRef]
4. Zhang, B.; Song, X.; Zhang, Y.; Ma, Y.; Tang, C.; Yang, L.; Wang, Z.L. The interaction between surface water and groundwater and its effect on water quality in the Second Songhua River basin, northeast China. *J. Earth Syst. Sci.* **2016**, *125*, 1495–1507. [CrossRef]
5. Xu, J.; Du, Y.; Sun, X.; Tian, H.; Zhu, S.; Gan, Y.; Wang, Y. Spatial variability of nutrient fluxes associated with lacustrine groundwater discharge in a typical oxbow lake, Central China. *Water Res.* **2025**, *281*, 123701. [CrossRef] [PubMed]
6. Paturel, J.E.; Servat, E.; Kouame, B.; Boyer, J.F. Manifestation de la sécheresse en Afrique de l'Ouest non sahélienne. Cas de la Côte d'Ivoire, du Togo et du Bénin. *Sécheresse* **1995**, *6*, 95–102.
7. Amoussou, E.; Camberlin, P.; Mahé, G. Impact de la variabilité climatique et du barrage Nangbéto sur l'hydrologie du système Mono-Couffo (Ouest Afrique). *Hydrol. Sci. J.* **2012**, *57*, 805–817. [CrossRef]
8. Kouanda, B. Modélisation Intégrée du Complexe Mouhoun Supérieur-Sourou Dans le Contexte des Changements Climatiques. Thèse en Sciences et Technologies de l'Eau, de l'Energie et de l'Environnement. 2019. 2iE, p. 250. Available online: http://documentation.2ie-edu.org/cdi2ie/opac_css/index.php?lvl=notice_display&id=30165 (accessed on 23 February 2024).
9. Mahe, G.; Paturel, J.-E.; Servat, E.; Conway, D.; Dezetter, A. The impact of land use change on soil water holding capacity and river flow modelling in the Nakambe River, Burkina-Faso. *J. Hydrol.* **2005**, *300*, 33–43. [CrossRef]
10. Mahe, G.; Lienou, G.; Descroix, L.; Bamba, F.; Paturel, J.E.; Laraque, A.; Meddi, M.; Habaieb, H.; Adeaga, O.; Dieulin, C.; et al. The rivers of Africa: Witness of climate change and human impact on the environment. *Hydrol. Process.* **2013**, *27*, 2105–2114. [CrossRef]
11. Dakouré, D. Etude Hydrogéologique et Géochimique de la Bordure Sud-Est du Bassin Sédimentaire de Taoudéni (Burkina Faso-Mali)—Essai de Modélisation. Ph.D. Thesis, Université Paris VI, Paris, France, 2003; p. 266.
12. Derouane, J. *Modélisation Hydrogéologique du Bassin Sédimentaire*; Programme de Valorisation des Ressources en Eau de l'Ouest (VREO); Direction Générale des Ressources en Eau. Ministère de l'Agriculture, de l'Hydraulique et des Ressources Halieutiques: Ouagadougou, Burkina Faso, 2008; p. 101.
13. Koussoube, Y. Hydrogéologie des Séries Sédimentaires de la Dépression Piézométrique du Gondo (Bassin du Sourou): Burkina Faso/Mali. Ph.D. Thesis, Université Pierre et Marie Curie—Paris VI, Paris, France, 2010.
14. Huneau, F.; Dakoure, D.; Celle-Jeanton, H.; Vitvar, T.; Ito, M.; Traore, S.; Compaore, N.F.; Jirakova, H.; Le Coustumer, P. Flow pattern and residence time of groundwater within the south-eastern Taoudeni sedimentary basin (Burkina Faso, Mali). *J. Hydrol.* **2011**, *409*, 423–439. [CrossRef]
15. Sauret, E. Etude des Potentialités Hydrogéologiques d'une Plaine Alluviale en Relation Avec Les Eaux Souterraines et de Surface Dans Un Contexte d'agriculture Irriguée (Burkina Faso). 2013. Available online: https://www.proquest.com/openview/b301b26b22e695d0e2c6678bace0a611/1.pdf?pq-origsite=gscholar&cbl=2026366&diss=y (accessed on 14 February 2024).
16. Tirogo, J.; Jost, A.; Biaou, A.; Valdes-Lao, D.; Koussoubé, Y.; Ribstein, P. Climate variability and groundwater response: A case study in Burkina Faso (West Africa). *Water* **2016**, *8*, 171. [CrossRef]

17. Derouane, J.; Dakouré, D. *Etude Hydrogéologique et Modélisation Mathématique du Système Aquifère du Bassin Sédimentaire de Taoudeni au Burkina Faso*; Colloque International—Gestion des Grands Aquifères: Dijon, France, 2006; p. 21. Available online: https://www.researchgate.net/publication/228467900_Etude_hydrogeologique_et_modelisation_mathematique_du_systeme_aquifere_du_bassin_sedimentaire_de_Taoudeni_au_Burkina_Faso (accessed on 14 February 2024).
18. Wang, J.; Xiao, Y.; Zhang, Y.; Hu, W.; Qi, Z.; You, X.; Gu, X.; Chen, H.; Han, J.; Zhang, Y.; et al. Hydrogeochemical sources and enrichment mechanisms of trace elements in river water of a typical endorheic headwater region on Tibetan Plateau. *Appl. Water Sci.* **2025**, *15*, 226. [CrossRef]
19. O'Connor, M.; Zabik, M.; Cady, C.; Cousens, B.; Chiarenzelli, J. Multi-Element Analysis and Geochemical Spatial Trends of Groundwater in Rural Northern New York. *Water* **2010**, *2*, 217–238. [CrossRef]
20. Doveri, M.; Mussi, M. Water Isotopes as Environmental Tracers for Conceptual Understanding of Groundwater Flow: An Application for Fractured Aquifer Systems in the "Scansano-Magliano in Toscana" Area (Southern Tuscany, Italy). *Water* **2014**, *6*, 2255–2277. [CrossRef]
21. Zhang, Y.; Xu, M.; Li, X.; Qi, J.; Zhang, Q.; Guo, J.; Yu, L.; Zhao, R. Hydrochemical characteristics and multivariate statistical analysis of natural water system: A case study in Kangding County, Southwestern China. *Water* **2018**, *10*, 80. [CrossRef]
22. Wu, Q.; Wang, G.; Zhang, W.; Cui, H.; Zhang, W. Estimation of groundwater recharge using tracers and numerical modeling in the North China Plain. *Water* **2016**, *8*, 353. [CrossRef]
23. Yuan, J.; Xu, F.; Deng, G.; Tang, Y.; Li, P. Hydrogeochemistry of shallow groundwater in a karst aquifer system of Bijie City, Guizhou Province. *Water* **2017**, *9*, 625. [CrossRef]
24. Peters, E.; Visser, A.; Esser, B.K.; Moran, J.E. Tracers Reveal Recharge Elevations, Groundwater Flow Paths and Travel Times on Mount Shasta, California. *Water* **2018**, *10*, 97. [CrossRef]
25. Sako, A.; Yaro, J.M.; Bamba, O. Impacts of hydrogeochemical processes and anthropogenic activities on groundwater quality in the Upper Precambrian sedimentary aquifer of northwestern Burkina Faso. *Appl. Water Sci.* **2018**, *8*, 88. [CrossRef]
26. Kumari, R.; Kumar, S.; Rao, M.S.; Mukherjee, S.; Sen, R. Isotopic Signature in Integration with Hydrochemistry to Infer the Groundwater Quality in Alluvial Aquifer, Jhajjar District, Haryana, NCR, India. 2021. Available online: https://assets-eu.researchsquare.com/files/rs-599293/v1/4633c7a3-6023-4072-bf23-60a821dbb046.pdf (accessed on 17 September 2024). [CrossRef]
27. Coplen, T.B.; Herczeg, A.L.; Barnes, C. Isotope engineering—Using stable isotopes of the water molecule to solve practical problems. In *Environmental Tracers in Subsurface Hydrology*; Cook, P.G., Herczeg, A.L., Eds.; Springer: Boston, MA, USA, 2000; pp. 79–110. [CrossRef]
28. Datta, P.S.; Bhattacharya, S.K.; Tyagi, S.K. 18O studies on recharge of phreatic aquifers and groundwater flowpaths of mixing in Delhi area. *J. Hydrol.* **1996**, *176*, 25–36. [CrossRef]
29. Gupta, S.K.; Deshpande, R.D. Groundwater isotopic investigations in India: What has been learned? *Curr. Sci.* **2005**, *89*, 825–835.
30. Jeelani, G.H.; Bhat, N.A.; Shivanna, K. Use of δ18O tracer to identify stream and spring origins of a mountainous catchment: A case study from Liddar watershed, Western Himalaya, India. *J. Hydrol.* **2010**, *393*, 257–264. [CrossRef]
31. Maurya, A.S.; Shah, M.; Deshpande, R.D.; Bhardwaj, R.M.; Prasad, A.; Gupta, S.K. Hydrograph separation and precipitation source identification using stable water isotopes and conductivity: River Ganga at Himalayan foothills. *Hydrol. Process.* **2011**, *25*, 1521–1530. [CrossRef]
32. Rina, K.; Singh, C.K.; Datta, P.S.; Singh, N.; Mukherjee, S. Geochemical modelling, ionic ratio and GIS based mapping of groundwater salinity and assessment of governing processes in Northern Gujarat, India. *Environ. Earth Sci.* **2013**, *69*, 2377–2391. [CrossRef]
33. Maurya, P.; Kumari, R.; Mukherjee, S. Hydrochemistry in integration with stable isotopes (δ18O and δD) to assess seawater intrusion in coastal aquifers of Kachchh district, Gujarat, India. *J. Geochem. Explor.* **2019**, *196*, 42–56. [CrossRef]
34. Breme, M. Caractérisation Physique du Bassin du Kou. Mémoire Master II, 2IE. 70 pp + Annexes. 2014. Available online: http://documentation.2ie-edu.org/cdi2ie/opac_css/doc_num.php?explnum_id=4066 (accessed on 23 January 2024).
35. Qu, S.; Wang, J.; Zhang, K.; Fan, M.; Zhao, Y.; Yang, X.; Wang, Z.; Gomes, H.I.; Gomes, R.L.; Duan, L.; et al. Fate and Risks of Potentially Toxic Elements Associated with Lacustrine Groundwater Discharge: Quantification, Modeling, and Biogeochemistry. *Environ. Int.* **2025**, *202*, 109707. [CrossRef] [PubMed]
36. Traoré, F.; Cornet, Y.; Denis, A.; Wellens, J.; Tychon, B. Monitoring the evolution of irrigated areas with Landsat images using backward and forward change detection analysis in the Kou watershed, Burkina Faso. *Geocarto Int.* **2013**, *28*, 733–752. [CrossRef]
37. Gourcy, L.; Aranyossy, J.F.; Olivry, J.C.; Zuppi, G.M. Space and time variations in the isotopic composition (d2H–d18O) of Niger inland delta water (Mali). *C. R. L'acad. Sci. Earth Planet. Sci.* **2000**, *331*, 701–707. [CrossRef]
38. Frappart, F.; Hiernaux, P.; Guichard, F.; Mougin, E.; Kergoat, L.; Arjounin, M.; Lavenu, F.; Koité, M.; Paturel, J.E.; Lebel, T. Rainfall regime across the Sahel band in the Gourma region, Mali. *J. Hydrol.* **2009**, *375*, 128–142. [CrossRef]
39. Taupin, J.D.; Gaultier, G.; Favreau, G.; Leduc, C.; Marlin, C. Isotopic variability of Sahelian rainfall at different time steps in Niamey (Niger, 1992–1999): Climatic implications. *C. R. Geosci.* **2002**, *334*, 43–50. [CrossRef]

40. Wellens, J.; Compaoré, N.F. *Renforcement de la Capacité de Gestion des Ressources en eau dans l'Agriculture Moyennant des Outils de Suivi-Évaluation–GEeau*; Rapport Annuel N1 (Décembre 2001–Novembre 2002); Direction Régionale de l'Agriculture, de l'Hydraulique et des Ressources Halieutiques des Hauts Bassins: Bobo-Dioulasso, Burkina Faso, 2003.
41. Ogou, C. *Etude des Potentialités Aquifères de la Plaine Alluviale du Kou Entre Nasso et Diaradougou*; Institut International D'ingénierie de L'eau et de L'environnement: Ouagadougou, Burkina Faso, 2008.
42. Ouedraogo, C. *Synthèse Géologique de la Région Ouest du Burkina Faso*; Programme VREO, SOFRECO-SAWES: Bobo-Dioulasso, Burkina Faso, 2006.
43. Bronner, G.; Roussel, J.; Trompette, R.A.; Clauer, N. Genesis and geodynamic evolution of the Taoudeni cratonic basin (Upper Precambrian and Paleozoic), Western Africa. *Dyn. Plate Inter.* **1980**, *1*, 81–90. [CrossRef]
44. Sogreah Ingénierie. *Notice Explicative de la Carte Hydrogéologique 1:50000 de la Région de Bobo-Dioulasso*; Etude des Ressources de la Zone Sédimentaire de la Région de Bobo-Dioulasso; Ministère de l'Agriculture de l'Hydraulique et des Ressources Halieutiques—Direction des Etudes et de la Planification—Direction Régionale de l'Eau des Hauts-Bassins: Bobo-Dioulasso, Burkina Faso, 1993; p. 25.
45. Sogreah Ingenierie. *Etude des Ressources en Eaux Souterraines de la Zone Sédimentaire de la Région de Bobo-Dioulasso*; Ministère de l'Agriculture de l'Hydraulique et des Ressources Halieutiques—Direction des Etudes et de la Planification—Direction Régionale de l'Eau des Hauts-Bassins: Bobo-Dioulasso, Burkina Faso, 1994.
46. Lorenzini, G. *Etude du Bassin D'alimentation des Sources de Nasso—Etat des Lieux des Ressources en Eaux Souterraines du Bassin du Kou, Département ArGEnCo*; Université de Liège: Liège, Belgique, 2007; p. 119.
47. Gombert, P. *Synthèse sur la Géologie et l'Hydrogéologie de la Série Sédimentaire du Sud-Ouest du Burkina Faso*; Programme RESO, IWACO-BURGEAP: Bobo-Dioulasso, Burkina Faso, 1998.
48. Sauret, E. Contribution à la Compréhension du Fonctionnement Hydrogeologique du Système Aquifère dans le Bassin du Kou (Burkina Faso) [Specialised Master, ULiège—Université de Liège]. ORBi-University of Liège. 2008. Available online: https://orbi.uliege.be/handle/2268/153604 (accessed on 14 February 2024).
49. Wellens, J.; Traoré, F.; Diallo, M.; Tychon, B. A framework for the use of decision-support tools at various spatial scales for the management of irrigated agriculture in West-Africa. *Agric. Sci.* **2013**, *4*, 9–15. [CrossRef]
50. Traoré, F. Optimisation de L'utilisation des Ressources en Eau du Bassin du Kou pour des Usages Agricoles. Ph.D. Thesis, Université de Liège, Liège, Belgium, 2012. p. 195. Available online: https://orbi.uliege.be/handle/2268/132698 (accessed on 14 February 2024).
51. RGPH. Cinquième Recensement Général de la Population et de l'Habitation du Burkina Faso. 2022. p. 136. Available online: https://www.insd.bf/sites/default/files/2023-08/INSD_Rapport_SYNTHESE%20DES%20RESULTATS%20DEFINITIFS_1.pdf (accessed on 19 August 2024).
52. Wellens, J. A framework for Using Decision-Support Tools at Various Spatial Scales for the Management of Irrigated Agriculture in Semi-Arid West-Africa. Ph.D. Thesis, Université de Liège, Liège, Belgium, 2014. Available online: https://orbi.uliege.be/handle/2268/163546 (accessed on 14 February 2024).
53. Wellens, J.; Sawadogo, I.; Diallo, M.; Dakouré, D.; Compaoré, N.F.; Traoré, F.; Tychon, B. *Recensement Exhaustif des Activités Hydro-Agricoles du Bassin du Kou*; GEeau: Bobo-Dioulasso, Burkina Faso, 2007.
54. Traoré, S.; Gombert, P. *Variations des Débits Interannuels de la Source de la Guinguette*; Rapport Interne; Direction Régionale de l'Hydraulique des Hauts-Bassins: Bobo-Dioulasso, Burkina Faso, 1997; p. 10.
55. Palm, P.M. Hydrological Investigation for Climate Change Investigation in the Kou Basin Burkina Faso: A Minor Field Study. TRITA LWR Degree Project 11:2. ISSN 1651-064X.. 2011. Available online: https://www.diva-portal.org/smash/record.jsf?pid=diva2%3A529737 (accessed on 23 January 2024).
56. Dione, M. Gestion Intégrée des Ressources en Eau de la Vallée du Kou: Profil de Pollution des Eaux de Surface en Rapport Avec les Usages. Master's Thesis, 2iE-Institut International d'Ingénierie de l'Eau et de l'Environnemen, Ouagadougou, Burkina Faso, 2005. p. 118. Available online: http://documentation.2ie-edu.org/cdi2ie/opac_css/doc_num.php?explnum_id=1008 (accessed on 23 January 2024).
57. Bieupoude, G.P. *Mapping Groundwater Intrinsic Vulnerability Using a New Physically Based Modeling in Kou Basin Bobo-Dioulasso/Burkina Faso*; Mémoire de Master en Hydraulique-Environnement et Infrastructure: Ouagadougou, Burkina Faso, 2008.
58. Ko, A. Influence des Rejets Industriels sur la Pollution des Eaux à Bobo Dioulasso (Burkina Faso). Master's Thesis, Université Polytechnique de Bobo-Dioulasso, Dindéresso, Burkina Faso, 2008. p. 78. Available online: https://beep.ird.fr/collect/upb/index/assoc/IDR-2008-KOH-INF/IDR-2008-KOH-INF.pdf (accessed on 23 January 2024).
59. Ben Moussa, A. Etude Hydrogéologique, Hydrochimique et Isotopique du Système Aquifère de Hammamet Nabeul, Cap Bon, Tunisie Nord Orientale. Ph.D. Thesis, Université de Sfax École Nationale d'Ingénieurs de Sfax, Sfax, Tunisia, 2011. p. 147. Available online: https://gnssn.iaea.org/main/NCP/Tunisia/LRAE/Documents/Student%20Programmes/Th%C3%A8se%20Ben%20Moussa%20Amor.pdf (accessed on 14 February 2024).

60. IAEA (International Atomic Energy Agency). *Introduction to Water Sampling and Analysis for Isotope Hydrology*; Non-Serial Publications; IAEA: Vienna, Austria, 2007.
61. Freeze, R.A.; Cherry, J.A. *Groundwater*; Prentice Hall: Englewood Cliffs, NJ, USA, 1979; p. 604.
62. Favreau, G. *Caractérisation et Modélisation d'une Nappe Phréatique en Hausse au Sahel: Dynamique et Géochimie de la Dépression Piézométrique Naturelle du Kori de Dantiandou (Sud-Ouest du Niger)*; Université Paris-Sud 11: Orsay, France; IRD: Montpellier, France, 2000; 348p.
63. Penna, D.; Stenni, B.; Šanda, M.; Wrede, S.; Bogaard, T.A.; Gobbi, A.; Borga, M.; Fischer, B.M.C.; Bonazza, M.; Chárówá, Z. On the reproducibility and repeatability of laser absorption spectroscopy measurements for δ2H and δ18O isotopic analysis. *Hydrol. Earth Syst. Sci.* **2010**, *14*, 1551–1566. [CrossRef]
64. Craig, H. Isotopic variations in meteoric waters. *Science* **1961**, *133*, 1702–1703. [CrossRef]
65. Dansgaard, W. Stable Isotope in Precipitation. *Tellus* **1964**, *16*, 436–468. [CrossRef]
66. Simler, R. DIAGRAMMES Software (Version 31-10-2014). Université d'Avignon. Available online: https://terre-et-eau.univ-avignon.fr/equipements-de-terrain-et-de-laboratoire/logiciels (accessed on 31 October 2024).
67. WHO(World Health Organization). *WHO—Guidelines on Drinking Water Quality*, 4th ed.; World Health Organization: Geneva, Switzerland, 2017.
68. IAEA (International Atomic Energy Agency). *(RAF/7/011) Gestion Intégrée et Durable des Systèmes Aquifères et des Bassins Partagés de la Région du Sahel, Bassin de Taoudéni*; l'AIEA: Vienna, Austria, 2017; p. 136. Available online: https://www.iaea.org/sites/default/files/18/02/raf7011_taoudeni_basin_fr.pdf (accessed on 19 January 2024).
69. Serge Gaëtan, S.É.; Marie Michelle, C.H.; Ouindinboudé Jacques, K.; Poulouma Louis, Y.; Idriss, S. Hydrogeochemistry of Shallow Groundwater and Suitability to Irrigation: The Case of the Karfiguéla Paddy Field in Burkina Faso. *Water* **2022**, *14*, 2574. [CrossRef]
70. Sheick, S. Water Nutrient Use Efficiency the Vertical Leaching Losses in Urban Vegetable Cropping Systems in Bobo–Dioulasso (Burkina Faso). Ph.D. Thesis, Université Catholique de Louvain, Louvain-la-Neuve, Belgium, 2012. p. 193. Available online: http://hdl.handle.net/2078.1/110879 (accessed on 14 February 2024).
71. Ouédraogo, R.A.; Kambiré, F.C.; Kestemont, M.P.; Bielders, C.L. Caractériser la diversité des exploitations maraîchères de la région de Bobo-Dioulasso au Burkina Faso pour faciliter leur transition agroécologique. *Cah. Agric.* **2019**, *28*, 20. [CrossRef]
72. Trabelsi, R.; Zouari, K.; Araguás Araguás, L.J.; Moulla, A.S.; Sidibe, A.M.; Bacar, T. Assessment of geochemical processes in the shared groundwater resources of the Taoudeni aquifer system (Sahel region, Africa). *Hydrogeol. J.* **2024**, *32*, 167–188. [CrossRef]
73. Huneau, F.; Travi, Y. The Miocene Aquifer of Valreas, France. In *Natural Groundwater Quality*; Edmunds, W.M., Shand, P., Eds.; Blackwell Publishing: London, UK, 2008; pp. 287–305. [CrossRef]
74. Yang, K.; Han, G.; Song, C.; Zhang, P. Stable HO isotopic composition and water quality assessment of surface water and groundwater: A case study in the Dabie Mountains, central China. *Int. J. Environ. Res. Public Health* **2019**, *16*, 4076. [CrossRef]
75. Sako, A.; Sawadogo, S.; Nimi, M.; Ouédraogo, M. Hydrogeochemical and Pollution Characterization of a Shallow Glauconitic Sandstone Aquifer in a Peri-Urban Setting of Bobo-Dioulasso, Southwestern Burkina Faso. *Environ. Earth Sci.* **2020**, *79*, 296. [CrossRef]
76. Xu, P.; Feng, W.; Qian, H.; Zhang, Q. Hydrogeochemical Characterization and Irrigation Quality Assessment of Shallow Groundwater in the Central-Western Guanzhong Basin, China. *Int. J. Environ. Res. Public Health* **2019**, *16*, 1492. [CrossRef]
77. Li, P.; Wu, J.; Qian, H.; Zhang, Y.; Yang, N.; Jing, L.; Yu, P. Hydrogeochemical Characterization of Groundwater in and around a Wastewater Irrigated Forest in the Southeastern Edge of the Tengger Desert, Northwest China. *Expo. Health* **2016**, *8*, 331–348. [CrossRef]
78. Subramani, T.; Rajmohan, N.; Elango, L. Groundwater Geochemistry and Identification of Hydrogeochemical Processes in a Hard Rock Region, Southern India. *Environ. Monit. Assess* **2010**, *162*, 123–137. [CrossRef] [PubMed]
79. Mahlangu, S.; Lorentz, S.; Diamond, R.; Dippenaar, M. Surface water-groundwater interaction using tritium and stable water isotopes: A case study of Middelburg, South Africa. *J. Afr. Earth Sci.* **2020**, *171*, 103886. [CrossRef]
80. Mathieu, R.; Bariac, T.; Fouillac, C.; Guillot, B.; Mariotti, A. Variations en isotopes stables dans les précipitations en 1988 et 1989 au Burkina Faso: Apports de la météorologie régionale. *Veill. Clim. Satellitaire* **1993**, *45*, 47–64.
81. Taupin, J.D.; Gallaire, R.; Arnaud, Y. Analyses isotopiques et chimiques des précipitations sahéliennes de la région de Niamey au Niger: Implications climatologiques. *IAHS Publ.-Ser. Proc. Rep.-Intern. Assoc. Hydrol. Sci.* **1997**, *244*, 151–164.
82. Song, F.; Nlend, B.; Boum-Nkot, S.N.; Huneau, F.; Ndondo, G.N.; Garel, E.; Leydier, T.; Celle, H.; Djieugoue, B.; Ntamak-Nida, M.-J.; et al. Groundwater Resources of the Transboundary Quaternary Aquifer of the Lake Chad Basin: Towards a Better Management via Isotope Hydrology. *Resources* **2023**, *12*, 138. [CrossRef]
83. Goni, I.B.; Taylor, R.G.; Favreau, G.; Shamsudduha, M.; Nazoumou, Y.; Nhatcha, B.N. Groundwater recharge from heavy rainfall in the southwestern Lake Chad Basin: Evidence from isotopic observations. *Hydrol. Sci. J.* **2021**, *66*, 1359–1371. [CrossRef]
84. Mamand, B.S.; Mawlood, D.K. Identifying sources of groundwater and recharge zone using stable environmental isotopes in the Erbil basin-northern Iraq. *Kuwait J. Sci.* **2024**, *51*, 100128. [CrossRef]

85. Mahamat Nour, A.; Vallet-Coulomb, C.; Gonçalves, J.; Sylvestre, F.; Deschamps, P. Rainfall-Discharge Relationship and Water Balance over the Past 60 Years within the Chari-Logone Sub-Basins, Lake Chad Basin. *J. Hydrol. Reg. Stud.* **2021**, *35*, 100824. [CrossRef]
86. Mahamat-Nour, A.; Huneau, F.; Mahamat, A.; Mahamat Saleh, H.; Ngo Boum-Nkot, S.; Nlend, B.; Djebebe-Ndjiguim, C.-L.; Foto, E.; Sanoussi, R.; Araguas-Araguas, L.; et al. Shallow Quaternary groundwater in the Lake Chad basin is resilient to climate change but requires sustainable management strategy: Results of isotopic investigation. *Sci. Total Environ.* **2022**, *851*, 158152. [CrossRef]
87. Beyerle, U.; Rueedi, J.; Leuenberger, M.; Aeschbach-Hertig, W.; Peeters, F.; Kipfer, R.; Dodo, A. Evidence for periods of wetter and cooler climate in the Sahel between 6 and 40 kyr BP derived from groundwater. *Geophys. Res. Lett.* **2003**, *30*, 1173. [CrossRef]
88. Liu, F.; Wang, S.; Wang, L.; Shi, L.; Song, X.; Yeh, T.C.J.; Zhen, P. Coupling hydrochemistry and stable isotopes to identify the major factors affecting groundwater geochemical evolution in the Heilongdong Spring Basin, North China. *J. Geochem. Explor.* **2019**, *205*, 106352. [CrossRef]
89. Böhlke, J.K. Groundwater recharge and agricultural contamination. *Hydrogeol. J.* **2002**, *10*, 153–179. [CrossRef]
90. Rawat, M.; Sen, R.; Onyekwelu, I.; Wiederstein, T.; Sharda, V. Modeling of Groundwater Nitrate Contamination Due to Agricultural Activities—A Systematic Review. *Water* **2022**, *14*, 4008. [CrossRef]
91. Lawniczak, A.E.; Sokołowska, K.; Dębska, B. Impact of Agriculture and Land Use on Nitrate Contamination of Groundwater and Surface Water in a Protected Area. *Environ. Monit. Assess.* **2016**, *188*, 172. [CrossRef]
92. Islam, A.R.M.T.; Pal, S.C.; Chowdhuri, I.; Salam, R.; Islam, M.S.; Rahman, M.M.; Zahid, A.; Idris, A.M. Application of novel framework approach for prediction of nitrate concentration susceptibility in coastal multi-aquifers, Bangladesh. *Sci. Total Environ.* **2021**, *801*, 149811. [CrossRef] [PubMed]
93. Rosillon, F.; Savadogo, B.; Kabore, A.; Bado-Sama, H.; Dianou, D. Attempts to Answer on the Origin of the High Nitrates Concentrations in Groundwaters of the Sourou Valley in Burkina Faso. *J. Water Resour. Prot.* **2012**, *4*, 663–673. [CrossRef]

water

MDPI

Article

Integrating Inverse Modeling to Investigate Hydrochemical Evolution in Arid Endorheic Watersheds: A Case Study from the Qaidam Basin, Northwestern China

Liang Guo [1], Yuanyuan Ding [1], Haisong Fang [1], Chunxue An [1], Wanjun Jiang [2] and Nuan Yang [1,]*

[1] College of Geology and Environment, Xi'an University of Science and Technology, Xi'an 710054, China; guoliang@xust.edu.cn (L.G.)

[2] Tianjin Center, China Geological Survey, Tianjin 300170, China; jiangwanjun0718@163.com

* Correspondence: yangnuan@xust.edu.cn

Abstract

The hydrochemical characteristics and evolution mechanisms of groundwater are critical for accurately understanding the input–output budget of hydrochemical constituents in pristine groundwater. However, few studies have analyzed the changes in mineral precipitation and dissolution equilibrium along the groundwater flow path, especially in arid regions. This study integrated hydrochemical analysis, stable isotopes, and inverse hydrochemical modeling to identify groundwater recharge sources, hydrochemical evolution, and controlling mechanisms in an arid endorheic watershed, northwestern China. A stable isotope signature indicated that groundwater is primarily recharged by high-altitude meteoric precipitation and glacial snowmelt. The regional hydrochemical type evolved from HCO_3·Cl-Ca·Mg·Na types in phreatic aquifers to more complex HCO_3·Cl-Ca·Mg Na and HCO_3·Cl-Na Mg types in confined aquifers and a Cl-Mg·Na type in high-salinity groundwater. The dissolution of halite, gypsum, calcite, K-feldspar, and albite was identified as the primary source of dissolved substances and a key factor controlling the hydrochemical characteristics. Meanwhile, hydrochemical evolution is influenced by cation exchange, mineral dissolution–precipitation, and carbonate equilibrium mechanisms. Inverse hydrochemical modeling demonstrated that high-salinity groundwater has experienced intensive evaporation and quantified the transfer amounts of associated minerals. This study offers deeper insight into hydrochemical evolution in the Golmud River watershed and elucidates mineral transport and enrichment mechanisms, providing a theoretical basis for investigating hydrochemical metallogenic processes.

Keywords: groundwater hydrochemical evolution; hydrochemical processes; inverse hydrochemical modeling; groundwater in arid regions

1. Introduction

Groundwater, as a stable source of freshwater supply, constitutes an indispensable component in the global response to escalating climate change challenges [1,2]. The hydrochemical characteristics of groundwater during recharge, runoff, and discharge processes are primarily governed by water–rock interactions, including mineral dissolution and precipitation, ion exchange, adsorption, and redox reactions. The processes and patterns of hydrochemical evolution are closely related to the spatial distribution and availability of freshwater resources [3,4]. To elucidate regional hydrochemical evolution and controlling

https://doi.org/10.3390/w17142074

mechanisms, various methods have been employed, including ratio relationships of hydrochemical parameters, graphical analysis, multivariate statistical methods, and inverse hydrochemical models [5]. Piper diagrams, Gibbs diagrams, and ionic ratios are integrated to analyze the hydrochemical distribution comprehensively and the controlling factors of both surface water and groundwater [6,7]. Hydrochemical components, stable isotopes (^{2}H, ^{18}O), and radiogenic isotopes (^{14}C, ^{222}Rn) enable the identification of groundwater recharge sources and the analysis of the impacts of processes such as evaporation and mixing on hydrochemical features, thereby revealing the spatial variability of surface water and groundwater interactions [8,9]. On the other hand, Phreeqc, a widely used geochemical modeling tool, facilitates the quantitative assessment of hydrochemical processes by calculating mineral saturation indices and simulating water–rock interactions between groundwater and the surrounding aquifer medium [10–12].

The Golmud River watershed, located in the middle of the southern margin of the Qaidam Basin, is a typical inland river system within the arid region of northwestern China. The study area is characterized by extremely high surface evaporation and a vulnerable ecological environment, where groundwater serves as a critical component in sustaining ecological stability [13,14]. Regional water balance and ecosystem functionality are strongly governed by the continuous exchange and interaction between surface water and groundwater [15,16]. Groundwater in the study area exhibits distinct spatial hydrochemical profiles, and previous studies have explored its geochemical evolution based on regional hydrochemical characteristics [15], isotopic tracing [8,17], and numerical modeling approaches [18,19]. However, the dissolution–precipitation equilibria of multiple mineral systems, including carbonates, sulfates, and silicates, during hydrochemical evolution in the study area remain poorly understood, hindering an accurate understanding of water–rock interactions in shaping regional hydrochemical characteristics [20].

To address these gaps, this study integrates inverse hydrochemical modeling to investigate hydrochemical evolution and its controlling mechanisms in the Golmud River watershed. The study objective is (1) to elucidate the hydrochemical evolution processes and (2) to evaluate the amounts of mineral transfer during water–rock interactions. Hydrochemical compositions and stable isotopic signatures ($\delta^{2}H$-$\delta^{18}O$) were used to trace groundwater recharge sources and identify the dominant hydrochemical evolution processes. This study enhances the understanding of the formation and evolution of hydrochemical characteristics in the Golmud River watershed and provides valuable scientific insights for the sustainable development of groundwater resources in the Qaidam Basin, as well as for preserving ecological balance.

2. Materials and Methods

2.1. Regional Hydrogeological Backgrounds

The Golmud River watershed is in the central part of the southern margin of the Qaidam Basin, originating from the Kunlun Mountains and flowing northward into Dabusun Lake, with elevations ranging from 2442 to 6212 m (Figure 1). The study area is characterized by intense evaporation, arid conditions with low precipitation, and large diurnal temperature variations. It exhibits a typical plateau continental climate, with a multi-year average precipitation of 42.3 mm and an average annual evaporation of 2626.9 mm [17]. Water resources in the study area are recharged primarily from snow and glacial meltwater in the Kunlun Mountains, as well as atmospheric precipitation. Groundwater and the Golmud River are closely interconnected, forming an endorheic groundwater–river system. The distribution of groundwater in the study area is influenced and constrained by regional stratigraphy, lithology, geological formations, and geomorphological conditions [21]. The Golmud River originates from the Kunlun Mountains, where the geological strata exhibit complex and diverse lithologies, and rock fractures is heterogeneous, leading to low overall permeability [22]. The piedmont alluvial fan is primarily composed of alluvial deposits, including gravel and sand. This area is characterized by thick, unconfined aquifers with broad distribution and high groundwater yield. In contrast, the central plain area mainly consists of fine-grained lacustrine and fluvial sediments, including clay, silt, and fine sand. These deposits form multi-layered heterogeneous confined aquifers whose hydraulic properties are governed by the texture and compaction of the deposits [23].

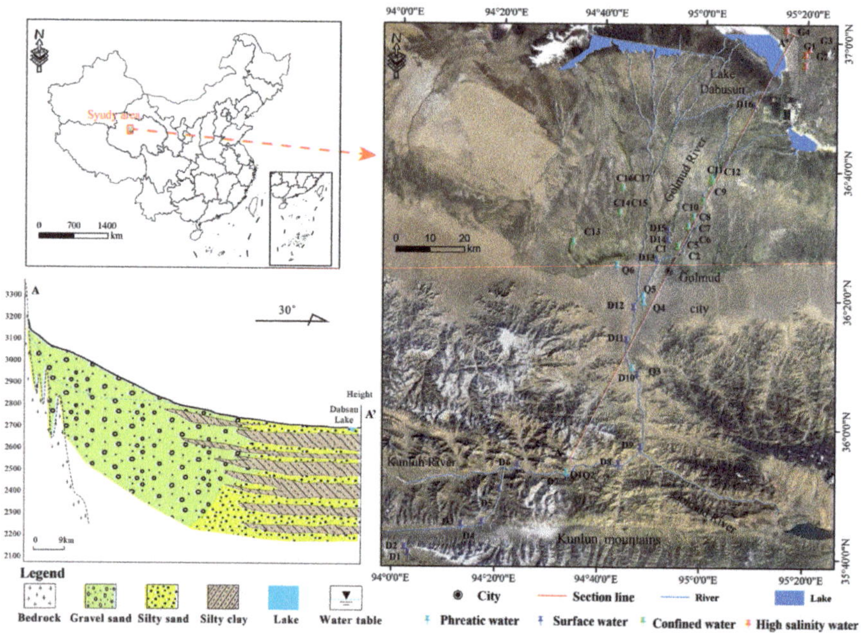

Figure 1. Topographical map of the Golmud River watershed and sampling locations along the groundwater flowpath.

2.2. Hydrochemical Sampling and Measurement

In this study, a total of 43 water samples were collected in August 2022, including 16 river samples, 6 phreatic water samples, 17 confined water samples, and 4 high salinity water samples. Surface water samples were collected along the Golmud River, while groundwater samples were obtained from boreholes or artesian wells (Figure 1). Groundwater samples were collected in field after at least 10 min of pumping or flowing until pH, dissolved oxygen (DO), and redox potential (ORP) had stabilized, ensuring that the samples accurately represent the native hydrochemical characteristics. Prior to sampling, all sample bottles were rinsed three times with the original water to be collected. Hydrogen and oxygen isotope samples were stored in 60 mL brown polyethylene bottles, while samples for cations, anions, and trace elements were stored in 500 mL polyethylene bottles. All water samples were filtered through 0.45 μm filter in the field. For cation analysis, 2–3 drops HNO_3 (analytical grade) were added to prevent precipitation. All sampling bottles were filled to avoid headspace.

pH, ORP, DO, and electrical conductivity (EC) were measured in the field with a set of portable hydrochemical testers (CLEAN, Shanghai ZhenMai Instruments, Shanghai, China). Cl^-, $SO_4{}^{2-}$, F^-, and $NO_3{}^-$ were analyzed using Ion Chromatography (ICS-90, Dionex

Thermo Scientific, Sunnyvale, CA, USA), while HCO_3^- and CO_3^{2-} were measured by Acid-base titration auto-analyzer (905 Titrando, Metrohm AG, Herisau, Switzerland). Ca^{2+}, Mg^{2+}, Na^+, and K^+, as well as trace elements including Sr and Li, were determined using an Inductively Coupled Plasma Optical Emission Spectrometer (ICP-OES, SPECTRO Analytical Instruments, Kleve, Germany). TDS was calculated as the sum of major ions minus half the concentration of HCO_3^-. Si and $HSiO_3^-$ were analyzed using spectrophotometry. Stable isotopes (^2H and ^{18}O) were measured using a gas isotope ratio mass spectrometry (LGR LWIA-24d, Los Gatos Research, Inc., Los Gatos, CA, USA) and reported with respect to the Vienna Standard Mean Ocean Water (VSMOW) standard. All analytical results met the charge balance requirements for hydrochemical cation–anion equilibrium, which was within the 5% error limit [24].

2.3. Hydrochemical Analyses

This study calculated the correlations among hydrochemical variables to assess the similarity of dissolved component sources in groundwater and to identify potential sources of dissolved minerals [25]. The correlation matrix, using color gradients and significance markers, was employed to visually display the magnitude and statistical significance of the relationships among hydrochemical variables. The Piper diagram was used to identify groundwater types and to analyze hydrochemical evolution processes along the groundwater flow path [26]. The Gibbs diagram was applied to identify the controlling mechanisms of major ionic compositions in groundwater by examining the relationships among $Na^+/(Na^++Ca^{2+})$, $Cl^-/(HCO_3^-+Cl^-)$, and TDS [27,28]. Based on the geochemical characteristic variations of weathering products from different rock types (silicate, carbonate, and evaporites), this study analyzed the distribution patterns of Mg^{2+}/Na^+ and HCO_3^-/Na^+ ratios relative to Ca^{2+}/Na^+ to identify the primary sources of dissolved components in groundwater [29]. In addition, the precipitation and dissolution processes of various minerals determine the quantitative relationships among hydrochemical parameters in groundwater. This study conducted a comprehensive analysis of ionic ratios to identify the potential mineral sources [30]. The evaporation–concentration process and cation exchange are also important mechanisms constraining the hydrochemical characteristics of groundwater [31]. The chlor-alkaline index (CAI) was used to assess whether cation exchange occurs in groundwater:

$$CAI - I = \frac{Cl^- - (Na^+ + K^+)}{Cl^-} \tag{1}$$

$$CAI - II = \frac{Cl^- - (Na^+ + K^+)}{HCO_3^- + SO_4^{2-} + CO_3^{2-} + NO_3^-} \tag{2}$$

Negative values of CAI-I and CAI-II indicate that Ca^{2+} and/or Mg^{2+} in groundwater have been exchanged with Na^+ and/or K^+ from the surrounding aquifer medium. Conversely, positive values suggest the opposite exchange direction [32].

2.4. Inverse Hydrochemical Modeling Approach

The inverse hydrochemical modeling approach applies the principle of mass conservation to calculate the possible water–rock interactions that may occur along a flow path between two points (a starting point and an endpoint), providing multiple potential solutions for the mineral reactions involved [33]. In detail, inverse modeling primarily involves the following key points. First, appropriate simulation paths are selected according to actual groundwater flow directions and hydrochemical characteristics. Next, the dissolution or precipitation status of potential minerals is identified based on groundwater hydrochemical data. This step includes quantifying the chemical compositions at the start and end points of the flow path and defining potential mineral and constraint phases based on the hydrogeological background and saturation index (SI) [30]. The saturation index is a quantitative measure to assess the equilibrium state of a specific mineral in hydrochemical samples, where

$$SI = \log(\frac{IAP}{K_{sp}}) \tag{3}$$

IAP is the ion activity product of the dissolved ions that form the specific mineral. K_{sp} is the solubility product constant of the corresponding mineral at the given temperature. Then, the inverse hydrochemical model is launched to generate possible reaction pathways. Inverse hydrochemical modeling was performed using Phreeqc software (version 3.3.5) [12]. The most plausible solution is evaluated and selected according to regional hydrogeological conditions, and the dissolution and precipitation processes and transfer amounts of minerals along the simulated path are interpreted accordingly [34]. Cation exchange was also considered, as determined by mole-balance modeling [12].

3. Results and Discussion

3.1. Hydrochemical Characterization in the Golmud River Watershed

The hydrochemical data of groundwater and river water in the study area are summarized in Table 1, and the results reveal pronounced variations in key hydrochemical parameters across surface river water, phreatic groundwater, confined groundwater, and high-salinity groundwater. The pH values ranged from 6.5 to 9.5, with phreatic groundwater (8.05) and confined groundwater (8.05) showing relatively small fluctuations and indicating weakly alkaline conditions. River water exhibited a slightly higher pH (8.48) compared to phreatic and artesian groundwater, while high-salinity groundwater had the lowest pH values (6.6), representing neutral to slightly acidic conditions. The total dissolved solids (TDS) of groundwater ranged from 230.19 mg/L to 302,909.31 mg/L, showing a continuous accumulation of dissolved ionic components along the groundwater flow path (Table 1). Similarly, surface water exhibited a comparable trend of accumulation, with TDS values ranging from 230.19 mg/L to 14,247.12 mg/L. Meanwhile, the concentration of DO decreased from unconfined groundwater (43.42%) to confined groundwater (26.94%) and ORP also showed a similar decline from an average of 208 mV in unconfined groundwater to −3.76 mV in confined groundwater, indicating a transition from oxidizing to reducing conditions along the groundwater flow path (Table 1). These systematic variations reflect distinct hydrochemical evolutionary pathways governed by regional hydrological and geochemical processes.

To investigate the interrelationships among different hydrochemical variables, a correlation analysis was conducted (Spearman, Figure 2). The results showed that the major cations (Na^+, K^+, and Mg^{2+}) and anions (Cl^- and $SO_4{}^{2-}$) in groundwater are significantly positively correlated ($p < 0.05$), suggesting that these ions likely originated from a relatively consistent source. Both Li^+ and Sr^{2+} also exhibited significant correlations with major cations (Na^+, K^+, and Mg^{2+}) and anions (Cl^- and $SO_4{}^{2-}$), hinting that these components may derive from common mineral sources. The primary sources of Li^+ and Sr^{2+} are largely released from the weathering and dissolution of carbonate and silicate minerals. Their enrichment is jointly controlled by water–rock interactions and evaporation concentration processes [35–37]. Studies have shown that groundwater in arid regions normally contains higher concentrations of Li^+ and Sr^{2+} [36]. The pH values displayed significant correlations with $HCO_3{}^-$ and $CO_3{}^{2-}$ ($p < 0.05$), indicating the control of carbonate equilibrium on $HCO_3{}^-$ and $CO_3{}^{2-}$ in the groundwater system. A positive correlation was also observed between $HSiO_3{}^-$ and $HCO_3{}^-$, suggesting that the carbonate equilibrium influences the dissolution of silicate minerals. This implies that multiple hydrochemical processes interact to form a complex hydrochemical environment, highlighting the need for further investigation into its evolution and controlling mechanisms. Additionally, NO_3^- exhibited weak correlations with other hydrochemical variables, and its concentration showed minimal variation among river water, unconfined groundwater, and confined groundwater, suggesting limited anthropogenic influence in the study area [38].

Table 1. Groundwater hydrochemical parameters in the Golmud River watershed.

Variables	Phreatic Water			Confined Water			High-Salinity Water			Surface Water		
	Maximum	Minimum	Average	Maximum	Minimum	Average	Maximum	Minimum	Average	Maximum	Minimum	Average
pH	8.37	7.61	8.05	8.57	7.44	8.05	6.66	6.55	6.60	9.35	8.19	8.48
DO	56.80	33.30	43.42	63.30	6.00	26.94	/	/	/	77.50	44.70	53.17
ORP	245.00	207.00	208.00	201.00	−234	−3.76	/	/	/	249.00	149.00	209.92
Alkalinity	198.00	168.00	185.17	311.90	173.00	211.70	263.00	178.00	217.75	217.00	97.10	159.57
TDS	755.03	420.59	522.99	1048.00	391.36	500.34	302,909.31	290,478.31	295,928.85	14,247.12	230.19	1201.58
Na	122.00	54.50	86.28	374.08	64.80	105.00	61,385.00	51,087.00	53,878.50	3338.00	14.20	250.56
K	5.38	3.01	4.32	6.99	2.68	5.51	16,739.00	2969.00	7238.75	200.00	1.80	15.66
Ca	68.30	38.70	51.97	73.00	4.93	36.80	2804.00	1052.00	1619.25	196.00	33.90	50.79
Mg	54.70	30.60	37.67	42.20	7.77	32.43	39,785.00	34,250.00	36,763.50	1186.00	15.80	97.84
Cl	209.00	77.20	123.77	200.56	79.70	102.63	194,240.00	189,970.00	192,451.25	8573.00	13.70	594.19
SO_4	188.00	73.30	102.27	139.96	62.60	78.97	4655.00	3336.00	3840.25	641.00	21.30	92.76
CO_3	0.00	0.00	0.00	9.36	0.00	1.07	0.00	0.00	0.00	58.60	0.00	4.71
HCO_3	241.00	205.00	225.50	404.77	211.00	258.46	320.00	218.00	265.75	264.00	108.00	184.94
NO_3	4.87	3.56	4.33	4.77	0.00	2.72	47.10	30.90	40.90	4.29	1.47	2.92
F	0.39	0.28	0.32	2.34	0.44	0.79	5.31	3.85	4.48	0.52	0.10	0.24
Li	0.07	0.04	0.06	0.06	0.02	0.04	66.48	1.76	45.04	59.71	0.02	4.02
Si	3.85	3.24	3.46	7.10	4.09	5.58	0.56	0.43	0.51	2.83	0.39	2.09
Sr	1.24	0.69	0.90	1.06	0.43	0.85	33.92	3.04	21.10	36.13	0.28	3.10
$HSiO_3$	9.58	9.46	9.53	20.10	11.50	15.30	9.93	7.44	8.86	8.71	6.61	7.32

Note: "/" represents data not available.

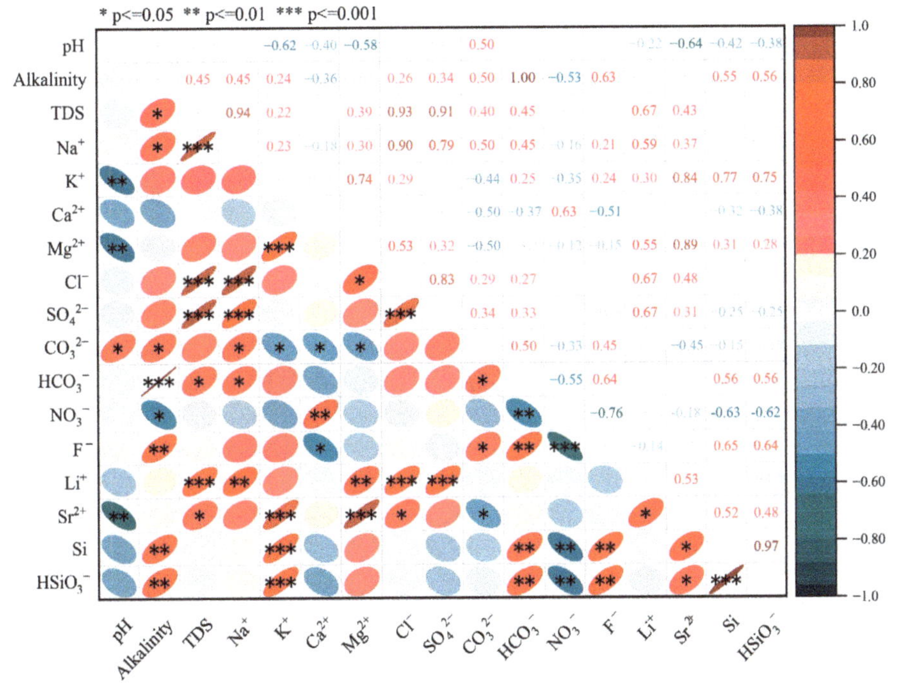

Figure 2. Correlation matrix of hydrochemical variables.

The Piper diagram revealed a systematic variation in hydrochemical types along the groundwater flow path (Figure 3). The hydrochemical type of phreatic groundwater was mainly composed of HCO_3·Cl-Ca·Mg Na, while confined groundwater exhibited more complex types, including HCO_3·Cl-Ca·Mg Na and HCO_3·Cl-Na Mg. In contrast, high-salinity groundwater was predominantly of the Cl-Mg·Na type. The hydrochemical evolution from phreatic to high-salinity groundwater followed a clear 'two-step' pattern [24]. In the first stage, while the proportions of major anions remained stable, the dominant cations shifted from Ca^{2+} and Mg^{2+} in the phreatic aquifer to Na^{+} and K^{+} in the confined aquifer. During this stage, the hydrochemical type experienced a significant transformation with little variation in TDS (Table 1). The second stage was characterized by a sharp increase in TDS along the groundwater flowpath, with Cl^{-} and Mg^{2+} emerging as the dominant ions in terms of ionic proportions. This pattern has also been identified in previous studies [21,24]. The complex hydrochemical evolution is controlled by multiple interacting mechanisms and merits further investigation [14]. In addition, surface river water was mainly characterized by HCO_3·Mg-Ca and HCO_3·Cl-Mg Na·Ca types and displayed evolutionary trends similar to those of low TDS groundwater.

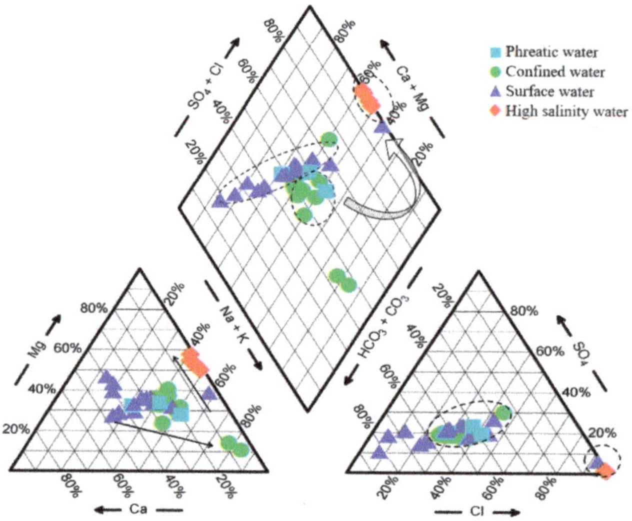

Figure 3. Piper diagram showing the hydrochemical characteristics of the study area.

3.2. Hydrogen and Oxygen Isotopes

The relationship between stable hydrogen (D) and oxygen (^{18}O) isotopes in groundwater and surface water can be used to identify groundwater recharge sources and the interactions between surface water and groundwater, thereby providing insights into the groundwater circulation and evolution processes [39]. In the study area, the $\delta^{18}O$ and δD values of phreatic groundwater ranged from $-10.74‰$ to $-9.32‰$ and from $-69.58‰$ to $-63.49‰$, respectively. For groundwater in confined aquifers, $\delta^{18}O$ values ranged from $-11.76‰$ to $-10.78‰$ and δD values from $-81.88‰$ to $-73.05‰$. In contrast, isotopic compositions of surface water ranged from of $-11.92‰$ to $0.09‰$ in $\delta^{18}O$ and from $-75.31‰$ to $-22.11‰$ in δD values (Figure 4). The $\delta^{18}O$ and δD values of all samples deviated significantly from the local meteoric water line of the Golmud River watershed (LMWL-GEM, $\delta D = 7.84\delta^{18}O-4.5$) [40], indicating that precipitation in the study area was not the primary recharge source for either groundwater or surface water [17]. Most river water, phreatic groundwater, and confined groundwater samples were distributed close to the atmospheric precipitation lines of high-altitude meteorological stations, including the Kunlun Mountain Pass station (LMWL-KMP, $\delta D = 8.5\delta^{18}O + 18.39$, >4700 m above sea level [m a.s.l.]) and the Nachitai station (LMWL-NCT, $\delta D = 7.4\delta^{18}O + 13.2$, 3552 m a.s.l.), as well as the glacier meltwater line ($\delta D = 7.32\delta^{18}O + 9.44$) [41,42]. This distribution illustrates that groundwater is mainly recharged by high-altitude atmospheric precipitation and glacier/snow meltwater. Compared to phreatic groundwater, the δD and $\delta^{18}O$ values of confined groundwater are more depleted and fall along the LMWL-KMP line, suggesting that the confined groundwater was recharged under colder climatic conditions, which led to lower isotopic values. In contrast, the isotopic signatures of river water and phreatic groundwater were closer to the glacier meltwater line, reflecting a significant contribution of glacier meltwater to river recharge [17]. The δD and $\delta^{18}O$ values of phreatic groundwater were close to those of river water, indicating hydraulic connectivity between river and unconfined aquifer (Figure 4). The $\delta^{18}O$ and δD values of high-salinity groundwater ranged from −0.44‰ to 0.85‰ and from −34.84‰ to −26.07‰, respectively, and deviated significantly from the local meteoric water line, suggesting that high-salinity groundwater was strongly affected by evaporation (Figure 4). In addition, there was a river sample (D16) which collected adjacent to the lake inlet, exhibiting enriched isotopic compositions similar to those of high-salinity groundwater, which was also attributed to strong surface evaporation and concentration processes.

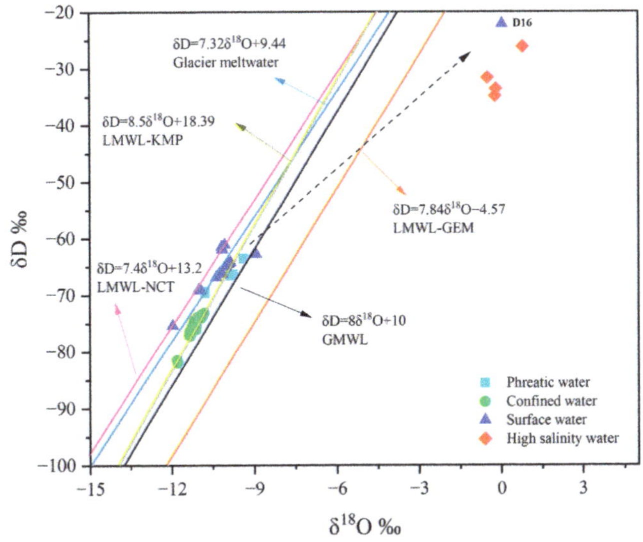

Figure 4. Stable isotope composition (δD and $\delta^{18}O$) of phreatic, confined, and high-salinity groundwater, as well as the river water. Reference lines in the plot consist of the Global Meteoric Water Line (GMWL) and various Local Meteoric Water Lines (LMWLs).

3.3. Hydrochemical Evolution and Its Controlling Mechanisms

3.3.1. Shift from Water–Rock Interaction to Evaporation Dominance

The dissolution–precipitation of evaporites and carbonate minerals, as well as the weathering of silicate minerals, significantly influenced the hydrochemical composition of groundwater [43,44]. The controlling mechanisms of hydrochemical evolution were assessed using the Gibbs diagram to identify rock weathering, evaporation, and atmospheric precipitation dominance in hydrochemical formation and evolution. The results showed that both phreatic and confined groundwater were mainly controlled by water–rock interactions (Figure 5). The release of Ca^{2+} and Mg^{2+} through mineral dissolution led to low $Na^+/(Na^+ + Ca^{2+})$ and $Cl^-/(Cl^- + HCO_3^-)$ ratios. In contrast, the hydrochemical composition of high-salinity groundwater was predominantly influenced by evaporation, resulting in elevated TDS levels and higher $Na^+/(Na^++Ca^{2+})$ and $Cl^-/(Cl^-+HCO_3^-)$ ratios. Along the groundwater flow path, the dominant hydrochemical processes gradually shift from water–rock interaction to evaporation dominance [14]. A similar hydrochemical process and evolutionary trend was also observed in river water (Figure 5). Evaporation in both surface river water and groundwater promotes TDS accumulation. This process elevates the proportions of Na^+ and Cl^-, leading to a higher $Na^+/(Na^++Ca^{2+})$ and $Cl^-/(Cl^-+HCO_3^-)$ ratios.

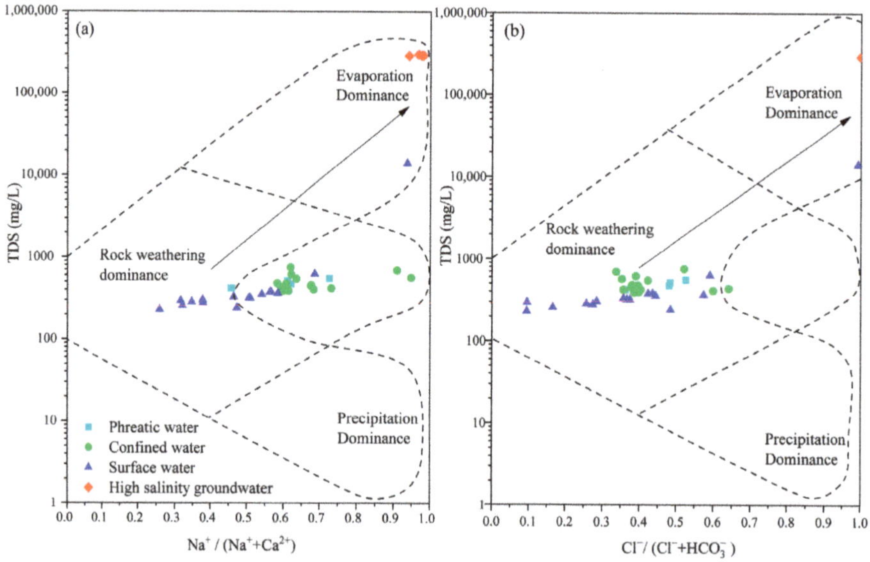

Figure 5. Gibbs diagram illustrates the controlling hydrochemical processes shifts in the study area. (**a**): TDS vs. $Na^+/(Na^++Ca^{2+})$, (**b**): TDS vs. $Cl^-/(Cl^-+HCO_3^-)$.

3.3.2. Major Sources of Hydrochemical Constituents

Water–rock interaction governs hydrochemical evolution and the major hydrochemical components in groundwater are primarily derived from interactions with host aquifer minerals [45]. Mineral dissolution typically occurs alongside precipitation, with the directions and rates of these reactions governed by equilibrium constants [46]. Analysis of major ionic stoichiometry in groundwater serves as a well-established methodology for identifying dissolved substrate sources and reconstructing hydrochemical processes [47]. The results showed that Na^++K^+ and Cl^- generally distributed along the 1:1 line in low TDS samples, consistent with the observed corresponding ionic correlations, indicating that Na^++K^+ and Cl^- primarily originated from halite dissolution (Figure 6a). However, high-salinity groundwater deviated significantly from the 1:1 line, with Na^++K^+ concentrations much higher than those of Cl^-, suggesting the possible involvement of cation exchange or additional input sources such as the weathering of carbonate or silicate minerals. The relationships between $(Ca^{2+}+Mg^{2+})$ and $(SO_4^{2-}+HCO_3^-)$ are widely used to identify the dissolution and precipitation of sulfate and carbonate minerals [24]. Similar to the relationship observed in Figure 6a, most low TDS groundwater and river water were distributed near the 1:1 line in Figure 6b, indicating that Ca^{2+} and Mg^{2+} largely originated from the dissolution of sulfate and carbonate minerals. In contrast, the high-salinity groundwater samples deviated from the 1:1 line and exhibited substantial enrichment in Ca^{2+} and Mg^{2+}, which may be attributed to cation exchange or weathering of other minerals such as silicates [6,7].

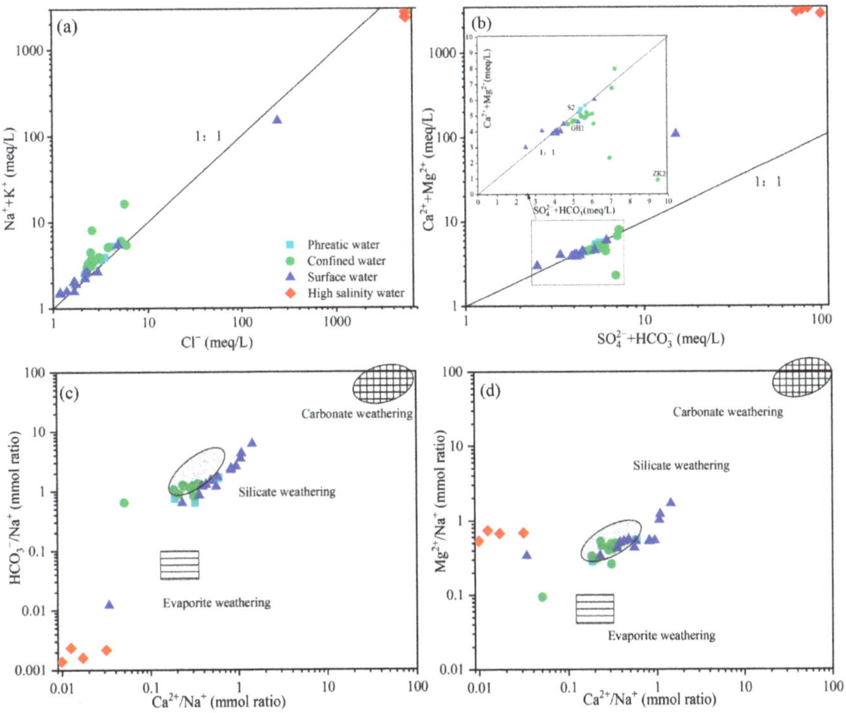

Figure 6. Scatter plot of hydrochemical variables and their ionic ratios. (**a**) Na^{+}+K^{+} vs. Cl^{-}, (**b**) Ca^{2+} + Mg^{2+} vs. HCO_3^{-} + SO_4^{2-}, (**c**) HCO_3^{-}/Na^{+} vs. Ca^{2+}/Na^{+}, and (**d**) Mg^{2+}/Na^{+} vs. Ca^{2+}/Na^{+}.

The dissolution–precipitation equilibria of silicate, carbonate, and evaporite minerals typically resulted in distinct characteristic ranges of Ca^{2+}/Na^{+}, Mg^{2+}/Na^{+} and HCO_3^{-}/Na^{+} molar ratios in groundwater, which can be used to infer the dominant hydrochemical processes governing water–rock interactions [48]. High-salinity groundwater samples in the study area mainly plotted adjacent to the evaporite weathering and dissolution area (Figure 6c,d), indicating that evaporite minerals (e.g., halite and potash) played a dominant role in controlling the hydrochemical characteristics. Elevated Na^{+} and K^{+} concentrations during evaporite weathering resulted in lower Ca^{2+}/Na^{+} and Mg^{2+}/Na^{+} ratios. The result suggests a significant contribution of evaporite dissolution to the hydrochemistry of high-salinity groundwater. However, the relatively elevated Mg^{2+}/Na^{+} ratios observed in high-salinity groundwater deviated from the typical evaporite dissolution signature, suggesting Mg^{2+} and Na^{+} concentrations were modulated by additional processes requiring further discrimination (Figure 6d). Low TDS samples were mostly located between the silicate, carbonate, and evaporite endmembers, with a tendency toward the silicate weathering domain, indicating that their hydrochemical composition was influenced by multiple mineral sources, particularly by silicate weathering.

3.3.3. Silicate Mineral Equilibrium Diagram

The results as mentioned above suggested that silicate minerals exerted notable influences on the hydrochemical evolution in the Golmud River watershed. To further examine the role of silicate weathering, the activities among Ca^{2+}/H^{+}, Mg^{2+}/H^{+}, Na^{+}/H^{+}, K^{+}/H^{+}, and H_4SiO_4 were analyzed. These relationships provide insight into the equilibrium state of silicate minerals, thereby clarifying the impact of silicate weathering on hydrochemical evolution in the study area [49–51]. All groundwater samples in the study area fell to the left of the quartz equilibrium line, indicating that groundwater was undersaturated with respect to quartz and exhibited a tendency for further dissolution. This result reflects the relatively low concentration of dissolved SiO_2 in groundwater. Under such conditions, additional soluble SiO_2 may be released through the weathering and dissolution of silicate minerals. In the equilibrium diagrams of anorthite, K-feldspar, and albite and their secondary minerals, the surface water, phreatic groundwater, and confined groundwater samples were all located within the stability fields of gibbsite, kaolinite, and Na-montmorillonite (Figure 7a,c,d). This indicates that during the hydrochemical evolution of the groundwater flow path, anorthite, K-feldspar, and albite tend to dissolve, contributing to the formation of corresponding secondary minerals [48]. In the chlorite equilibrium diagram, almost all groundwater samples were located within the stability range of chlorite, while only a few fell into the gibbsite field, indicating a state of equilibrium between chlorite and groundwater (Figure 7b). High-salinity groundwater exhibited relatively low concentrations of dissolved SiO_2. Due to elevated TDS and low pH, these samples consistently fell within the equilibrium fields of aluminosilicate minerals relative to groundwater (Figure 7), which reflects that related aluminosilicate minerals existed in the study area are generally in a state of equilibrium and, thus, exert limited influences on hydrochemical evolution in the high-salinity groundwater of the study area.

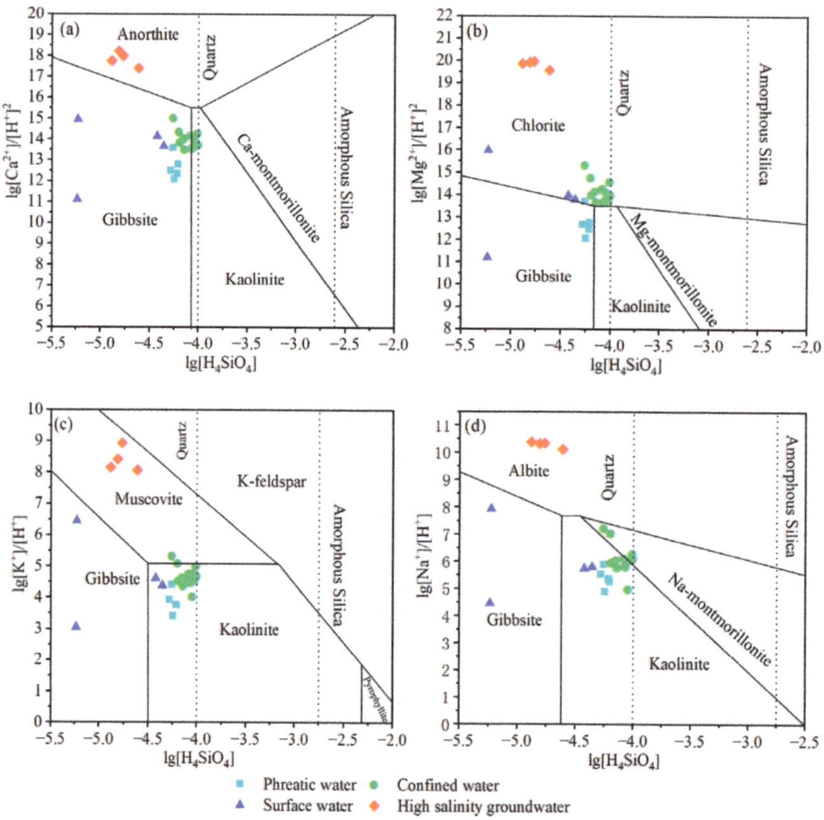

Figure 7. Log activity diagrams of (**a**) $CaO-Al_2O_3-SiO_2-H_2O$, (**b**) $MgO-Al_2O_3-SiO_2-H_2O$, (**c**) $K_2O-Al_2O_3-SiO_2-H_2O$, and (**d**) $Na_2O-Al_2O_3-SiO_2-H_2O$ versus H_4SiO_4, illustrating the thermodynamic stability fields of relevant silicate minerals and the chemical equilibrium states of different water types in relation to mineral dissolution or precipitation.

3.3.4. Cation Exchange in the Groundwater Flow System

In the study area, some hydrochemical anomalies could not be fully explained by mineral dissolution–precipitation and evaporation processes or silicate mineral equilibrium states. A notable example is the enrichment of Mg^{2+} ratios in high-salinity groundwater. According to the previous results, we propose that cation exchange has exerted a significant influence on hydrochemical evolution. This study employed chlor-alkaline indices (CAI-I and CAI-II) to evaluate the occurrence of cation exchange processes in groundwater [28,46]. The results showed that the CAI-I and CAI-II values of river water and phreatic groundwater samples clustered together with near zero values, suggesting a negligible cation exchange process in these hydrological compartments (Figure 8). Further, confined groundwater samples located in the third quadrant showed negative CAI values. This trend suggests an increase in Na^+ and K^+ relative to Cl^-, implying that the replacement of Na^+ and K^+ in the aquifer medium by dissolved Ca^{2+} and Mg^{2+} occurred in the confined aquifer, ultimately leading to the enrichment of Na^+ and K^+ in confined groundwater. High-salinity groundwater samples were plotted in the first quadrant with positive CAI values, indicating a profoundly reverse cation exchange process, where dissolved Na^+ and K^+ exchange mineral-bound Ca^{2+} and Mg^{2+} in aquifer medium. It is noteworthy that, due to the limited solubility of carbonate minerals, the reverse exchange in high-salinity groundwater does not result in a substantial release of Ca^{2+}., in line with the hydrochemical dataset. Instead, Mg^{2+} is predominantly exchanged and released into groundwater, which explains the continuous increase in Mg^{2+} concentrations while Ca^{2+} remains relatively stable in high-salinity groundwater.

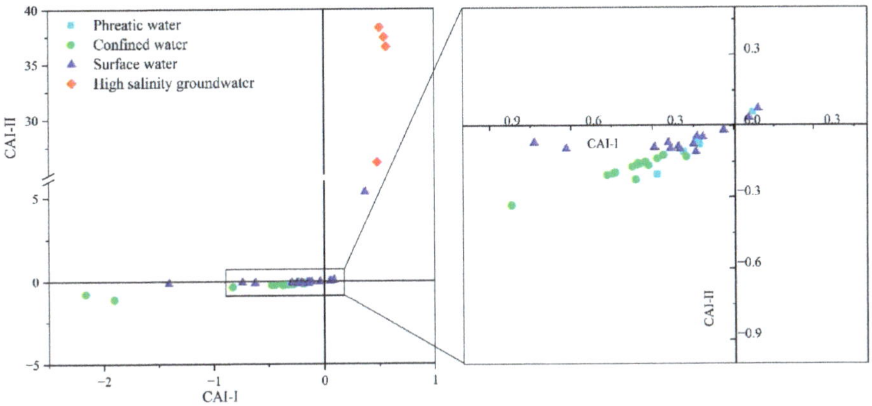

Figure 8. Chlor-alkaline indices (CAI-I and CAI-II) of different water types, indicating ion exchange processes between groundwater and aquifer medium.

3.4. *Inverse Hydrochemical Modeling*

The preceding results qualitatively delineated the hydrochemical variations and their controlling mechanisms in the Golmud River watershed. However, the accuracy of these interpretations lacked validation through quantitative simulations. To address this limitation, representative groundwater flow paths were selected for inverse hydrochemical modeling based on the regional groundwater flow system. The simulation route included Q3-Q5, C1-C3, C5-C11, and C11-G3 to capture the evolution from phreatic groundwater to high-salinity groundwater. The modeling framework identified dominant water–rock interactions during hydrochemical evolution by tracking hydrochemical variations along the groundwater flow path.

3.4.1. Possible Mineral Phases and Constraints

Considering the aforementioned results and lithological features of the Golmud River watershed, silicates, evaporites, and carbonates represent the dominant mineral assemblages in the study area [52]. The saturation indices (SI) of representative minerals were calculated using Phreeqc to assess their potential dissolution and precipitation equilibrium states (Figure 9). The SI values of gypsum ranged from −6 to −2, indicating that it was undersaturated and tended to dissolve in groundwater. Calcite and quartz showed SI values between −2 and 2, suggesting near-saturation and a dynamic equilibrium between precipitation and dissolution. Dolomite exhibited SI values of 4 and 6, reflecting a supersaturated state and a tendency to precipitate. Albite and K-feldspar had negative SI values, indicating undersaturation to near-saturation states. Combined with the earlier results of silicate mineral equilibrium state (Figure 7), albite and K-feldspar were inferred to dissolve in groundwater. In all, the selected mineral phases for inverse hydrochemical modeling include gypsum, dolomite, calcite, halite, K-feldspar, albite, and quartz. The effect of carbonate equilibrium CO_2 (g) on hydrochemical evolution is also considered, reflecting its role in regulating dissolution–precipitation equilibrium along the groundwater flow path [53].

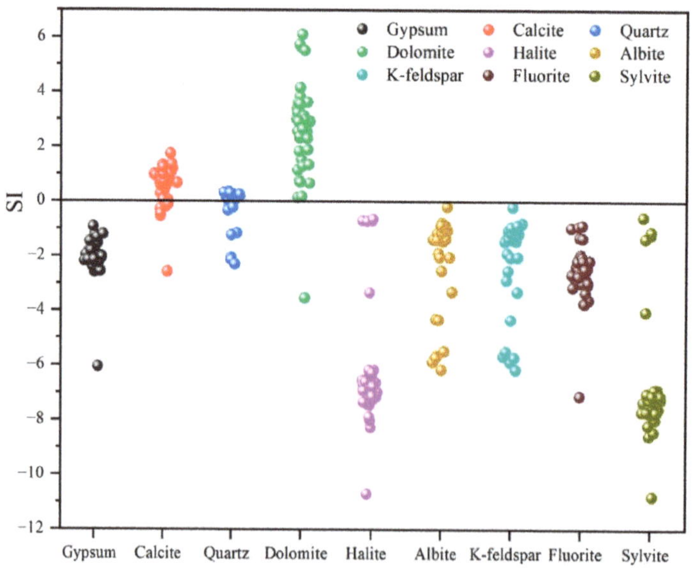

Figure 9. Saturation index (SI) of potential minerals in the study area. The saturation index is characterized by a threshold value of 0 (SI > 0: supersaturation; SI < 0: undersaturation; SI = 0: equilibrium).

Previous results have shown that water–rock interactions and evaporation concentration, with additional influence from processes such as cation exchange, primarily control hydrochemical evolution [54,55]. Based on these findings, the potential mineral-groundwater reactions were selected accordingly (Table 2).

Table 2. Chemical reaction of possible minerals in the groundwater flow system.

Mineral Phase	Chemical Formula	Chemical Reaction
Calcite	$CaCO_3$	$CaCO_3 \rightleftharpoons Ca^{2+} + CO_3{}^{2-}$
Gypsum	$CaSO_4 \cdot 2H_2O$	$CaSO_4 \cdot 2H_2O \rightleftharpoons Ca^{2+} + SO_4{}^{2-} + 2H_2O$
Quartz	SiO_2	$SiO_2 + 2H_2O \rightleftharpoons H_4SiO_4$
Dolomite	$CaMg\ (CO_3)_2$	$CaMg\ (CO_3)_2 \rightleftharpoons Ca^{2+} + Mg^{2+} + 2CO_3{}^{2-}$
Halite	$NaCl$	$NaCl \rightarrow Na^+ + Cl$
Sylvite	KCl	$KCl \rightarrow K^+ + Cl^-$
Albite	$NaAlSi_3O_8$	$NaAlSi_3O_8 + 8H_2O \rightleftharpoons Na^+ + Al\ (OH)^{4-} + 3H_4SiO_4$
K-feldspar	$2KAlSi_3O_8$	$2KAlSi_3O_8 + 11H_2O + 2CO_2 \rightleftharpoons 2K^+ + 2HCO_3{}^- + 4H_4SiO_4 + Al_2Si_2O_5(OH)_4$
Cation exchange	CaX_2 MgX_2	$Ca^{2+} + 2NaX \rightleftharpoons 2Na^+ + CaX_2$ $Mg^{2+} + 2NaX \rightleftharpoons 2Na^+ + MgX$

3.4.2. Results of Inverse Modeling

Due to the non-uniqueness solutions of inverse hydrochemical modeling, the optimal solution requires a comprehensive consideration of the hydrogeological conditions, saturation indices, and hydrochemical analysis of the study area. The results of the mass balance simulation are listed below (Table 3).

Table 3. Groundwater inverse simulation results (in mmol/L).

Modeling Path	Q3-Q5	C1-C3	C5-C11	C11-G3
Gypsum	/	0.0595	0.4149	16.81
Calcite	0.3478	0.4783	2.508	/
Quartz	/	/	/	/
Dolomite	−0.1852		−0.7159	
Halite	0.6455	0.1441	0.5851	7268
Sylvite	0.0038	0.0311	/	188.6
Albite	/	/	0.0096	/
K-Feldspar	/	0.0152	−0.0053	/
Fluorite	−0.0023	−0.0023	/	/
kaolinite	/	−0.0076	−0.0021	/
CaX2	−0.3949	−0.7426	−2.864	73.53
NaX	0.7899	1.197	5.727	−4400
MgX2	/	0.1442	/	2127
CO2 (g)	0.0228	0.5226	2.09	/

Note: /: data not available, +: dissolve, −: precipitation.

In the phreatic aquifer of the piedmont alluvial fan, hydrochemical evolution was primarily driven by the dissolution of calcite (0.348 mmol/L) and halite (0.645 mmol/L) and the precipitation of dolomite (−0.182 mmol/L), coupled with minor dissolution of sylvite (0.004 mmol/L) and CO_2 (0.023 mmol/L). The dissolution of calcite, halite, and potash minerals served as the dominant mechanism in shaping the hydrochemical characteristics of phreatic groundwater. Continuous input of Ca^{2+} led to a small amount of dolomite precipitation. Meanwhile, cation exchange between dissolved Ca^{2+} and mineral-bound Na^+ limited Ca^{2+} concentrations in groundwater. These processes collectively explained the deviation of phreatic groundwater from the 1:1 stoichiometric line of (Ca^{2+}+Mg^{2+}) and ($SO_4{}^{2-}$+$HCO_3{}^-$) towards the anion-dominant side (Figure 6b).

In the transitional zone from the piedmont alluvial fan to the upper part of the alluvial-lacustrine plain, changes in aquifer composition and structure led to the shift from phreatic to confined groundwater. The confined groundwater along the C1–C3 modeling path was primarily influenced by the dissolution of calcite (0.478 mmol/L), gypsum (0.059 mmol/L), halite (0.144 mmol/L), and sylvite (0.031 mmol/L). Due to the intrinsic features of confined aquifer, increased CO_2 dissolution enhanced the weathering of silicate minerals such as K-feldspar (0.015 mmol/L). The formation of kaolinite (−0.008 mmol/L) from K-feldspar dissolution is consistent with the results of the silicate equilibrium state (Figure 7c). The alluvial-lacustrine plain was the main distribution zone of confined groundwater in the study area, particularly for artesian groundwater. Followed the modeling path from C5 to C11, the hydrochemical processes were generally similar to those observed upstream (C1–C3), including the dissolution of calcite (2.508 mmol/L), gypsum (0.415 mmol/L), halite (2.508 mmol/L), and albite (0.01 mmol/L) and the precipitation of dolomite (−0.716 mmol/L). Throughout the hydrochemical evolution of the confined aquifer, a continuous cation exchange process, where dissolved Ca^{2+} replaced mineral-bound Na^+, modified the spatial distribution of Ca^{2+}, Na^+, and Mg^{2+}, serving as an important mechanism that constrains hydrochemical evolution [56].

C11-G3 was chosen to simulate and quantify the hydrochemical evolutionary process from confined groundwater to high-salinity groundwater. The modeling results suggested that the initial groundwater sample (C11) would need to undergo a 32.7-fold concentration to reproduce the hydrochemical characteristics observed in the high-salinity groundwater at the terminal point, confirming that evaporation concentration is the dominant process controlling the formation of high-salinity groundwater. The dissolution of gypsum (0.514 mmol/L), halite (222.263 mmol/L), and sylvite (5.768 mmol/L) was the primary source of dissolved ions during the modeling path. This model also revealed an intense reversed cation exchange process, wherein dissolved Na^+ in groundwater exchanges mineral-bound Mg^{2+} and Ca^{2+} in aquifer minerals. However, due to the limited solubility of carbonate minerals, the exchange primarily involves Na^+ replacing Mg^{2+}, which supports the observed increase in Mg^{2+} concentrations from confined to high-salinity groundwater (stage 2). The substantial precipitation of H_2O in the simulation corresponds to water loss caused by evaporation.

Based on the modeling results, mass balance equations (in mmol/L) were obtained for each modeling path.

Q3 + 0.348 Calcite + 0.646 Halite + 0.004 Sylvite + 0.790 NaX + 0.023 CO_2 (g) − 0.185 Dolomite − 0.002 Fluorite − 0.395 CaX_2 = Q5

C1 + 0.059 Gypsum + 0.478 Calcite + 0.144 Halite + 0.031 Sylvite + 0.01 K-Feldspar + 1.197 NaX + 0.144 MgX_2 + 0.523 CO_2 (g) − 0.002Fluorite − 0.008 kaolinite = C3

C5 + 0.415 Gypsum + 2.508 Calcite + 0.585 Halite + 0.01 Albite + 5.727 NaX +2.090 CO_2 (g) − 0.716 Dolomite − 0.005 K-Feldspar − 0.002 kaolinite − 2.864 CaX_2 = C11

C11 + 0.514 Gypsum + 222.263 Halite + 5.768 Sylvite + 2.249 CaX_2 + 65.046 MgX_2 − 134.557 NaX = G3/32.7

4. Conclusions

This study integrated hydrochemical analysis, hydrogen and oxygen isotopes, and inverse hydrochemical modeling to investigate hydrochemical evolution in a typical arid endorheic watershed from the Qaidam Basin, northwestern China. The δD and $\delta^{18}O$ values indicated that the primary recharge sources of groundwater are from high-altitude atmospheric precipitation and glacier/snowmelt. Regionally, groundwater types evolved from HCO_3·Cl-Ca·Mg·Na types in phreatic aquifers to HCO_3·Cl-Ca·Mg Na and HCO_3·Cl-Na Mg types in confined aquifers, with terminal high-salinity groundwater dominated by Cl-Mg·Na type, characterized by progressive TDS enrichment along the groundwater flow path. The hydrochemical evolution of regional groundwater in the Golmud River watershed is predominantly controlled by water–rock interactions and evaporation concentration processes, with the dominant mechanism shifting from water–rock interaction to evaporation concentration along the groundwater flow path. The dissolution of halite, gypsum, calcite, K-feldspar, and albite provided the primary source of dissolved components in shaping hydrochemical characteristics. Additionally, precipitation dissolution equilibrium, cation exchange, and carbonate equilibrium influenced hydrochemical evolution. Inverse hydrochemical modeling confirmed the findings of the hydrochemical analysis and further demonstrated that high-salinity groundwater has experienced intense evaporation concentration. The enrichment of Mg^{2+} in high-salinity groundwater was mainly attributed to reverse cation exchange, highlighting that hydrochemical evolution processes in the Golmud River watershed act as a major driving force for regional geochemical cycling. The study area hosts abundant strategic mineral resources (e.g., Li, K). Integrating the multidisciplinary approaches employed at the present study, future investigations could elucidate mineral transport processes and enrichment mechanisms from a hydrochemical evolution perspective, providing a theoretical basis for understanding hydrochemical metallogenic processes.

Author Contributions: Conceptualization, L.G., W.J. and N.Y.; Data curation, Y.D.; Formal analysis, Y.D., H.F. and C.A.; Funding acquisition, L.G. and N.Y.; Investigation, L.G. and N.Y.; Methodology, C.A.; Project administration, L.G.; Software, Y.D. and H.F.; Supervision, W.J. and N.Y.; Validation, L.G. and C.A.; Visualization, H.F.; Writing—original draft, L.G. and Y.D.; Writing—review and editing, L.G., W.J. and N.Y. All authors have read and agreed to the published version of the manuscript.

Funding: This work was financially supported by grants from the National Natural Science Foundation of China (42107101), the China Postdoctoral Science Foundation (2022MD723822), and the Natural Science Basic Research Program of Shaanxi Province (2023-JC-QN-0317).

Data Availability Statement: The data presented in this study are available on request from the corresponding author due to their complexity and ongoing use in further analyses.

Acknowledgments: We sincerely thank Guangcai Wang and Hairu Mao for the valuable discussion and Xinming Song, Luoyao Xiong, Chenyu Wang for their assistance during the field investigation and sampling collection.

Conflicts of Interest: The authors declare that the research was conducted without any commercial or financial relationships that could be construed as potential conflicts of interest.

References

1. Kuang, X.; Liu, J.; Scanlon, B.R.; Jiao, J.J.; Jasechko, S.; Lancia, M.; Biskaborn, B.K.; Wada, Y.; Li, H.; Zeng, Z.; et al. The Changing Nature of Groundwater in the Global Water Cycle. *Science* **2024**, *383*, eadf0630. [CrossRef] [PubMed]
2. Scanlon, B.R.; Fakhreddine, S.; Rateb, A.; de Graaf, I.; Famiglietti, J.; Gleeson, T.; Grafton, R.Q.; Jobbagy, E.; Kebede, S.; Kolusu, S.R.; et al. Global Water Resources and the Role of Groundwater in a Resilient Water Future. *Nat. Rev. Earth Environ.* **2023**, *4*, 87–101. [CrossRef]
3. Elango, L.; Kannan, R. Rock–Water Interaction and Its Control on Chemical Composition of Groundwater. *Dev. Environ. Sci.* **2007**, *5*, 229–243.
4. Yuan, R.; Li, Z.; Guo, S. Hydrochemical evolution of groundwater in a river corridor: The compounded impacts of various environmental factors. *Discov. Water* **2024**, *4*, 32. [CrossRef]
5. Hou, Z.; Huang, L.; Zhang, S.; Han, X.; Xu, J.; Li, Y. Identification of Groundwater Hydrogeochemistry and the Hydraulic Connections of Aquifers in a Complex Coal Mine. *J. Hydrol.* **2024**, *628*, 130496. [CrossRef]
6. Li, P.; Qian, H.; Wu, J.; Zhang, Y.; Zhang, H. Major Ion Chemistry of Shallow Groundwater in the Dongsheng Coalfield, Ordos Basin, China. *Int. J. Mine Water* **2013**, *32*, 195–206. [CrossRef]
7. Zhang, B.; Zhao, D.; Zhou, P.; Qu, S.; Liao, F.; Wang, G. Hydrochemical Characteristics of Groundwater and Dominant Water–Rock Interactions in the Delingha Area, Qaidam Basin, Northwest China. *Water* **2020**, *12*, 836. [CrossRef]
8. Ma, B.; Jin, M.; Liang, X.; Li, J. Groundwater Mixing and Mineralization Processes in a Mountain–Oasis–Desert Basin, Northwest China: Hydrogeochemistry and Environmental Tracer Indicators. *Hydrogeol. J.* **2018**, *26*, 233–250. [CrossRef]

9. Yang, N.; Zhou, P.; Wang, G.; Zhang, B.; Shi, Z.; Liao, F.; Li, B.; Chen, X.; Guo, L.; Dang, X.; et al. Hydrochemical and Isotopic Interpretation of Interactions between Surface Water and Groundwater in Delingha, Northwest China. *J. Hydrol.* **2021**, *598*, 126243. [CrossRef]
10. Mao, H.; Wang, C.; Shen, Q.; Fu, L.; Wang, G.; Shi, Z. Source and Evolution of Sulfate in the Multi-Layer Groundwater System in an Abandoned Mine—Insight from Stable Isotopes and Bayesian Isotope Mixing Model. *Sci. Total Environ.* **2023**, *859*, 160368. [CrossRef]
11. Belkhiri, L.; Boudoukha, A.; Mouni, L.; Baouz, T. Application of Multivariate Statistical Methods and Inverse Geochemical Modeling for Characterization of Groundwater—A Case Study: Ain Azel Plain (Algeria). *Geoderma* **2010**, *159*, 390–398. [CrossRef]
12. Plummer, L.N.; Parkhurst, D.L. Application of the Pitzer Equations to the PHREEQE Geochemical Model. In *Chemical Modeling of Aqueous Systems II*; Chapter 10; ACS: Washington, DC, USA, 1990; pp. 128–137.
13. Sun, Y.; Wang, Y.; Zhang, W.; Ye, S. Regional water resources carrying capacity in China based on analytic hierarchy process and system dynamics model: A case study of Golmud City. *Front. Environ. Sci.* **2024**, *12*, 1450747. [CrossRef]
14. Xiao, Y.; Shao, J.; Frare, S.; Cui, Y.; Dang, X.; Wang, S.; Ji, Y. Groundwater Origin, Flow Regime and Geochemical Evolution in Arid Endorheic Watersheds: A Case Study from the Qaidam Basin, Northwestern China. *Hydrol. Earth Syst. Sci.* **2018**, *22*, 4381–4400. [CrossRef]
15. Xiao, Y.; Liu, K.; Yan, H.; Zhou, B.; Huang, X.; Hao, Q.; Zhang, Y.; Zhang, Y.; Liao, X.; Yin, S. Hydrogeochemical Constraints on Groundwater Resource Sustainable Development in the Arid Golmud Alluvial Fan Plain on Tibetan Plateau. *Environ. Earth Sci.* **2021**, *80*, 750. [CrossRef]
16. Guo, L.; Wang, G.; Sheng, Y.; Shi, Z.; Sun, X. Groundwater Microbial Communities and Their Connection to Hydrochemical Environment in Golmud, Northwest China. *Sci. Total Environ.* **2019**, *695*, 133848. [CrossRef]
17. Yang, N.; Wang, G.; Hou, E.; Guo, L.; Xiong, L.; Song, X. Triple Isotopes (δD, δ18O, δ17O) Characteristic of River Water and Groundwater in an Arid Watershed from Qaidam Basin, Northwestern China: Implications for Hydrological Cycle. *Sci. Total Environ.* **2024**, *927*, 172229. [CrossRef] [PubMed]
18. Xiao, Y.; Liu, K.; Zhang, Y.; Yang, H.; Wang, S.; Qi, Z.; Hao, Q.; Wang, L.; Luo, Y.; Yin, S. Numerical Investigation of Groundwater Flow Systems and Their Evolution Due to Climate Change in the Arid Golmud River Watershed on the Tibetan Plateau. *Front. Earth Sci.* **2022**, *10*, 943075. [CrossRef]
19. Yi, L.; Chen, F.; Han, J.; Liu, X.; Yang, J.; Zhou, Z.; Lu, X.; Ma, Z.; Wei, H.; Han, F. SWAT-based Distributed Hydrologic Simulation and Runoff Prediction in the Upper Golmud River Watershed. *Int. J. Sediment Res.* **2024**, *32*, 509. [CrossRef]
20. Hou, W.; Li, X.; Yang, L.; Ta, W.; Jia, X. Carrying Capacity of Water Resources for Renewable Energy Development in Arid Regions in Northwest China: A Case Study of Golmud, Qinghai. *Front. Environ. Sci.* **2022**, *10*, 892414. [CrossRef]
21. Jiang, W.; Sheng, Y.; Shi, Z.; Guo, H.; Chen, X.; Mao, H.; Liu, F.; Ning, H.; Liu, N.; Wang, G. Hydrogeochemical Characteristics and Evolution of Formation Water in the Continental Sedimentary Basin: A Case Study in the Qaidam Basin, China. *Sci. Total Environ.* **2024**, *957*, 177672. [CrossRef]
22. Wang, J.; Huang, J.; Fang, T.; Song, G.; Sun, F. Relationship of Underground Water Level and Climate in Northwest China's Inland Basins under the Global Climate Change: Taking the Golmud River Catchment as an Example. *China Geol.* **2021**, *4*, 402–409. [CrossRef]
23. Hu, X.; Sun, Q.; Huang, J.; Geng, J.; Wang, J. Pore Structure and Radon Emission Characteristics of Shallow Soil in the Golmud River Basin of the Qinghai–Tibet Plateau. *Phys. Chem. Earth Parts A/B/C* **2025**, *139*, 103911. [CrossRef]
24. Guo, L.; Ding, Y.; Fang, H.; An, C.; Wang, G.; Mao, H.; Yang, N. Regional Variations and Hydrochemical Evolution in Golmud River Watershed (Qaidam Basin, China): An Integration of Self-Organizing Maps and Multi-Statistic Approaches. *Environ. Earth Sci.* **2025**, *84*, 143. [CrossRef]
25. Wang, Y.; Jiao, J.J. Origin of Groundwater Salinity and Hydrogeochemical Processes in the Confined Quaternary Aquifer of the Pearl River Delta, China. *J. Hydrol.* **2012**, *438–439*, 112–124. [CrossRef]
26. Piper, A.M. A Graphic Procedure in the Geochemical Interpretation of Water-analyses. *Neurochem. Int.* **1984**, *6*, 27–39.
27. Gibbs, R.J. Mechanisms Controlling World Water Chemistry. *Science* **1970**, *170*, 1088–1090. [CrossRef]
28. Huang, X.; Wang, G.; Liang, X.; Cui, L.; Ma, L.; Xu, Q. Hydrochemical and Stable Isotope (δD and δ18O) Characteristics of Groundwater and Hydrogeochemical Processes in the Ningtiaota Coalfield, Northwest China. *Mine Water Environ.* **2018**, *37*, 119–136. [CrossRef]
29. Liu, Z.; Wang, X.; Wan, X.; Jia, S.; Mao, B. Evolution Origin Analysis and Health Risk Assessment of Groundwater Environment in a Typical Mining Area: Insights from Water-Rock Interaction and Anthropogenic Activities. *Environ. Res.* **2024**, *252*, 118792. [CrossRef]
30. Yu, F.; Zhou, D.; Li, Z.; Li, X. Hydrochemical Characteristics and Hydrogeochemical Simulation Research of Groundwater in the Guohe River Basin (Henan Section). *Water* **2022**, *14*, 1461. [CrossRef]

31. Dong, F.; Yin, H.; Cheng, W.; Li, Y.; Qiu, M.; Zhang, C.; Tang, R.; Xu, G.; Zhang, L. Study on Water Inrush Pattern of Ordovician Limestone in North China Coalfield Based on Hydrochemical Characteristics and Evolution Processes: A Case Study in Binhu and Wangchao Coal Mine of Shandong Province, China. *J. Clean. Prod.* **2022**, *380*, 134954. [CrossRef]
32. Liu, J.; Hao, Y.; Gao, Z.; Wang, M.; Liu, M.; Wang, Z.; Wang, S. Determining the Factors Controlling the Chemical Composition of Groundwater Using Multivariate Statistics and Geochemical Methods in the Xiqu Coal Mine, North China. *Environ. Earth Sci.* **2019**, *78*, 364. [CrossRef]
33. Lecomte, K.L.; Pasquini, A.I.; Depetris, P.J. Mineral Weathering in a Semiarid Mountain River: Its Assessment through PHREEQC Inverse Modeling. *Aquat. Geochem.* **2005**, *11*, 173–194. [CrossRef]
34. Jia, Z.; Zang, H.; Hobbs, P.; Zheng, X.; Xu, Y.; Wang, K. Application of Inverse Modeling in a Study of the Hydrogeochemical Evolution of Karst Groundwater in the Jinci Spring Region, Northern China. *Environ. Earth Sci.* **2017**, *76*, 312. [CrossRef]
35. Li, D.; Gan, S.; Li, J.; Dong, Z.; Long, Q.; Qiu, S.; Zhou, Y.; Lu, C. Hydrochemical Characteristics and Formation Mechanism of Strontium-Rich Groundwater in Shijiazhuang, North China Plain. *J. Chem.* **2021**, *2021*, 5547924. [CrossRef]
36. Lindsey, B.D.; Belitz, K.; Cravotta, C.A.; Toccalino, P.L.; Dubrovsky, N.M. Lithium in Groundwater Used for Drinking-Water Supply in the United States. *Sci. Total Environ.* **2021**, *767*, 144691. [CrossRef]
37. Qiu, Y.; Zhou, A.; Gao, L.; Wang, Z.; Hu, X.; Li, Y.; Zhang, F.; Ma, C. Cation Exchange and Leakage as Dominant Processes in Controlling Salinity and Strontium in Sandy and Argillaceous Coastal Aquifer: Insights from Hydrochemistry and Multi-Isotopes. *J. Hydrol.* **2024**, *638*, 131529. [CrossRef]
38. Guo, W.; Zhang, D.; Zhang, W.; Li, S.; Pan, K.; Jiang, H.; Zhang, Q. Anthropogenic Impacts on the Nitrate Pollution in an Urban River: Insights from a Combination of Natural-Abundance and Paired Isotopes. *J. Environ. Manag.* **2023**, *333*, 117458. [CrossRef] [PubMed]
39. Chung, C.H.; You, C.F.; Yeh, Y.L. Strontium Isotopes and Rare Earth Elements as Tracers of Water–Rock Interactions in Taiwan Hot Springs. *Water* **2024**, *17*, 71. [CrossRef]
40. Zhu, J.; Chen, H.; Gong, G. Hydrogen and Oxygen Isotopic Compositions of Precipitation and Its Water Vapor Sources in Eastern Qaidam Basin. *Huan Jing Ke Xue* **2015**, *36*, 2784–2790.
41. Yang, Y.; Wu, Q.; Jin, H. Evolutions of Water Stable Isotopes and the Contributions of Cryosphere to the Alpine River on the Tibetan Plateau. *Environ. Earth Sci.* **2016**, *75*, 49. [CrossRef]
42. Rao, W.; Li, Y. Stable Hydrogen-Oxygen Isotope Composition and Atmospheric Moisture Sources of Precipitation in an Arid-Alpine Region: A Case Study of the Golmud River Watershed on the North Slope of the Kunlun. *J. Hydraul. Eng.* **2021**, *52*, 1116–1125.
43. Xu, J.; Gui, H.; Chen, J.; Li, C.; Li, Y.; Zhao, C.; Guo, Y. Hydrogeochemical Characteristics and Formation Mechanisms of the Geothermal Water in the Qingdong Coal Mine, Northern Anhui Province, China. *Mine Water Environ.* **2022**, *41*, 1015–1026. [CrossRef]
44. Hao, Q.; Li, Y.; Xiao, Y.; Yang, H.; Zhang, Y.; Wang, L.; Liu, K.; Liu, G.; Wang, J.; Hu, W.; et al. Hydrogeochemical Fingerprint, Driving Forces and Spatial Availability of Groundwater in a Coastal Plain, Southeast China. *Urban Clim.* **2023**, *51*, 101611. [CrossRef]
45. Luo, A.; Wang, G.; Dong, S.; Wang, H.; Shi, Z.; Ji, Z.; Xue, J. Effect of Large-Scale Mining Drainage on Groundwater Hydrogeochemical Evolution in Semi-Arid and Arid Regions. *Front. Environ. Sci.* **2022**, *10*, 926866. [CrossRef]
46. Li, P.; Wu, J.; Qian, H. Assessment of Groundwater Quality for Irrigation Purposes and Identification of Hydrogeochemical Evolution Mechanisms in Pengyang County, China. *Environ. Earth Sci.* **2013**, *69*, 2211–2225. [CrossRef]
47. Chen, Z.; Zhou, X.; Du, J.; Xie, C.; Liu, L.; Li, Y.; Yi, L.; Liu, H.; Cui, Y. Hydrochemical Characteristics of Hot Spring Waters in the Kangding District Related to the Lushan MS Combining Double Low Line 7.0 Earthquake in Sichuan, China. *Nat. Hazards Earth Syst. Sci.* **2015**, *15*, 1149–1156. [CrossRef]
48. Mao, H.; Wang, G.; Rao, Z.; Liao, F.; Shi, Z.; Huang, X.; Chen, X.; Yang, Y. Deciphering spatial pattern of groundwater chemistry and nitrogen pollution in Poyang Lake Basin (Eastern China) using self-organizing map and multivariate statistics. *J. Clean. Prod.* **2021**, *329*, 129697. [CrossRef]
49. Xing, L.; Guo, H.; Zhan, Y. Groundwater Hydrochemical Characteristics and Processes along Flow Paths in the North China Plain. *J. Asian Earth Sci.* **2013**, *70*, 250–264. [CrossRef]
50. Raju, N.J.; Patel, P.; Reddy, B.C.S.R.; Suresh, U.; Reddy, T.V.K. Identifying Source and Evaluation of Hydrogeochemical Processes in the Hard Rock Aquifer System: Geostatistical Analysis and Geochemical Modeling Techniques. *Environ. Earth Sci.* **2016**, *75*, 1157. [CrossRef]
51. Werner, S.; Morgan, J.J. *Aquatic Chemistry: Chemical Equilibria and Rates in Natural Waters*, 3rd ed.; John Wiley & Sons, Inc.: Hoboken, NJ, USA, 1996.
52. Xiao, L.; Wang, J.; Dang, Y.; Cheng, Z.; Huang, T.; Zhao, J.; Xu, Y.; Huang, J.; Xiao, Z.; Komatsu, G. A new terrestrial analogue site for Mars research: The Qaidam Basin, Tibetan Plateau (NW China). *Earth Sci. Rev.* **2017**, *164*, 84–101. [CrossRef]

53. Gaillardet, J.; Dupré, B.; Louvat, P.; Allègre, C.J. Global Silicate Weathering and CO2 Consumption Rates Deduced from the Chemistry of Large Rivers. *Chem. Geol.* **1999**, *159*, 3–30. [CrossRef]
54. Jiang, W.; Sheng, Y.; Wang, G.; Shi, Z.; Liu, F.; Zhang, J.; Chen, D. Cl, Br, B, Li, and noble gases isotopes to study the origin and evolution of deep groundwater in sedimentary basins: A review. *Environ. Chem. Lett.* **2022**, *20*, 1497–1528. [CrossRef]
55. Stober, I.; Zhong, J.; Bucher, K. From freshwater inflows to salt lakes and salt deposits in the Qaidam Basin, W China. *Swiss J. Geosci.* **2023**, *116*, 5. [CrossRef]
56. Möller, P.; De Lucia, M. The impact of Mg^{2+} ions on equilibration of Mg-Ca carbonates in groundwater and brines. *Geochemistry* **2020**, *80*, 125611. [CrossRef]

water

MDPI

Article

Research on Ecological–Environmental Geological Survey and Evaluation Methods for the Kundulun River Basin in Baotou City

Yi Hao [1,2], Junwei Wan [3,*], Yihui Xin [4,*], Wenhui Zhou [1,2], Yongli Li [1,2], Lei Mao [1,2], Xiaomeng Li [1,2], Limei Mo [1,2] and Ruijia Li [1,2]

1 Hohhot General Survey of Natural Resources Center, China Geological Survey, Hohhot 010010, China; haoyi@mail.cgs.gov.cn (Y.H.); zhouwenhui@mail.cgs.gov.cn (W.Z.); liyongli@mail.cgs.gov.cn (Y.L.); maolei@mail.cgs.gov.cn (L.M.); lixiaomeng@mail.cgs.gov.cn (X.L.); molimei@mail.cgs.gov.cn (L.M.); liruijia@mail.cgs.gov.cn (R.L.)

2 Innovation Base for Water Resource Exploration and EcoEnvironmental Effects in the Daheihe Basin of the Yellow River, China Geological Survey, Hohhot 010010, China

3 School of Resources, China University of Geosciences (Wuhan), Wuhan 430074, China

4 School of Architecture and Art, Hebei University of Architecturez, Zhangjiakou 075000, China

* Correspondence: wanjw@cug.edu.com (J.W.); xinyihui0809@163.com (Y.X.)

Abstract

The Kundulun River Basin is the most prominent branch of the Yellow River system within the jurisdiction of Baotou City. As an important water source and ecological barrier, its ecological quality is directly related to the ecological security and sustainable development of the surrounding areas. This study selected the Kundulun River Basin in Baotou City as the research area. On the basis of collecting relevant information, a field investigation was conducted on the ecological and geological conditions of the atmospheric surface subsurface Earth system, using the watershed as the survey scope and water as the carrier for the transfer and conversion of materials and energy in the watershed. This study selected the main factors that affect the ecological geological quality of the watershed and established an evaluation model using the analytic hierarchy process, the coefficient of variation method, and the comprehensive analysis method. We have established an ecological geological quality evaluation index system for the Kundulun River Basin. We conducted quantitative evaluation and comprehensive analysis of regional ecological and geological environment quality. The results indicate that ecological environment indicators contribute the most to the ecological quality of the study area, while the impact of human activities on ecological quality is relatively small. From the perspective of evaluation indicators, grassland has the highest weight, followed by precipitation, groundwater depth, forest land, and cultivated land. Approximately 30.26% of the land in the research area is in a state of high or relatively high ecological and geological–environmental quality risk. It can be seen that the overall quality of the ecological geological environment is not optimistic and needs further protection.

Keywords: ecological environment; geological environment; ecological quality assessment; evaluation indicators; Kundulun River Basin

1. Introduction

Under the dual pressures of global climate change and intensified human activities, the ecological environment quality in arid and semi-arid regions is facing unprecedented major challenges [1,2]. Due to their inherently fragile ecosystems, these areas are particularly sensitive to environmental changes, and issues such as the frequent occurrence of extreme weather events, water resource shortages, and land degradation further exacerbate ecosystem instability [3]. The state of the ecological environment not only affects the conservation of biodiversity and the sustainable use of natural resources but also directly affects the quality of life for local residents and the stable development of regional socioeconomics [4,5]. Against this background, it has become especially important to adopt scientific and effective measures to assess and enhance ecological conditions in arid and semi-arid regions [6,7]. Therefore, conducting systematic evaluations of the ecological–environmental quality in these areas can not only provide a solid theoretical foundation for local ecological civilization construction but also promote harmonious coexistence between humans and the natural environment, laying the groundwork for achieving long-term sustainable development goals [8,9]. Thus, carrying out geological assessments of ecological and environmental quality in arid and semi-arid regions can offer scientific theoretical support for the construction of ecological civilization and the harmonious development of the relationship between humans and the natural environment.

In recent years, with the continuous advancement and integration of GIS (geographic information system), GPS (global positioning system), and RS (remote sensing) technologies, research on ecological environment quality assessment has experienced significant development both domestically and internationally [10–13]. The analytic hierarchy process (AHP), proposed by Saaty in the 1970s, is a practical and effective method for determining weights [14], which effectively divides complex problems into hierarchical levels, making them more organized. This method remains mainstream today; however, some scholars have explored alternative approaches to enhance the scientific rigor of assessments by introducing objective evaluation methods such as the coefficient of variation, neural network method, and principal component analysis, aiming to achieve more objective determination of factor weights [15–17]. However, completely abandoning subjective experience may lead to the neglect of actual information specific to the study area. Therefore, evaluation models that combine both subjective and objective approaches have, to some extent, balanced the contradictions between subjective judgment and objective reality [18,19].

The natural environment of the Kundulun River Basin exhibits typical characteristics of arid and semi-arid regions, with scarce and unevenly distributed precipitation and low vegetation coverage, resulting in a relatively fragile ecological environment [20]. Water scarcity limits vegetation growth, leading to severe land degradation and desertification. In addition, the soil in the basin is mostly infertile chestnut calcic soil, which has poor water retention capacity and is easily affected by weathering and erosion [21]. Recent studies on the basin have mainly focused on climatic conditions [22,23], hydrogeological features [24], and the distribution characteristics of soil pollution [25], while research specifically targeting its ecological and geological–environmental conditions is still limited. This study aims to address the lack of research on ecogeological–environmental assessments in arid and semi-arid regions, particularly within the specific context of the Kundulun River Basin in Baotou City. We propose the following hypotheses: grassland coverage is significantly positively correlated with ecological quality; precipitation plays a crucial role in vegetation growth and groundwater recharge. Therefore, this study selects the Kundulun River Basin in Baotou City as the research area, conducting an in-depth investigation into its ecological environment, natural conditions, geological features, and humanistic factors. Based on GIS technology and using methods such as analytic hierarchy process, coefficient of variation, and comprehensive analysis, this study comprehensively evaluates the quality of the ecological and geological environment within the basin through resampling at 50 × 50 m grid intervals, reclassification, and raster calculation with weighted overlay techniques, providing scientific and rational support for the protection, sustainable development, and utilization of the basin's ecological environment.

2. The Study Area

The study area is located in the central-western region of Inner Mongolia, primarily encompassing the administrative districts of Jiuyuan, Qingshan, Kundulun, Shiguai, Guyang, Damu Banner, and parts of Wulateqian Banner in Baotou City [26], with its specific location shown in Figure 1. The Kundulun River has a total length of 134 km and is accompanied by 23 tributaries of varying sizes. The basin features a complex topography with mountains and hills crisscrossing the region: to the east lies the towering Chun Kun Mountain, to the north is the Ata Mountain, to the west is the Mingan Mountain, and to the south and southwest are the Wula Mountain and Daqing Mountain, respectively. The climate is classified as a temperate continental arid to semi-arid monsoon climate, characterized by significant continental features, with windy and dry springs, hot summers, and cold winters. The annual average precipitation is 254.9 mm, with 80% occurring between July and September [27], and the annual evaporation rate ranges from 2200 to 2800 mm, which is 8 to 17 times the precipitation.

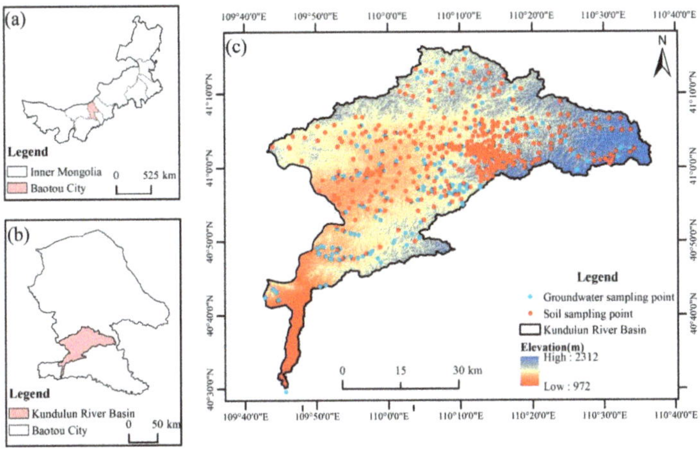

Figure 1. Study area location map. (**a**) Location of Baotou City within Inner Mongolia Autonomous Region; (**b**) Location of the Kundulun River Basin within Baotou City; (**c**) Sampling sites in the Kundulun River Basin.

The study area is situated on the northern margin of the North China Platform, with the central and southern parts dominated by extensive Proterozoic metamorphic rock formations. Relatively complete Paleozoic, Mesozoic, and Cenozoic strata are also exposed within the region [28]. The plains belong to the Hetao geological sub-region, where Holocene deposits are widely distributed. These deposits are especially common in the Yellow River alluvial plain and the foothills of the Daqing Mountains, where alluvial and fluvioglacial sediments dominate. Additionally, large areas of Holocene deposits are found on the second terraces of rivers and lakes, as well as on the surfaces of the Wula Mountains and Daqing Mountains, primarily characterized by aeolian sand accumulation [29]. The study area spans two groundwater subsystems: the inland water system of the Yinshan Mountains' northern hills and the Yellow River water system of the Yinshan Mountains' southern hills [30]. Based on the aquifer medium and storage conditions, the regional groundwater can be categorized into three main types: Quaternary unconsolidated porous water, fractured porous water in clastic rocks, and fractured water in bedrock [31].

In the southern part of the Kundulun River Basin, the topography is mainly characterized by mid-mountain terrain, with elevations mostly exceeding 1600 m. The annual average precipitation is 346 mm, and the dominant rock types include Archean to Proterozoic gneiss, plagioclase amphibolite, schist, and granulite. The soil types are mainly stony soil and grey-cinnamon soils, with vegetation dominated by trees and shrubs, while herbaceous cover is relatively sparse. The primary tree species are poplar, elm, and pine, and the dominant shrub is small-leafed chickenweed. The main herbaceous species are dogtail grass, foxtail grass, Salsola collina, and Artemisia, with low vegetation coverage [32].

In the northern part of the Kundulun River Basin, the topography is mainly characterized by low mountains, hills, and high plains [33], with elevations mostly below 1600 m. The annual average precipitation ranges from 170.2 to 248.5 mm, and the dominant rock types include Silurian to Cretaceous sedimentary rocks, metamorphic rocks, and intrusive rocks, with relatively complex lithological combinations. The soil types are primarily calcic cambisols and brown calcic soils, dominated by herbaceous plants, with extensive distribution of shrub vegetation and scattered tree distributions. Trees are mainly distributed around towns, farmlands, and roadsides, primarily consisting of poplar, elm, and pine, with scattered jujube trees. The dominant shrub is small-leafed chicken weed, and the herbaceous vegetation is diverse, including common species such as Stipa krylovii, Stipa bungeana, Salsola collina, Artemisia, Potentilla anserina, and *Leymus secalinus*.

3. Materials and Methods

3.1. Establishment of the Evaluation Indicator System

To address the specific ecological and geological issues of the Kundulun River Basin in Baotou City, this study integrated multi-dimensional data from geological background, meteorological records, and field surveys. A systematic analysis was conducted to identify the core factors influencing the environmental quality of the region. The key elements affecting the ecological and geological environment were categorized into four main classes: natural geographic background, climatic conditions, geological environment background, and ecological environment background. Further, geological environment conditions and human activities were further identified as critical factors impacting the ecological environment.

To construct a comprehensive evaluation system, this study selected five primary evaluation criteria: natural geography, climatic conditions, geological environment, ecological environment background, and human activities. Under these primary criteria, more specific secondary evaluation indicators were established, totaling 15 sub-indicators. To verify the stability of the model, sensitivity analysis was conducted using the single-factor variation method. These include, but are not limited to, vegetation cover, precipitation, groundwater quality, arable land status, and others. The detailed evaluation indicator system is illustrated in Figure 2.

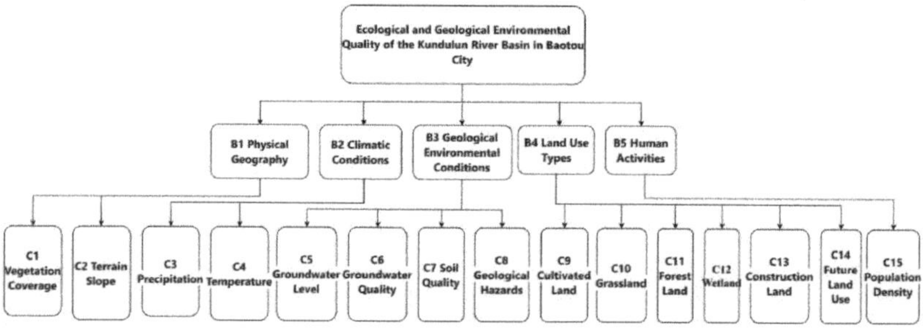

Figure 2. Evaluation index system.

3.2. Selection of Evaluation Indicators

(1) Vegetation Coverage

When estimating vegetation coverage using remote sensing, it is common to use vegetation indices as a proxy to reflect the condition of plant growth. In the southern part of the Kundulun River Basin, vegetation is mainly composed of trees and shrubs, with relatively sparse herbaceous vegetation. The primary tree species include poplar, elm, and pine, while the main shrub species is Caragana microphylla. Herbaceous vegetation primarily consists of Setaria viridis, Chloris virgata, Salsola collina, and Artemisia scoparia. The basin features predominantly herbaceous plants alongside extensive shrubbery, with scattered distribution of trees. Among these, Astragalus membranaceus is widely distributed in Guyang County. By comparing vegetation coverage across different regions, we can clearly observe the close relationship between vegetation coverage and ecological–environmental quality. Therefore, vegetation coverage should be considered a key reference indicator when constructing an ecological environment quality evaluation system.

(2) Terrain Slope

Terrain slope is a crucial indicator for measuring the inclination of surface units, significantly influencing the redistribution process of water resources. Specifically, slope controls the flow and accumulation of organic matter and water within soil, thereby profoundly impacting vegetation growth and its spatial distribution. Terrain slope determines the speed and direction of surface runoff, affecting water resource allocation and flood routing. Steep slopes may lead to rapid surface runoff, increasing the risk of soil erosion, whereas gentle slopes may facilitate water accumulation and infiltration. The study area exhibits significant terrain characteristics, especially with strong topographic dissection, forming complex and varied landscapes. Notably, some areas in the south have steep slopes exceeding 50° (as shown in Figure 3(C2)). This pronounced variation in terrain slope not only affects surface hydrological processes but also deeply reflects the migration of beneficial elements in the soil and local spatial aggregation of heavy metals. Thus, terrain slope is not only an important parameter in geomorphology but also a critical factor to consider when assessing ecological safety and environmental quality in the study area.

(3) Precipitation

Precipitation influences natural processes such as surface water cycles, vegetation growth, soil erosion, and deposition, as well as human activities like agriculture, water conservancy, and urban planning. Appropriate rainfall is beneficial for vegetation growth and recovery, while insufficient or excessive precipitation can negatively impact vegetation. Heavy rainfall may cause soil erosion and threaten soil quality. Therefore, precipitation is a necessary consideration when evaluating ecological–environmental quality, predicting environmental changes, and formulating ecological protection strategies. The Kundulun River Basin has scarce precipitation and abundant sunlight, with an annual average precipitation of 291.1 mm, mostly concentrated from June to August, accounting for 64% of the annual total. The historical maximum daily precipitation was 175.3 mm on 19 July 2018 (previously, the highest daily precipitation was 114.9 mm on 7 August 1958), with an annual average evaporation of 1941.4 mm. Precipitation decreases from east to west in the Kundulun River Basin area of Baotou City (Figure 3(C3)).

(4) Temperature

Temperature is one of the key factors affecting vegetation growth and distribution. Suitable temperature conditions are conducive to vegetation growth and reproduction, while excessively high or low temperatures may stress or limit vegetation. Changes in temperature also affect soil microbial activity, thus influencing the decomposition of soil organic matter and nutrient transformation. Therefore, temperature is an important factor in ecological environment evaluation. The Kundulun River Basin has a typical arid and semi-arid continental climate characterized by low temperatures and large temperature differences. The frost-free period lasts 69–177 days annually. The annual average temperature is 5.5 °C, with the extreme maximum temperature reaching 38.6 °C on 22 June 2005, and the extreme minimum temperature dropping to −36.1 °C on 28 December 1967. Annual average temperatures increase from east to west in the Kundulun River Basin area of Baotou City (Figure 3(C4)).

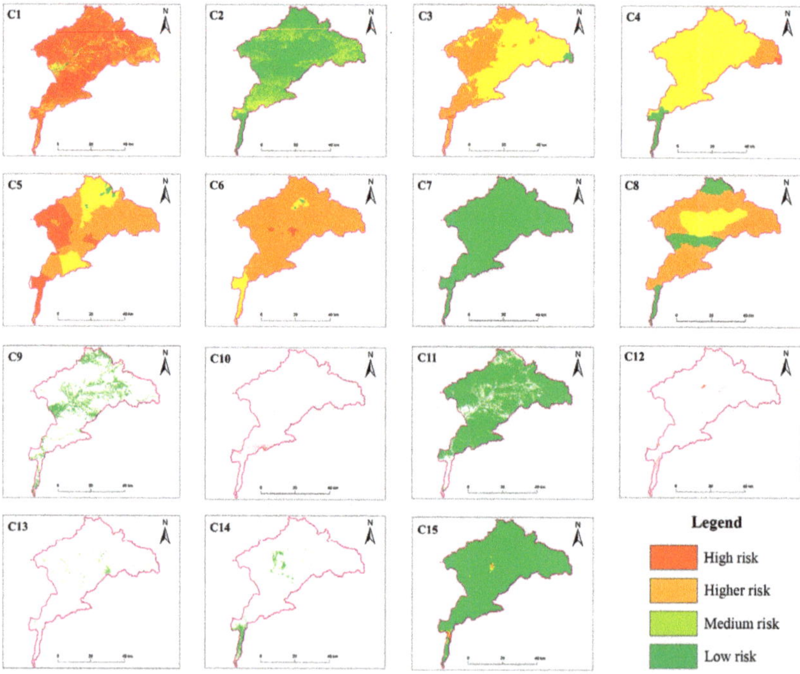

Figure 3. Risk assessment chart for indicators. (**C1**): Vegetation Coverage Distribution Map; (**C2**): Topographic Slope Distribution Map; (**C3**): Rainfall Distribution Map; (**C4**): Temperature Distribution Map; (**C5**): Groundwater Level Distribution Map; (**C6**): Groundwater Quality Distribution Map; (**C7**): Soil Quality Distribution Map; (**C8**): Geological Hazard Distribution Map; (**C9**): Cultivated Land Distribution Map; (**C10**): Grassland Distribution Map; (**C11**): Forest Land Distribution Map; (**C12**): Wetland Distribution Map; (**C13**): Construction Land Distribution Map; (**C14**): Future Land Use Distribution Map; (**C15**): Population Density Distribution Map.

(5) Groundwater Depth

The level of groundwater can influence the moisture conditions of surface water and soil, thereby affecting vegetation growth and distribution. Excessively high groundwater levels may lead to soil salinization, adversely affecting vegetation growth, while excessively low levels may result in soil drought, hindering normal vegetation growth.

(6) Groundwater Quality

Groundwater quality assessment typically involves multiple aspects, including chemical composition (such as pH, total dissolved solids, ion concentrations), physical properties (such as water temperature, color, turbidity), and biological characteristics (such as microbial content). Variations in these indicators may directly or indirectly affect the usability value of groundwater and ecological–environmental health.

(7) Soil Quality

Soil geochemical properties are core elements in assessing land quality in the study area, directly related to nutrients, heavy metals, organic pollutants, and the physicochemical properties of the soil. Phenomena such as the activation and migration of certain soil elements, like the enrichment of heavy metals and the loss of nutrients, are closely related to soil geochemical factors. These changes not only reveal the health status of the land but also serve as important criteria for assessing ecological safety and stability.

The quality of soil geochemical properties directly impacts the ecological functions and resource value of the land. Healthy soil ensures crop growth and yield in agricultural production, affecting productivity. Additionally, soil quality is a decisive factor in optimizing land resource allocation, influencing land development methods and sustainability. Lastly, research on soil geochemical properties is crucial for ecological–environmental protection efforts, aiding in identifying and controlling potential pollution sources.

(8) Geological Hazards

Geological hazards are an essential factor in ecological environment evaluation, involving geological structure stability, changes in topography, and risks associated with natural disasters. The occurrence of geological hazards can severely damage and impact the ecological environment, such as land destruction, water source pollution, and loss of biodiversity habitats. Geological hazards commonly include earthquakes, landslides, debris flows, and ground subsidence. These events may be influenced by various factors such as geological structures, topography, climatic conditions, and human activities. Monitoring and assessing geological hazards can help understand their risk levels, frequency, and potential impacts on the ecological environment.

(9) Land Use Types

Land resources are fundamental for human survival, and their utilization directly reflects the degree of human intervention in the natural environment, impacting ecosystem structure and function. In the Kundulun River Basin area of Baotou City, land use types are categorized into six classes: cultivated land, grassland, forest land, wetlands, construction land, and unused land.

(10) Population Density

Population density directly affects land resource utilization and cover. High-population-density areas often experience more urbanization and industrial activities, leading to over-exploitation and utilization of land resources, resulting in issues such as land degradation, water shortages, and biodiversity loss. Population density is also closely related to energy consumption and environmental pollution. High-density areas generally have higher energy consumption and pollutant emissions, such as vehicle exhaust and industrial wastewater, all of which negatively impact the ecological environment. Moreover, population density influences biodiversity conservation and restoration. High-density areas face greater pressures on biodiversity loss, necessitating increased attention to biodiversity protection and restoration efforts. In the Kundulun River Basin area of Baotou City, population clusters are mainly concentrated in the Kunqu District of Baotou City, Guyang County, and surrounding villages.

3.3. Data Collection and Processing

The data primarily include vegetation, topography, meteorology, land use, field measurements, socioeconomic, and population data for the Kundulun River Basin in Baotou City in 2020. Among these, vegetation data mainly consist of vegetation cover; slope gradient data mainly include slope and elevation; meteorological data primarily comprise precipitation and temperature; land use data encompass categories such as cultivated land, grassland, forest land, water bodies, construction land, and unused land; field measurement data comprise groundwater depth, groundwater quality, soil quality, and types and scales of geological hazards; and socioeconomic and population data are primarily based on population density. The sources of the data used are listed in Table 1.

Table 1. Data source and description.

Data Name	Data Source
Vegetation Cover	Multispectral remote sensing images from Landsat 8 OLI_TIRS satellite digital products available on the Geospatial Data Cloud
Slope Gradient	30 m resolution ASTER GDEM (Digital Elevation Model) data from the Geospatial Data Cloud
Precipitation	1 km raster data of 2020 precipitation from the National Earth System Science Data Center, part of the National Science and Technology Infrastructure Platform
Temperature	1 km raster data of 2020 temperature from the National Earth System Science Data Center, part of the National Science and Technology Infrastructure Platform
Groundwater Depth	Field measurements conducted in July 2020 under the China Geological Survey project (ZD20220217)
Groundwater Quality	Field measurements conducted in July 2020 under the China Geological Survey project (ZD20220217)
Soil Quality	Field measurements conducted from June to August 2020 under the China Geological Survey project (ZD20220217)
Geological Hazards	Field measurement results from the First Hydrogeological and Engineering Geological Survey Institute of Inner Mongolia Autonomous Region
Cultivated Land Grassland Forest Land Water Bodies Construction Land Unused Land	China Land Cover Dataset, CLCD (30 m Fine Land Cover Dataset of Inner Mongolia in 2020)
Population Density	1 km resolution population spatial distribution raster data from the LandScan dataset for 2020

The collection of soil samples followed the surface soil sampling methods and requirements specified in DZ/T 0295-2016 Specifications for Geochemical Survey of Land Quality [34] and DZ/T 0167-2012 Specifications for Regional Geochemical Exploration [35]. The soil quality standard refers to DZ/T 0295-2016 Specification for Geochemical Survey of Land Quality. The unified groundwater measurement work was conducted in accordance with the standards outlined in Specifications for 1:50,000 Hydrogeological Survey (DD2019-03) [36] and Technical Specifications for Unified Groundwater Measurement (Trial Version 20200831) [37]. Groundwater quality sampling and analysis were carried out in reference to the Specifications for Multi-Purpose Geochemical Survey (1:250,000) [38]. The groundwater quality standard refers to GB/T 14848-2017 Groundwater Quality Standard [39]. All samples were collected within the Kundulun River Basin area (109°30′–110°45′ E, 40°40′–41°40′ N).

3.4. Calculation of Indicator Weights

3.4.1. Determination of Indicator Weights Based on the Analytic Hierarchy Process (AHP)

The analytic hierarchy process (AHP) is a systematic analysis method that combines qualitative and quantitative analysis. By clearly defining the problem, establishing a hierarchical structure model, constructing a fuzzy consistency matrix, and calculating the weights of each indicator, AHP can quantify uncertain problems that are difficult to solve quantitatively using the theory and methods of fuzzy mathematics. In the evaluation process, quantified weight values are used to obtain final quantified evaluation results [40].

According to the principles of AHP, a judgment matrix is established, and consistency is evaluated using Formulas (1) and (2). When the CR value is less than 0.1 [14], the judgment matrix is considered to have reasonable consistency.

$$CI = (\lambda_{max} - n)/(n - 1) \tag{1}$$

$$CR = CI/RI \quad (2)$$

CR is the consistency ratio, CI is the consistency index, RI is the random index, λ_{max} is the maximum eigenvalue of the judgment matrix, and n is the number of indicators.

In this study, the consistency ratio (CR) values of the judgment matrices for each criterion layer relative to the goal layer were all less than 0.1 [14]. The results of the consistency check for the overall hierarchy ranking were as follows: consistency index (CI) = 0.015; random index (RI) = 0.756; and consistency ratio (CR) = 0.020. Since $CR < 0.1$, this indicates that the measurement is reasonable and passes the consistency test. After passing the consistency check, the components of the eigenvector corresponding to the maximum eigenvalue (λ_{max}) represent the weights of the indicators at that level.

3.4.2. Determination of Indicator Weights Based on the Coefficient of Variation Method

In the evaluation of ecological and environmental quality, the coefficient of variation (CV) method can be used to determine the weights of different evaluation indicators. Its core principle lies in quantifying the relative variability among data to determine the weight of each indicator [41]. Compared to the entropy weight method, the CV method is more suitable for scenarios with significant data dispersion. For example, the standard deviation of precipitation is σ = 120 mm, with a mean value of x = 380 mm, resulting in a coefficient of variation (vi) of σ/x = 0.316. This indicates a high sensitivity of precipitation to the ecogeological environment; thus, it is assigned a higher weight (0.107).

Given that the units and scales of various indicators differ, this directly affects the feasibility of establishing a unified evaluation model. To accurately assess the ecological and environmental quality of the Kundulun River Basin in Baotou City, this study employed the natural breaks classification method as an analytical tool, with the classification results presented in Table 2. The ecological quality status of the study area was divided into four risk levels, as detailed in Table 2.

Table 2. Ecological quality risk level table.

Risk Level	Assigned Value
Low risk	9
Medium risk	7
Higher risk	5
High risk	3

To achieve a more precise evaluation of ecological and environmental quality, 15 industry experts were convened to independently score multiple key evaluation indicators within this evaluation unit, constructing a weight table for each indicator. The 15 experts who participated in the AHP weight scoring are all from the fields of ecology, hydrogeology, and remote sensing applications, including 7 with senior professional titles and 8 who have experience in ecological restoration projects in arid areas. The consistency ratio of the judgment matrix was calculated using Formulas (1) and (2), resulting in a CR value of 0.02, which is less than 0.1, indicating a high level of consistency in the expert scoring. Using Excel software (2019), the average scores and standard deviations for each evaluation indicator were calculated, and the coefficient of variation for each indicator was further computed using Formula (3). Finally, through normalization processing (Formula (4)). We normalized the coefficients of variation into weight values for each evaluation indicator.

$$v_i = \frac{\sigma_i}{x_i} \quad (3)$$

$$W_i = \frac{V_i}{\sum_{i=1}^{n} V_i} \quad (4)$$

V_i is the coefficient of variation for the i-th indicator, σ_i is the standard deviation of the i-th indicator, x_i is the mean of the i-th indicator, and W_i is the weight of the i-th indicator.

3.4.3. Comprehensive Analysis

An equal-weighted average (AHP weight × 0.5 + CV weight × 0.5) was used to integrate the subjective and objective weighting results, balancing expert experience with data characteristics. The stability of the weight allocation was verified using the single-factor variation method (Table 3). The results show that adjustments to the weights of highly sensitive indicators such as grassland coverage (C10) and precipitation (C3) significantly affect the comprehensive score (change rate > 8%), proving that the current weight allocation is reasonable.

Table 3. Research area evaluation index division standard reference table.

Goal Layer	Criterion Layer	Indicator Layer	Classification Standards			
			Low Risk (9)	Medium Risk (7)	Higher Risk (5)	High Risk (3)
A	B1	C1	0.63~1	0.32~0.63	0.12~0.32	<0.12
		C2	0~12°	12~32°	32~43°	43~75°
	B2	C3	>440 mm	405~440 mm	370~405 mm	<370 mm
		C4	>6 °C	3.5~6 °C	1~3.5 °C	<1 °C
	B3	C5	0.4~3 m	3~5 m	5~10 m	>10 m
		C6	Class I, Class II	Class III	Class IV	Class V
		C7	Class I, Class II	Class III		
		C8	Rare	Low Probability	Moderate Probability	High Probability
	B4	C9	<30%	30~50%	50~80%	>80%
		C10	>80%	50~80%	30~50%	<30%
		C11	>80%	50~80%	30~50%	<30%
		C12	>80%	50~80%	30~50%	<30%
		C13	<30%	30~50%	50~80%	>80%
		C14	<30%	30~50%	50~80%	>80%
	B5	C15	<1000 people/km^2	1000~3000 people/km^2	3000~5000 people/km^2	>5000 people/km^2

3.4.4. Sensitivity Analysis

To verify the model's sensitivity to weight allocation and ensure the reliability of the results, a sensitivity analysis of the weight parameters was conducted using the single-factor variation method. Particular attention was given to the sensitivity of five indicators: grassland (C10), precipitation (C3), groundwater depth (C5), forest land (C11), and cultivated land (C9).

(1) Single-Factor Variation Method

Based on the comprehensive weights listed in Table 4, each of the five indicators was unilaterally adjusted by ±10%, while the other weights remained unchanged. The comprehensive scores were then recalculated.

(2) Analysis Results:

Highly Sensitive Indicators: Grassland (C10), precipitation (C3), and forest land (C11) showed the greatest impact on the comprehensive score (score change rate > 8%). For example, a 10% increase in grassland weight led to a 7.2% improvement in the comprehensive score, indicating that vegetation coverage plays a significant role in regulating ecogeological–environmental quality.

Moderately Sensitive Indicator: Groundwater depth (C5) exhibited a score change rate of 3.1–3.4%, suggesting that weight adjustments have some influence on the model results but are not dominant factors.

Low-Sensitivity Indicator: Cultivated land (C9) had a score change rate between 2.5% and 2.8%, indicating that weight adjustments have minimal impact on the comprehensive score.

Table 4. Results of sensitivity analysis for five key indicators.

Indicator	Original Weight	Score Change Rate After +10% Weight	Score Change Rate After −10% Weight
Grassland (C10)	0.083	+7.2%	−7.5%
Precipitation (C3)	0.109	+9.3%	−10.1%
Groundwater Depth (C5)	0.093	+3.1%	−3.4%
Forest Land (C11)	0.114	+8.7%	−9.6%
Cultivated Land (C9)	0.077	+2.5%	−2.8%

3.5. Comprehensive Evaluation of Ecological and Geological–Environmental Quality

We used the comprehensive analysis method to perform a weighted overlay of 15 indicators—vegetation cover, slope gradient, precipitation, temperature, groundwater depth, groundwater quality, soil quality, geological hazards, cultivated land, grassland, forest land, wetlands, construction land, unused land, and population density—was performed using the ArcGIS (10.8) Raster Calculator. The weight values of these indicators were applied in the weighted overlay process. The ecological and environmental quality of each indicator is shown in Figure 3. Based on the evaluation grade table (Table 5), the evaluation results for the study area were classified into different levels.

Table 5. Evaluation table of ecological geological environment quality in the study area.

Evaluation Results	Poor Ecological and Geological–Environmental Quality	Fair Ecological and Geological–Environmental Quality	Good Ecological and Geological–Environmental Quality	Excellent Ecological and Geological–Environmental Quality
Evaluation Grades	3.26~4.08	4.08~4.45	4.45~4.74	4.74~5.53

4. Results and Discussion

4.1. Analysis of Indicator Weights

Based on the weights obtained from the analytic hierarchy process (AHP) and the coefficient of variation (CV) method, the final weights for the criteria and indicators in this evaluation are summarized in Table 6. From the perspective of the criterion layer, it is evident that the land use type evaluation indicator has the highest weight (0.368), while the human activity evaluation indicator has a relatively lower weight of 0.100. This result indicates that ecological indicators play a dominant role in assessing the ecological quality of the study area, with their contribution significantly outweighing that of human activities. Although some areas within the study region (e.g., the eastern part) have experienced significant local ecological degradation due to activities such as mineral extraction, these high-intensity human activities are relatively concentrated in spatial distribution and have limited impact coverage, accounting for only a small proportion of the total study area. Therefore, from the perspective of the entire watershed, the influence of human activities on the comprehensive evaluation results of ecogeological environment quality is relatively weak. Based on this, the weight assigned to human activities in this study, 0.100, is considered reasonable. In addition, land use changes (e.g., expansion of construction land) and irrigation practices (e.g., excessive groundwater extraction) also have significant impacts on ecological quality.

Table 6. Comprehensive weight of evaluation indexes of Kundulun River Basin in Baotou.

Goal Layer	Criterion Layer	AHP Weight	CV Weight	Comprehensive Weight	Indicator Layer	AHP Weight	CV Weight	Comprehensive Weight
A	B1	0.097	0.141	0.119	C1	0.065	0.043	0.054
					C2	0.032	0.047	0.04
	B2	0.160	0.166	0.163	C3	0.12	0.093	0.107
					C4	0.04	0.075	0.058
	B3	0.263	0.236	0.250	C5	0.123	0.063	0.093
					C6	0.025	0.089	0.057
					C7	0.042	0.04	0.041
					C8	0.073	0.064	0.069
	B4	0.418	0.318	0.368	C9	0.106	0.045	0.076
					C10	0.156	0.068	0.112
					C11	0.068	0.093	0.081
					C12	0.043	0.098	0.071
					C13	0.027	0.045	0.036
					C14	0.018	0.04	0.029
	B5	0.061	0.139	0.100	C15	0.061	0.04	0.051

From the perspective of the indicator layer, the grassland evaluation indicator has the highest weight (0.112), followed by precipitation (0.107), groundwater depth (0.093), forest land (0.081), and cultivated land (0.076). This indicates that grasslands contribute most significantly to the ecological quality of the study area. Grasslands are the largest land use type in the study area, covering 81.15% of the total area, and their health status directly affects soil conservation, water retention, climate regulation [42], and other critical aspects. Healthy grasslands help stabilize soil, reduce surface runoff, and minimize erosion caused by rainfall and surface water flow. Additionally, grasslands prevent wind erosion, enhance the ability to block wind-blown sand, and protect soil from being carried away by wind [43]. Moreover, grasslands play a crucial role in climate regulation, improving the regional environment and contributing significantly to soil formation and conservation [44]. The weight of forest land is 0.081, which is relatively low, mainly due to two limiting factors: first, its coverage accounts for only 0.20% of the study area (much lower than grassland at 81.15%), with a scattered spatial distribution and low dispersion (CV weight of 0.093), leading to a dilution of its overall contribution; second, in the AHP scoring, experts in arid regions prioritized high-coverage grassland (weight of 0.112) and key climatic factors (such as precipitation with a weight of 0.107), while forests, due to their small area and significant local effects but limited overall impact, were assigned a relatively lower weight.

Precipitation is also a vital factor influencing the ecological quality of the study area. First, adequate precipitation supports plant photosynthesis and nutrient absorption, promoting lush vegetation growth [45]. Second, precipitation is essential for replenishing surface water and groundwater, which is crucial for maintaining the water quantity and quality of wetlands, rivers, and lakes [46]. Third, soil moisture, which is influenced by precipitation, helps maintain soil structure and fertility, providing favorable conditions for plant growth [47].

Groundwater is an essential component of the water cycle. In arid and semi-arid regions, a decline in groundwater levels can lead to vegetation degradation, soil exposure, and increased susceptibility to wind erosion, ultimately causing land desertification [48]. Reduced or vanished river flows supplied by groundwater can result in diminished surface runoff, leading to comprehensive ecological degradation [49]. When groundwater extraction causes a drop in water levels, pore water pressure decreases while total stress remains constant, resulting in increased effective stress. This can lead to consolidation and dewatering of the geotechnical skeleton, potentially triggering geological issues such as land subsidence [50].

Forests and cultivated land cover 0.20% and 13.98% of the study area, respectively. Forests, as a crucial part of ecosystems, play a significant ecological role. They help retain soil and water, prevent soil erosion, and maintain land health and productivity [51]. Additionally, forests regulate climate by absorbing carbon dioxide and releasing oxygen, mitigating global warming trends. Cultivated land has a dual impact on ecological quality: improper use and management can degrade ecological quality, while scientific management and protection can enhance its positive ecological functions. Unsustainable agricultural practices can expose soil surfaces, making them more vulnerable to erosion by water, wind, and gravity, leading to soil loss and reduced fertility and productivity [52]. Furthermore, agricultural activities can cause the decomposition and oxidation of soil organic matter, reducing its content and leading to soil impoverishment and quality decline [53]. Long-term application of chemical fertilizers and specific crop choices can also cause soil acidification, negatively impacting soil ecology [54]. However, cultivated land also possesses many positive ecological functions. Agricultural ecosystems can absorb waste, maintaining ecological and economic balance, protecting soil, and preventing pollution from being amplified through crops, food chains, and production chains [55]. Crops also provide functions such as windbreaks, soil stabilization, and the regulation of soil pH [56]. Therefore, the role of cultivated land in ecological quality is also significant.

4.2. *Analysis of the Comprehensive Evaluation Results for Ecological and Geological–Environmental Quality*

Based on the weighted overlay of 15 indicators, the weighted formula is as follows: Vegetation Cover × 0.055 + Slope Gradient × 0.041 + Precipitation × 0.109 + Temperature × 0.060 + Groundwater Depth × 0.093 + Groundwater Quality × 0.057 + Soil Quality × 0.041 + Geological Hazards × 0.068 + Cultivated Land × 0.077 + Grassland × 0.083 + Forest Land × 0.114 + Wetlands × 0.073 + Construction Land × 0.030 + Unused Land × 0.038 + Population Density × 0.052. According to the ecological quality evaluation table (Table 4), the ecological and geological–environmental quality grades were determined. The final evaluation map of the ecological and geological–environmental quality for the Kundulun River Basin in Baotou City is shown in Figure 4. Figure 4 shows that most areas of the study region have good or excellent ecological and geological–environmental quality, with the quality improving from the northwest to the southeast.

Areas with poor ecogeological–environmental quality are mainly located in the central, southern, and eastern parts of the study area, covering a total area of 127.86 km^2, accounting for 6.01% of the total study area. Areas with relatively poor ecogeological–environmental quality are primarily found in the western part of the study area, covering 836.19 km^2, or 39.30%, of the total study area. Areas with better ecogeological–environmental quality are mainly located in the mountainous valleys of the eastern and central regions, with a total area of 1072.72 km^2, accounting for 50.41% of the total study area. Areas with good ecogeological–environmental quality are concentrated in the central region where vegetation growth conditions are favorable and groundwater is shallow (with a burial depth of 0.4–3 m), covering an area of 91.07 km^2, or 4.28% of the total study area. The specific proportion of the evaluation results is shown in Figure 5.

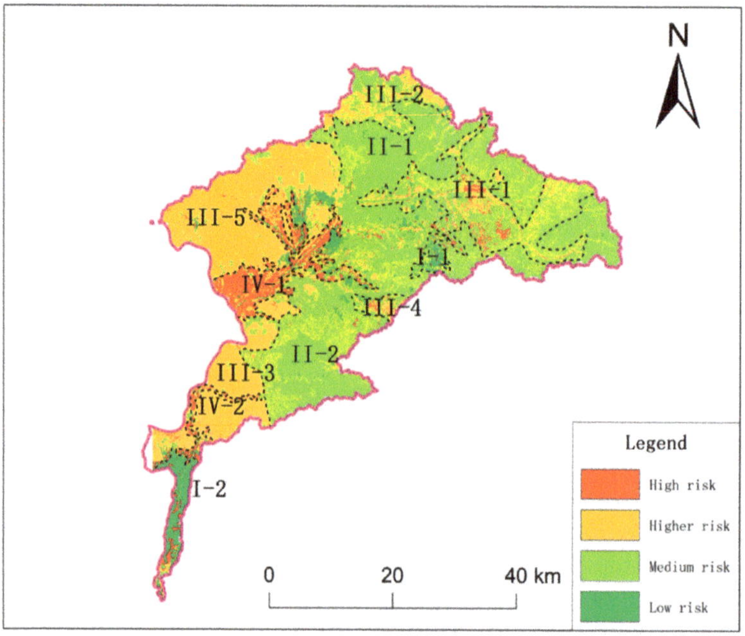

Figure 4. Ecological environment quality assessment map of Kundulun River basin in Baotou City.

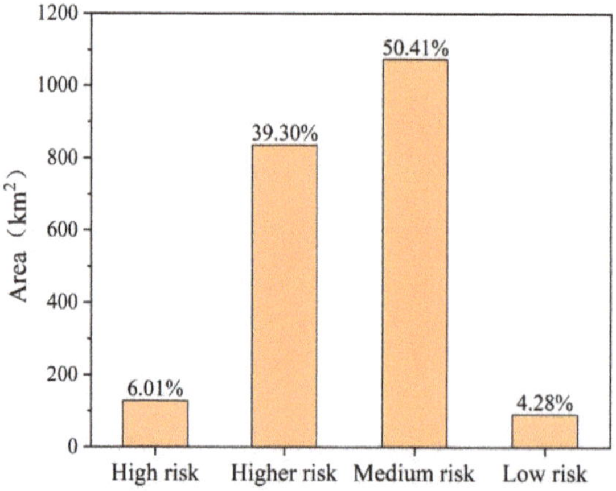

Figure 5. Histogram of ecological quality evaluation results in the study area.

In summary, the total area of regions with poor and relatively poor ecogeological–environmental quality is 964.05 km^2, accounting for 45.31% of the total study area. These areas are mainly ecologically fragile zones characterized by deep and poor-quality groundwater, harsh natural conditions, single vegetation types, and low vegetation coverage. The soil is infertile and has low productivity, and precipitation is relatively lower compared to the eastern part of the study area. A small number of areas are located within mineralization zones, and the overall ecological environment faces certain challenges.

In summary, the total area of regions with poor and relatively poor ecological and geological–environmental quality in the study area is 647.94 km^2, accounting for 30.26% of the total area of the study area. These areas are predominantly ecologically fragile zones, characterized by deep groundwater levels, poor water quality, harsh natural conditions, limited vegetation types, and low coverage. The soil is barren, with low productivity, and precipitation is relatively scarce compared to the eastern part of the study area. These regions belong to the agropastoral ecotone, significantly affected by human activities and irrigation. Some areas are located within mineral concentration zones, leading to vegetation destruction, reduction in forest and grassland areas, and deterioration of the ecological environment.

4.3. Zoning of Ecological and Geological–Environmental Quality

According to the characteristics of ecological and geological–environmental quality zoning, the ecological and geological–environmental quality of the county study area is divided into 11 sub-regions (Figure 4) based on specific ecological, environmental, and geological issues and geological–environmental conditions.

4.3.1. Areas with Good Ecogeological–Environmental Quality (Category I)

(1) I-1 Area: Located in the eastern part of the study area, this region has shallow groundwater (0.4–3 m) with excellent water quality, mainly classified as Class I and II. Vegetation coverage is high, primarily consisting of grasslands with a small amount of forest land. The terrain slope is gentle (0–12°), which facilitates water infiltration and retention. Annual precipitation is moderate (370–405 mm), temperatures are suitable, averaging between 1 °C and 3.5 °C, providing favorable conditions for vegetation growth.

(2) I-2 Area: Situated near the Kundulun River in the southern part of the study area, this region also features shallow groundwater with good water quality, mainly Class III. It contains abundant grassland and a small amount of forest land, with overall good soil quality and minimal heavy metal pollution. The terrain slope is moderate, without severe geological hazards. Climate conditions are favorable, with annual precipitation of 370–405 mm and average temperatures ranging from 3.5 °C to 6 °C, contributing to a stable natural ecosystem.

4.3.2. Areas with Better Ecogeological–Environmental Quality (Category II)

(1) II-1 Area: Located in the northeastern part of the study area, this region has relatively high vegetation coverage but slightly lower than Category I areas. The terrain slope is moderate, ranging from 5°to 15°, with no severe geological hazards. Predominantly covered by chestnut calcic soils, the vegetation coverage is 50–60%, with annual precipitation of 405–440 mm and an average temperature of 3.5–6 °C. Groundwater depth is 3–5 m, with Class III water quality, making it suitable for agriculture.

(2) II-2 Area: Located in the mountainous valley areas of the study region, such as Liufa Gully, the area features diverse vegetation types and rich herbaceous plant coverage. Groundwater depth is shallow with good water quality, mainly Class III. Climatic conditions are favorable, with temperatures and precipitation conducive to vegetation growth, annual precipitation of about 405–440 mm, and an average temperature around 5.5 °C. Land use is reasonable, mostly comprising grasslands and cultivated land. This area is predominantly covered by grey-brown soils, which have good water and nutrient retention capabilities. The terrain slope is moderate, ranging from 5°to 15°.

4.3.3. Areas with Relatively Poor Ecogeological–Environmental Quality (Category III)

(1) III-1, III-2, and III-4 Areas: Located in the eastern and northern parts of the study area, Zone III-1 has a small-scale concentrated mining area in its central part, primarily

consisting of iron ore, with sporadically distributed construction granite and quartzite mines. The groundwater level is deep, and the water quality is relatively poor, mainly classified as Class IV and V. Vegetation coverage is low, dominated by sparse grasslands. Terrain slopes are steep, exceeding 15° in some areas, posing a risk of geological hazards. Climatic conditions are harsh, with annual precipitation below 365 mm, unfavorable for vegetation growth. These areas are mainly covered by chestnut calcic soils, which are infertile and prone to wind erosion, with lower temperatures.

(2) III-3, III-5 Areas: Located in the western part of the study area, including the northwest and southwest of Guyang County, these regions have low vegetation coverage and steep terrain slopes, with low annual precipitation and average temperatures of 1–3.5 °C. Groundwater depth is significant, and water quality is poor. Geological hazards occur at a moderate frequency, and population density is higher.

4.3.4. Areas with Poor Ecogeological–Environmental Quality (Category IV)

(1) IV-1, IV-2 Areas: Located in Guyang County and its surrounding areas in the central part of the main study area, this region is characterized by deep groundwater levels and poor water quality, mainly classified as Class IV and V. Vegetation types are limited and show low coverage. In addition, frequent geological hazards further deteriorate the ecological environment. Although the area features diverse soil types, soil salinization and wind erosion are commonly observed. Annual precipitation is less than 365 mm, and the climate is cold, with an average annual temperature below 2 °C.

4.4. Discussion and Comparison

Through a systematic assessment of the ecogeological–environmental quality in the Kundulun River Basin of Baotou City, we have reached several important conclusions. To better understand these findings and verify the effectiveness of our research methods, we compared the results from the Kundulun River Basin with those from Li Hang et al. (2022) [57] in Ertaizhen, Zhangbei County. Additionally, we compared them with recent international cutting-edge research on the assessment of ecogeological environments in arid and semi-arid regions.

(1) Comparison with Li Hang et al. (2022) [57] in Ertaizhen, Zhangbei County

Comparing our findings with those of Li Hang et al. (2022) [57] in Ertaizhen, Zhangbei County, it was found that both the Kundulun River Basin and Ertaizhen are geographically close, located in central-western Inner Mongolia, with similar annual precipitation levels (346 mm for the Kundulun River Basin and approximately 400 mm for Ertaizhen). Despite differences in geomorphological features (the Kundulun River Basin consists mainly of medium and low mountains and hills, while Ertaizhen is characterized by plateau hills), both areas exhibit common ecological, geological, and environmental issues. For example, both regions have a significant proportion of land at high risk of desertification, primarily concentrated in the central and eastern areas.

Further comparisons reveal that the evaluation index system of the Kundulun River Basin is more comprehensive, covering key factors such as groundwater depth and water quality, which significantly impact the ecological quality in arid and semi-arid regions. Specifically, the Kundulun River Basin not only considers vegetation coverage and climatic conditions but also deeply analyzes changes in groundwater levels and their impact on ecosystems, making its evaluation results more accurate and comprehensive. In contrast, although Li Hang et al.'s study in Ertaizhen also focuses on soil texture and vegetation coverage, it places less emphasis on geological–environmental aspects. Therefore, the Kundulun River Basin study can better reveal the complexity of the ecogeological environment and provide more targeted management recommendations.

(2) Comparison with International Cutting-Edge Research

1. Comparison with Sun et al. (2020) [17] using the PCA-Disaster Theory Method

Sun et al. (2020) [17] used a combination of principal component analysis (PCA) and disaster theory to assess the overall ecological–environmental quality in the mid-Atlantic region of the United States. This method reduced data complexity through dimensionality reduction techniques and effectively identified major ecological risk factors. In contrast, the Kundulun River Basin study combined the analytic hierarchy process (AHP) with coefficient of variation (CV), considering expert experience while reflecting the importance of each indicator through objective data. This subjective–objective integrated approach provides a more comprehensive assessment of ecogeological–environmental quality, especially in data-scarce arid and semi-arid regions.

2. Comparison with Zhang et al. (2021)'s [18] AHP-GPCA Model

Zhang et al. (2021) [18] proposed an integrated assessment method for the Qinghai–Tibet Plateau's ecogeological environment based on RS/GIS and the AHP-GPCA model. This method uses geographic information systems (GIS) and remote sensing (RS) technologies to obtain multi-source data and determine weight values through AHP. Although both studies use AHP, the Kundulun River Basin study further incorporates CV and validates the stability of weight allocation through the single-factor variation method. Results show that adjustments in the weights of highly sensitive indicators such as grassland (C10) and precipitation (C3) significantly affect the comprehensive score (change rate > 8%), proving the current weight allocation is reasonable and highly reproducible.

(3) Uniqueness of the Kundulun River Basin Study

Through comparisons with the aforementioned international cutting-edge research and Li Hang et al.'s study in Ertaizhen, we find that the Kundulun River Basin study has several unique aspects and advantages:

Comprehensive Index System: The evaluation index system of the Kundulun River Basin not only includes conventional factors like vegetation coverage and climatic conditions but also delves into groundwater depth and its impact on ecosystems, making its evaluation results more accurate and comprehensive.

Methodological Innovation: By combining AHP with CV, this study considers expert knowledge while objectively reflecting the importance of each indicator, thereby enhancing the rationality and reliability of weight allocation.

Targeted Recommendations: Addressing specific problems in arid and semi-arid regions, such as declining groundwater levels and land degradation, this study provides more targeted management recommendations, offering scientific support for ecological protection in similar regions.

5. Conclusions and Recommendations

(1) From the Evaluation Criteria Layer

The evaluation index weight for land use type is the highest at 0.368, while the human activity evaluation index has the lowest weight at 0.100. This indicates that land use type contributes most significantly to the ecological quality of the study area, whereas human activities have a relatively smaller impact. Spatial distribution analysis of human activity evaluation indices shows that population density is sparse overall, resulting in only a minor influence.

(2) From the Evaluation Indicators Layer

The grassland evaluation indicator has the highest weight (0.112), with its coverage directly impacting regional ecological quality. Grasslands enhance soil stability through their extensive root networks, reducing the risk of soil erosion and water loss, which is especially crucial in arid and semi-arid regions. High grassland coverage effectively retains moisture, reduces surface runoff, prevents soil erosion, and maintains ecosystem stability and water conservation functions. Therefore, increasing grassland coverage is one of the key measures to improve the ecological quality of this region.

Precipitation, as an important factor affecting ecological quality (weight: 0.107), not only directly promotes vegetation photosynthesis and nutrient absorption but also indirectly affects the replenishment of surface water and groundwater. Adequate precipitation helps increase vegetation coverage, improving soil structure and fertility. Moreover, precipitation is essential for maintaining water levels and quality in wetlands and rivers. However, in the Kundulun River Basin, low and unevenly distributed annual precipitation limits vegetation growth in some areas, leading to declining groundwater levels and exacerbating–environmental degradation.

Groundwater depth (weight: 0.093) is closely related to ecological quality, particularly in arid and semi-arid regions, where a decline in groundwater levels can lead to vegetation degradation and land desertification. Groundwater extraction can reduce pore water pressure, causing soil compaction, and potentially triggering geological problems like land subsidence. Thus, managing groundwater resources rationally and avoiding over-extraction are crucial for protecting local ecosystems.

(3) Ecological Environment Quality Assessment Results

The assessment results show that the ecological–environmental quality around Guyang County in the central part of the study area is poor, while the western part has relatively poorer ecological conditions. The total area of regions with poor and relatively poor ecological quality is 964.05 km^2, accounting for 45.31% of the total study area. The main reasons include low precipitation, deep groundwater levels with poor water quality, harsh natural conditions, single vegetation types, infertile soils, and parts of the area being located in mining zones, leading to vegetation destruction, reduced forest and grassland areas, and environmental degradation.

(4) Sensitivity Analysis Using Single-Factor Variation Method

Vegetation coverage (C1), precipitation (C3), and forest land (C11) are highly sensitive indicators (score change rate > 8%), with precipitation having the most significant impact. Groundwater depth (C5) and cultivated land (C9) are moderately sensitive indicators (change rate: 2.5–3.4%). Population density (C15), geological disasters (C8), etc., are low-sensitivity indicators (change rate: < 1.5%). These findings highlight the critical role of water resource factors in the evaluation of ecogeological–environmental quality.

(5) Overall Status and Recommendations

Approximately 45.31% of the land in the study area is in a high- or relatively high-risk state in terms of ecogeological–environmental quality, and the overall status of the ecogeological–environmental quality is not optimistic, requiring further strengthening of protection measures. It is recommended that in Zone I (areas with good ecogeological environment), strict protection should be the main approach, establishing ecological red lines to prohibit mineral exploitation and expansion of construction land, promoting water-saving agricultural technologies, and setting up grassland health monitoring stations; in Zone II (better areas), grassland coverage should be increased to 70% through measures such as converting farmland back to grassland (e.g., planting drought-resistant forage like Stipa tenacissima) and constructing rainwater harvesting facilities; in Zone III (poorer areas), efforts should focus on mine reclamation (e.g., increasing vegetation coverage from <30% to 60% in mining areas within the Kundulun River Basin) and groundwater management (e.g., constructing artificial wetlands to intercept mine water pollution); in Zone IV (areas with poor conditions), urgent interventions should be implemented, such as improving saline soils through soil replacement and microbial remediation in Zone IV-2 around Baotou City, while implementing grazing bans and constructing forage bases in the agropastoral ecotone

(Zone IV-1). Regarding implementation and safeguards, priority should be given to the treatment of Zone IV (within 5 years) and Zone III (within 3–5 years), combined with remote sensing/GIS dynamic monitoring indicators (e.g., vegetation coverage, groundwater depth), and special national funding should be applied for to support restoration projects in Zone IV.

(6) Research Limitations and Future Prospects

This study constructed an ecogeological–environmental quality evaluation model for the Kundulun River Basin based on static data from 2020, with a focus on quantitative assessment of current ecological quality and zoning-based management strategies. Due to limitations in research duration, data availability, and technical methods, the dynamic trends under climate change or human activity disturbances were not thoroughly explored. For example, an ecological response model under drought intensification scenarios (e.g., a 15% reduction in precipitation) was not developed, nor was the long-term impact of grassland restoration projects quantitatively assessed. This limitation restricts the forward-looking application of the research findings in policy-making. Future work will integrate IPCC climate projections and MODFLOW groundwater modeling to conduct multi-scenario simulations (e.g., precipitation changes, land-use transitions) to predict the spatiotemporal evolution of ecological quality indicators. Additionally, the latest remote sensing data (e.g., Landsat 9) and field monitoring data will be used to verify the stability of the current model and supplement dynamic weight allocation methods.

Author Contributions: Y.H., writing—original draft preparation, software, and formal analysis; J.W., methodology and writing—review and editing; Y.X., writing—review and editing; W.Z., validation, investigation, and supervision; Y.L., conceptualization, methodology, and data curation; L.M. (Lei Mao), investigation, validation, and resources; X.L. responsible for data organization; L.M. (Limei Mo) investigation and funding acquisition; R.L. investigation and supervision. All authors have read and agreed to the published version of the manuscript.

Funding: This research was funded by the Survey of Lakes in the Mengxin Plateau Lake Region (DD20230510). It was further Supported by Natural Science Foundation of Inner Mongolia Autonomous Region of China (2024QN04003) and Innovation Base for Water Resource Exploration and Eco-environmental Effects in the Daheihe Basin of the Yellow River.

Data Availability Statement: The data presented in this study are available on request from the corresponding author.

Acknowledgments: We appreciate the anonymous reviewers for their invaluable comments and suggestions on this manuscript.

Conflicts of Interest: The authors declare no conflicts of interest.

References

1. Jiang, W.; Wang, G.; Sheng, Y.; Shi, Z.; Zhang, H. Isotopes in groundwater (2H, 18O, 14C) revealed the climate and groundwater recharge in the Northern China. *Sci. Total Environ.* **2019**, *666*, 298–307. [CrossRef] [PubMed]
2. Jiang, W.; Wang, G.; Liu, F.; Zhang, J.; Chen, D. Cl, Br, B, Li, and noble gases isotopes to study the origin and evolution of deep groundwater in sedimentary basins: A review. *Environ. Chem. Lett.* **2022**, *20*, 1497–1528. [CrossRef]
3. Jiang, W.; Sheng, Y.; Shi, Z.; Guo, H.; Chen, X.; Mao, H.; Liu, F.; Ning, H.; Liu, N.; Wang, G. Hydrogeochemical characteristics and evolution of formation water in the continental sedimentary basin: A case study in the Qaidam Basin, China. *Sci. Total Environ.* **2024**, *957*, 177672. [CrossRef]
4. Sheng, Y.; Baars, O.; Guo, D.; Whitham, J.; Srivastava, S.; Dong, H. Mineral-bound trace metals as cofactors for anaerobic biological nitrogen fixation. *Environ. Sci. Technol.* **2023**, *57*, 7206–7216. [CrossRef] [PubMed]
5. Zhou, X.; Sheng, Y.; Zheng, Y.; Jiang, M.; Wang, M.; Zhu, Z.; Li, G.; Baars, O.; Dong, H. Bioavailability of molybdenite to support nitrogen fixation on early Earth by an anoxygenic phototroph. *Earth Planet. Sci. Lett.* **2024**, *647*, 119056. [CrossRef]
6. Bernardino, P.N.; De Keersmaecker, W.; Fensholt, R. Global-scale characterization of turning points in arid and semi-arid ecosystem functioning. *Glob. Ecol. Biogeogr.* **2020**, *29*, 1230–1245. [CrossRef]

7. Fan, X.; Yu, H.; Tiando, D.S.; Rong, Y.; Luo, W.; Eme, C.; Ou, S.; Li, J.; Liang, Z. Impacts of Human Activities on Ecosystem Service Value in Arid and Semi-Arid Ecological Regions of China. *Int. J. Environ. Res. Public Health* **2021**, *18*, 11121. [CrossRef]
8. Jiang, W.; Liu, H.; Sheng, Y.; Ma, Z.; Zhang, J.; Liu, F.; Chen, S.; Meng, Q.; Bai, Y. Distribution, source apportionment, and health risk assessment of heavy metals in groundwater in a multi-mineral resource area, North China. *Expo. Health* **2022**, *14*, 807–827. [CrossRef]
9. Chen, X.; Sheng, Y.; Wang, G.; Zhou, P.; Liao, F.; Mao, H.; Zhang, H.; Qiao, Z.; Wei, Y. Spatiotemporal Successions of N, S, C, Fe, and As Cycling Genes in Groundwater of a Wetland Ecosystem: Enhanced Heterogeneity in Wet Season. *Water Res.* **2024**, *251*, 121105. [CrossRef]
10. Tudes, S.; Ceryan, S.; Bulut, F. Geoenvironmental evaluation for planning: An example from Gumushane City, close to the North Anatolia Fault Zone, NE Turkey. *Bull. Eng. Geol. Environ.* **2012**, *71*, 679–690. [CrossRef]
11. Ren, B.Y.; Cheng, J.; Shi, L.P.; Liu, H.; Guo, Y.; Jackson, T.J.; Chen, J.M.; Gong, P.; Liang, S. *Eco-Geological Environment Assessment of Datong Basin Using Satellite Remote Sensing*; SPIE: Bellingham, WA, USA, 2014; p. 92604E.
12. Wang, Z.F.; Wang, Y.J.; Wang, L.; Zhang, T.; Tang, Z. Research on the comprehensive evaluation system of eco-geological environmental carrying capacity based on the analytic hierarchy process. *Clust. Comput.* **2019**, *22*, 5347–5356. [CrossRef]
13. Duan, J.S.; Huang, J.; Chen, Y.F.; Du, K. Research on Eco-geological Environment Carrying Capacity Based on GIS Technology. *IOP Conf. Ser. Earth Environ. Sci.* **2021**, *651*, 042003. [CrossRef]
14. Saaty, T.L.; Bennett, J.P. A theory of analytical hierarchies applied to political candidacy. *Syst. Res. Behav. Sci.* **1977**, *22*, 237–245. [CrossRef]
15. Shi, J.T.; Liu, J.J.; Zhang, J.C.; Wang, J.Y.L.; Jiang, Y.G.; Wang, M.; Li, H.F.; Yang, W.H.; Yan, X.J. Analysis of soil heavy metal influencing factors and sources in typical small watersheds in shallow mountainous area. *Geophys. Geochem. Explor.* **2024**, *48*, 834–846. (In Chinese)
16. Long, X.J.; Zheng, J.S.; Li, X.J.; He, Z.W.; Liu, Y.S. An evaluation system of ecogeology environment. *Sci. Surv. Mapp.* **2018**, *43*, 65–70.
17. Sun, X.F.; Shao, H.Y.; Xiang, X.Y.; Yuan, L.; Zhou, Y.; Xian, W. A Coupling Method for Eco-Geological Environmental Safety Assessment in Mining Areas Using PCA and Catastrophe Theory. *Nat. Resour. Res.* **2020**, *29*, 4133–4148. [CrossRef]
18. Zhang, J.C.; Pan, X.C. Comprehensive Assessment for Changes of Ecological Environment in Qinghai-Tibet Plateau Based on RS/GIS and AHP-GPCA Model. *J. Earth Sci. Environ.* **2011**, *33*, 434–440.
19. Zhao, F.F.; He, M.C.; Wang, Y.T.; Tao, Z.-G.; Li, C. Eco-geological environment quality assessment based on multi-source data of the mining city in red soil hilly region, China. *J. Mt. Sci.* **2021**, *19*, 253–275. [CrossRef]
20. Du, Y.F.; Guo, H.; Wang, S.; Sun, J.; Siqi, Z.; Zeshi, C. Hydrochemical Characteristics and Formation Mechanism of Groundwater in the West of Kundulun River in Baotou City. *Environ. Sci. Technol.* **2024**, *47*, 121–130.
21. Chang, F.; Xing, X.F.; Wang, F. Study on Dynamic Monitoring of Soil and Water Loss in Baotou City in 2019. *Soil Water Conserv. China* **2021**, *7*, 56–58.
22. Chen, J. Simulation of Rainfall-Runoff in the Kundulun River, Baotou City, Inner Mongolia. *J. Sichuan Univ. (Eng. Sci. Ed.)* **2003**, *1*, 31–33.
23. Li, M.Z.; Guo, X.; Du, Z. Correlation Analysis of Precipitation and Runoff in Kundulun Reservoir. *Inn. Mong. Water Resour.* **2015**, *1*, 12–13.
24. Zhao, J.; Geng, D.J. Analysis of Hydrogeological Characteristics in the Area West of Kundulun River, Baotou City, Inner Mongolia. *Ground Water* **2016**, *38*, 240–241+265.
25. Han, X.D.; Li, H.J.; Su, M.X.; An, P. Spatial network analysis of surface soil pollution from heavy metals and some other elements: A case study of the Baotou region of China. *J. Soils Sediments* **2019**, *19*, 629–640. [CrossRef]
26. Zhang, M.Y.; Rong, L.H.; Li, Y.T.; Hui, D. Land use function transformation in the agro-pastoral ecotone based on ecological-production-living spaces and associated eco-environment effects:A case of Baotou City. *Arid Land Geogr.* **2023**, *46*, 958–967.
27. Yang, B.; Ning, X.L. Study on Construction of Ecological City's Evaluation of Ecological Environment in Baotou City. *J. Inn. Mong. Norm. Univ. (Nat. Sci. Ed.)* **2015**, *44*, 104–107+112.
28. Deng, J.X.; Liu, Z.H.; Xu, Z.Y. Subdivision and correlation of the late Pleistocene-Holocene strata in the Baotou area and its paleoenvironment variation. *J. Stratigr.* **2007**, *31*, 133–140.
29. Liao, Z.L.; Long, Y.H.; Wei, Y.F.; Guo, Z.; Jiao, R.; Son, Y.; Cui, Y. Responses of the sustainable yield of groundwater to annual rainfall and pumping patterns in the Baotou Plain. *Desalination Water Treat.* **2018**, *131*, 96–106. [CrossRef]
30. Bai, L.P.; Wang, Y.Y.; Guo, Y.L.; Zhou, Y.; Liu, L.; Yan, Z.; Li, F.; Xie, X. Health Risk Assessment Research on Heavy Metals Ingestion through Groundwater Drinking Pathway for the Residents in Baotou, China. *J. Environ. Health* **2016**, *78*, 84–90.
31. Zhang, C.M. Discussion on the Development and Utilization of Groundwater Resources in Baotou City, Inner Mongolia. *Inn. Mong. Sci. Technol. Econ.* **2021**, *15*, 84–85.
32. Xin, S.Q. The research on cold resistance of several kinds of cover plant in Baotou area. *J. North. Agric.* **2016**, *44*, 83–85.

33. Zhao, Y.; Yu, Z.C.; Chen, F. Spatial and temporal patterns of Holocene vegetation and climate changes in arid and semi-arid China. *Quat. Int.* **2009**, *194*, 6–18. [CrossRef]
34. *DZ/T 0295-2016*; Specifications for Geochemical Survey of Land Quality. Ministry of Land and Resources of the People's Republic of China: Beijing, China, 2016.
35. *DZ/T 0167-2012*; Specifications for Regional Geochemical Exploration. Ministry of Land and Resources of the People's Republic of China: Beijing, China, 2012.
36. *DD2019-03*; Specifications for 1:50,000 Hydrogeological Survey. China Geological Survey: Beijing, China, 2019.
37. *Technical Specifications for Unified Groundwater Measurement (Trial Version 20200831)*; Institute of Hydrogeology and Environmental Geology, Chinese Academy of Geological Sciences: Shijiazhuang, China, 2020.
38. *Specifications for Multi-Purpose Geochemical Survey (1:250,000)*; China Geological Survey: Beijing, China, 2016.
39. *GB/T 14848-2017*; Groundwater Quality Standard. State Administration for Market Regulation and Standardization Administration of China: Beijing, China, 2017.
40. Wind, Y.; Saaty, T.L. Marketing Applications of the Analytic Hierarchy Process. *Manag. Sci.* **1980**, *26*, 641–658. [CrossRef]
41. Chen, H.G.; Li, X.N.; Li, C.Y. Resilience Evaluation of Water Resource System Based on Coefficient of Variation Entropy Weight Method: A Case Study of Water Resources in Heilongjiang Province from 2007 to 2016. *Ecol. Econ.* **2021**, *37*, 179–184.
42. Cao, D.; Yu, F.; Zhu, W.Q.; Xie, G.X.; Song, C.Y. Evaluation of economic losses caused by grassland ecosystem degradation based on remote sensing technology. *Acta Sci. Circumstantiae* **2011**, *31*, 1799–1807.
43. Yang, Y.; Wang, G.X.; Li, Y.; Wang, Z.W. Spatial-Temporal Changes in the Critical Vegetation Coverage of Grasslands with Functions of Soil-Water Conservation, Wind-Breaking and Sand-Fixing in the Three-Rivers Headwater Region, China. *Mt. Res.* **2024**, *42*, 143–153.
44. Su, J.S.; Xu, F.W.; Wang, Y.; Zhao, Y.; Bai, Y. Collaborative restoration between vegetation and soil attributes in degraded grassland of different grassland types. *Chin. Sci. Bull.* **2024**, *70*, 1519–1536. [CrossRef]
45. Yu, Z.X.; Wang, T.Y.; Wang, P.; Yu, J. The Spatiotemporal Response of Vegetation Changes to Precipitation and Soil Moisture in Drylands in the North Temperate Mid-Latitudes. *Remote Sens.* **2022**, *14*, 3511. [CrossRef]
46. Venkatesan, G.; Subramani, T.; Karunanidhi, D.; Sathya, U.; Li, P. Correction to: Impact of precipitation disparity on groundwater fluctuation in a semi-arid region (Vellore district) of southern India using geospatial techniques. *Environ. Sci. Pollut. Res.* **2021**, *28*, 18552. [CrossRef]
47. Rahul, W.S.; Arnab, B.; Aditi, B. Assessment of impact of climate change on streamflow and soil moisture in Pare watershed of Arunachal Pradesh, India. *Sustain. Water Resour. Manag.* **2024**, *10*, 82.
48. Wang, P.S.; Dong, S.G.; Zang, X.C.; Yang, X.; Ji, Y.; Li, L.; Han, X.; Hou, F. Effects of groundwater level changes on soil characteristics and vegetation response in arid and semiarid coal mining areas. *Environ. Geochem. Health* **2024**, *46*, 441. [CrossRef]
49. Zhang, J.T.; Xi, H.Y. Spatiotemporal dynamics of groundwater levels in a desert riparian forest and its response to surface runoff. *Arid Land Geogr.* **2020**, *43*, 388–397.
50. Tran, T.V.; Vuong, H.M.T.; Ram, A.; Kumar, P.; Van Hiep, H.; Kurasaki, M. Spatiotemporal variations in groundwater levels and the impact on land subsidence in CanTho, Vietnam. *Groundw. Sustain. Dev.* **2021**, *15*, 100680.
51. Shen, H.; Jiang, F.Q.; Du, X.J.; Lu, T. Evaluation on soil anti-erodibility of soil and water conservation forest. *Chin. J. Appl. Ecol.* **2000**, *11*, 345–348.
52. Song, Y.; Zhang, Z.X. The Effect of Different Tillage Measures on Soil Erosion in Slope Farmland in Black Soil Region. *Res. Soil Water Conserv.* **2011**, *18*, 14–16+25.
53. Li, J.J.; Zhao, X.; Pan,T.H.; Yan, J.X.; Li, H.J. Effects of Different Land-use Types on Labile Organic Matter. *J. Soil Water Conserv.* **2011**, *25*, 147–151.
54. Zhou, X.Y.; Xu, M.G.; Zhou, S.W.; Gilles, C. Soil acidification characteristics in southern China's croplands under long-term fertilization. *J. Plant Nutr. Fertil.* **2015**, *21*, 1615–1621.
55. Yang, Z.X. Study on Integrated Assessment of Positive and Negative Value of Beijing Cropland Ecological System. Ph.D. Thesis, China Agricultural University, Beijing, China, 2006.
56. Mo, H.W.; Ren, Z.Y.; Wang, X. Study on the Dynamic Change of Value of Vegetation Sand-fixing Effect—A Case Study in the Yuyang Region. *Arid. Zone Res.* **2006**, *2*, 56–59.
57. Li, H. Eco-Geological Survey and Assessment: A Case Study of Ertaizhen Town, Zhangbei County. *Chin. Acad. Geol. Sci.* **2022**, *43*, 665–675.

water

MDPI

Article

Characterization of Groundwater Dynamics and Their Response Mechanisms to Different Types of Compound Stress in a Typical Hilly Plain Area

Qian Zhang [1,2,†], Meng Zhang [1,2,†], Wanjun Jiang [3,*], Yang Hao [4], Feiwu Chen [1,2] and Mucheng Zhang [1,2]

1 College of Water Conservancy Engineering, Tianjin Agricultural University, Tianjin 300392, China; cathy_zhang@tjau.edu.cn (Q.Z.); 2408028119@stu.tjau.edu.cn (M.Z.); 2208028105@stu.tjau.edu.cn (F.C.)
2 Tianjin Agricultural University-China Agricultural University Joint Smart Water Conservancy Research Center, Tianjin 300392, China
3 Tianjin Center, China Geological Survey, Tianjin 300170, China
4 College of New Energy and Environment, Jilin University, Changchun 130021, China; haoyang@jlu.edu.cn
* Correspondence: jiangwanjun0718@163.com
† These authors contributed equally to this work.

Abstract

Groundwater is a crucial source of water supply and an important ecological element globally. Research on the dynamic characteristics of groundwater and their causative mechanisms is fundamental to objectively evaluating groundwater resources and their sustainable utilization. Based on the large amount of hydrogeological data collected and analyzed in typical hilly plain areas, a multi-factor weighted comprehensive evaluation system (MFWCES) based on GIS was used to evaluate the response of groundwater dynamics to combined stress elements in Tangshan City. The study area is located in the plains and hilly regions of Tangshan City. The evaluation system was based on seven influencing factors, including hydraulic conductivity, soil media, aquifer thickness, depth of groundwater, land use type, extraction intensity of groundwater, and groundwater evaporation. The results of groundwater dynamics in the study area were obtained by weighted comprehensive evaluation, with their score size ranging from 2.4 to 12.7. The spatial distribution of groundwater dynamics was classified into four categories: rapid response (10.3–12.7), dual response to precipitation and anthropogenic extraction (9.6–10.3), delayed response (7.6–9.6), and strong superimposed response to human activities (2.4–7.6). The related conclusions will provide key references for regional water resource planning, ecological protection, and the development of differentiated groundwater management strategies under compound stress.

Keywords: groundwater system; groundwater dynamic; mathematical statistics; multi-factor weighted comprehensive evaluation system (MFWCES); typical hilly plain area

1. Introduction

In recent years, rapid social and economic growth coupled with deteriorating ecological conditions have steadily reduced global water availability, while the exploitation of inland water resources has intensified [1,2]. Because of shifting climate patterns and increasing human activities, coastal regions are increasingly threatened by groundwater-related environmental issues [3]. Groundwater is an essential resource that supports the livelihoods of over two billion people worldwide [4]. Due to its widespread distribution and protection beneath the vadose zone, it is relatively resistant to contamination [5], making it a vital water source for urban life, industry, and agriculture [6]. As an indispensable component of the hydrological cycle, groundwater recharge is primarily driven by precipitation, which encapsulates various factors influencing groundwater dynamics over time. In order to make more reasonable use of valuable groundwater resources and to protect groundwater, it is particularly important to master the dynamic laws of groundwater and study their dynamic characteristics and causes of formation.

The main sources of groundwater recharge include atmospheric precipitation infiltration, surface water infiltration, overflow recharge of adjacent aquifers, and agricultural irrigation water recharge, among which precipitation is the main factor affecting groundwater dynamics in most areas. The dynamics of groundwater in response to precipitation are crucial for regional groundwater balance and the storage of groundwater [7]. Analyzing the fluctuation patterns of groundwater in response to precipitation forms the basis for assessing and efficiently utilizing freshwater resources [8]. Relevant studies in Baotou [9], Salento [10], western Japan [11], and other typical areas with drought or lack of surface fresh water showed that the response of groundwater levels to variation in precipitation possesses a complicated correlation and a lag effect, and could not be well described by analyzing only macro annual variation trends in precipitation. Cui et al. [9] and Yin et al. [12] used the wavelet coherence method to investigate the response of groundwater depth to precipitation and found that the lag in groundwater depth changes in response to precipitation was affected by long-term precipitation. Russo et al. [13] and Wang et al. [14] developed a long-term evaluation framework from several years to several decades and revealed that the annual groundwater recharge reflected by groundwater level changes was significantly correlated with seasonal rainfall, such as precipitation and snowmelt, but that the response lag time might vary spatially and temporally due to aquifer lithology and groundwater exploitation intensity. Qi et al. [15] and Kellner et al. [16] conducted correlation and regression analyses based on decades of observational data, and the results indicated that, under intense precipitation patterns, the sensitivity of groundwater dynamics was higher, while its hysteresis was influenced by agricultural irrigation. Infiltration and groundwater recharge are mainly affected by the underlying surface conditions, such as the thickness of the infiltration zone, vegetation cover, and the modernization of agriculture and animal husbandry. Analyzing the dynamic patterns of groundwater and its response to precipitation is an important aspect of water resource utilization and protection. Previous studies on groundwater response to precipitation have often been limited to analyzing the relationship between precipitation and groundwater level changes in small areas or based on several decades of precipitation data. Rarely have these studies explored, from a macroscopic perspective, the correlation between precipitation infiltration and groundwater dynamics affected by combined geological conditions and anthropogenic disturbances. This study just makes up for this direction.

Groundwater dynamic patterns are not only related to precipitation, but also to other geographic and human activity factors. This study initially focused on analyzing the correlation between precipitation and groundwater dynamics. However, examining groundwater dynamics solely from the perspective of precipitation may lead to misleading conclusions regarding water resource utilization. Subsequently, this study integrates natural and anthropogenic factors into a unified evaluation framework by considering the combined effects of groundwater recharge and discharge. We investigate groundwater dynamic characteristics and underlying mechanisms under the joint influence of multiple factors—including hydraulic conductivity, soil media, aquifer thickness, groundwater depth, land use type, groundwater extraction intensity, and groundwater evaporation [17]. A multi-factor weighted comprehensive evaluation system (MFWCES) has been established to address this evaluation. Considering the impact of recent human activities, this comprehensive study examined the causes of groundwater dynamics under compound stresses. This investigation of groundwater dynamics in a typical hilly plain area lays the foundation for rational and sustainable utilization of regional groundwater resources [18], while also providing insights for ecological remediation and the prevention of geological disasters.

2. Study Area and Data

2.1. Location of Study Area

As shown in Figure 1, Tangshan City is situated in central Bohai Bay in China, between longitude 117°31′–119°19′ East and latitude 38°55′–40°28′ North. It is situated in the northeast of the North China Plain, with Hebei Province to the east, making Tangshan City a choke point of the North China–Northeast China corridor. Tangshan City covers a land area of 13,472 km^2, consisting of 5131 km^2 of mountainous terrain (38.09%) and 8341 km^2 of plains (61.91%). The sea area is 4472 km^2, with a coastline of 229.72 km. This study focuses on the plain and hilly areas of Tangshan City (the transitional region between plains and mountains), covering a total area of 12,925 km^2.

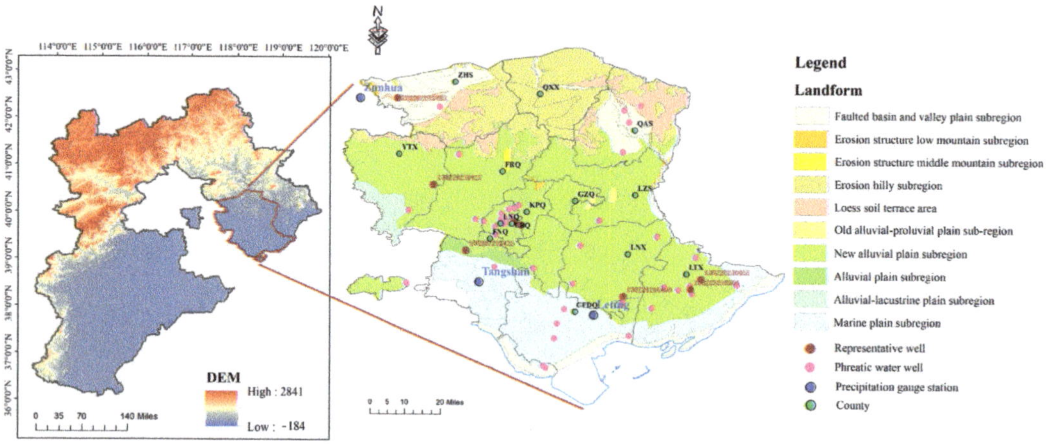

Figure 1. The location of the study area. The figure on the right depicts the hydrogeologic zoning of the study area. Green dots indicate the locations of administrative districts (All the county in the Figure 1 are presented in their abbreviations. For the correspondence between the full names and the abbreviations, please refer to Table A1), red and pink dots denote the observation wells from which water level data were collected, and the blue line delineates the coastline of the study area.

2.2. Hydrogeological and Meteorological Characteristics

Based on storage conditions and aquifer medium pore characteristics, groundwater in Tangshan City can be categorized into three types: pore water in loose rock, karst water in carbonate rock, and fissure water in bedrock. Pore water is mainly distributed in the Yanshan pre-mountain plains, coastal plains, mountain basins, and valley areas of the hilly area. The water is continuous and uniform in space, and the hydraulic connection within this system is better. Its main recharge sources include rainfall infiltration and river recharge, and the dynamic changes in shallow groundwater levels with the annual distribution of precipitation are obvious. Karst water-bearing rock groups are mainly distributed in the southern piedmont of Yanshan Mountain and hidden in the sloping plains in front of the mountains, intermontane basins, and other places. The water yield property of karst aquifers is generally strong but extremely uneven; in addition, there is obvious anisotropy in hydraulic connection [19]. The dynamic changes in karst water levels are characterized by large amplitude and rapid change. Fissure water is distributed in the hilly areas of northern Tangshan, and its abundance varies significantly across locations due to factors such as the nature of the fractures, their development characteristics, and recharge conditions. In the hilly and plain regions of Tangshan, fault structures are densely distributed; these fault zones not only act as barriers to water flow, but also serve as preferential conduits, resulting in a highly heterogeneous groundwater system [20]. The hydrogeological cross-section diagrams of FNQ and FRQ are presented in Figure 2.

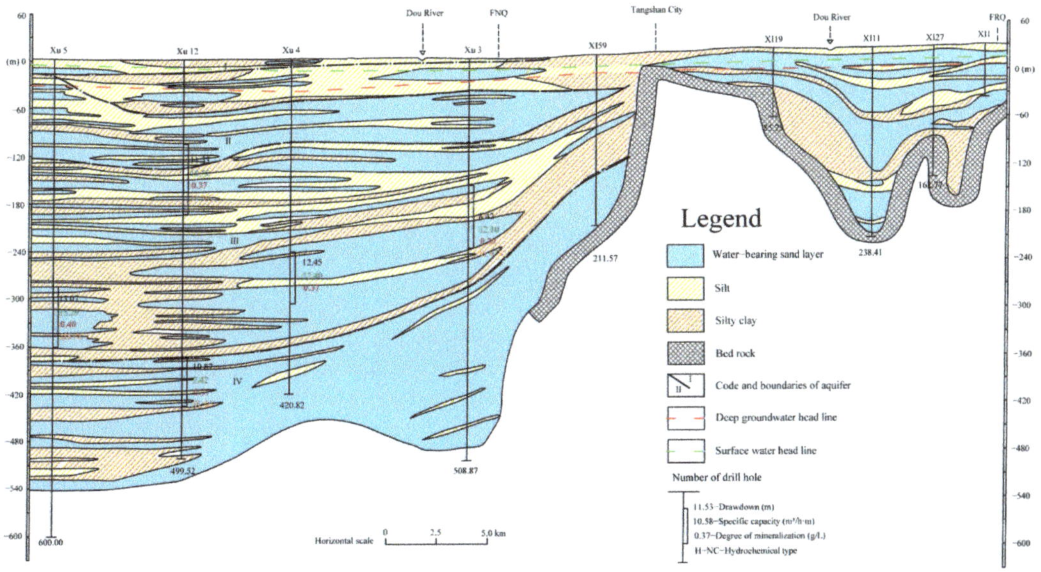

Figure 2. Aquifer profile hydrogeological cutaway view of FNQ and FRQ.

The study area exhibits a diverse range of soil types, with loamy sand and cohesive clay being the most widely distributed. The stratigraphy reflects a complex depositional and tectonic history, progressing from ancient metamorphic rocks to shallow-marine carbonates of the early Paleozoic, followed by Mesozoic fluvial-lacustrine deposits, and finally transitioning into Quaternary unconsolidated sediments. Furthermore, Tangshan is located within the North China Plain, where long-term agricultural irrigation and industrial water demands have led to significant groundwater overextraction and land subsidence, causing pronounced hydrogeological disturbances [21]. The central part of the North China Plain has a flat topography with good stratigraphic continuity and a relatively homogeneous groundwater system, whereas Tangshan has a complex groundwater runoff due to the presence of fracture zones and earthquakes having occurred several times in history. In the southern coastal plain, there may be seawater intrusion into freshwater areas of the aquifer.

In the study area, there are three major river systems—the Luan River, the Jiyun River, and the Jidong Coastal Water System—with over 70 rivers in total. Among these, rivers with drainage areas exceeding 200 km^2 include the Luanhe River, the Huanxiang River, the Dou River, the Sha River, and the Jiyun River.

Tangshan has a warm, temperate, semi-humid continental monsoon climate. Winters are relatively cold, with scarce precipitation and more northerly winds; summers are hot and humid, with concentrated rainfall and more southerly winds [22]. The highest temperature throughout the year reaches 40 °C (July), the lowest temperature reaches −26 °C (January), and the average temperature for many years is 10.6 °C. The average annual precipitation is 608.7 mm, while evaporation reaches 1509 mm.

3. Data and Methods

3.1. Data Sources

This study mainly analyzes the response characteristics of groundwater dynamics under the influence of precipitation and other compound factors. Precipitation data were primarily obtained from three precipitation gauge stations located in Zunhua, Tangshan and Leting (spanning January 2018 to August 2020). In the groundwater dynamics investigation, all 52 monitoring wells with complete basic information and containing long series of observation data for well-hole location projection were projected on the map. In each district and county-level administrative area, 1–2 monitoring points were selected as the object of analysis to ensure that the selected points were evenly distributed within the study area and were controlled and representative, and that the data were complete (spanning January 2018 to August 2020). Since the main objective of this study is to analyze the relationship between groundwater dynamics and precipitation, unconfined aquifer wells that directly receive recharge from precipitation were selected as the study object, and the main type of groundwater monitored was unconfined aquifers. A confined aquifer well was selected as the study object in the Xiaobeihai area, where there are no unconfined aquifer wells. In addition, the selected monitoring wells are shown in the topographic and geomorphic maps, indicating the geomorphic units in which the wells are located, trying to ensure that there are representative points in each geomorphic feature subdistrict. Some monitoring wells were selected as research objects according to the above principles, and their distribution locations are shown in Figure 1.

In evaluating the groundwater response to multiple compound factors, data collection was conducted for the study area based on information from hydrogeology reports, well drilling, and meteorological stations. The collected data included parameters such as precipitation, groundwater level, hydraulic conductivity, soil media, aquifer thickness, and groundwater depth. Land use data were obtained from the first Landsat-derived annual land cover product of China (CLCD); groundwater extraction data were calculated based on the Tangshan City Water Use Comprehensive Table; and phreatic evaporation data were sourced from the Tangshan evaporation stations.

3.2. Correlation Analysis (Auto-Correlation and Cross-Correlation)

The auto-correlation function (ACF) $r(k)$ can be used to identify repeating patterns in a sequence. It describes the correlation of a random signal at different times, essentially performing cross-correlation on the signal itself. The formula is as follows [19,23]:

$$r(k) = \frac{C(k)}{C(0)} \tag{1}$$

$$C(k) = \frac{1}{n}\sum_{t=1}^{n-k}(x_t - \overline{x})(x_{t+k} - \overline{x}) \tag{2}$$

$$C(0) = \frac{1}{n}\sum_{t=1}^{n}(x_t - \overline{x})^2 \tag{3}$$

where x_t is the t-th observation, $\overline{x}$ is its sample mean, n is the length of the time series, k is the time lag, $C(k)$ is the covariance function at lag k, $C(0)$ is the zero-lag covariance, and $r(k)$ is the auto-correlation coefficient. Equation (4) can be used to test whether the auto-correlation at lag k is significant [19,24].

$$|r_k| > \frac{2}{\sqrt{n}} \tag{4}$$

The cross-correlation function (CCF) is used to measure the similarity between independent signals x_t and y_t at any two different times. It provides an indicator of whether the two signals are correlated in the frequency domain, linking the cross-spectrum between two measurement points with their respective auto-spectra, thereby determining the extent to which the output signal is derived from the input signal. The cross-correlation function is not an even function, and $r_{xy}(k) \neq r_{yx}(k)$.

$$r_{xy}(k) = \frac{C_{xy}(k)}{\sqrt{C_x^2(0)C_y^2(0)}} \tag{5}$$

$$C_{xy}(k) = \frac{1}{n}\sum_{t=1}^{n-k}(x_t - \overline{x})(y_{t+k} - \overline{y}) \tag{6}$$

$$C_x(0) = \frac{1}{n}\sum_{t=1}^{n}(x_t - \overline{x})^2 \tag{7}$$

$$C_y(0) = \frac{1}{n}\sum_{t=1}^{n}(y_t - \overline{y})^2 \tag{8}$$

where x_t is the t-th observation of the input sequence, y_t is the t-th observation of the output sequence, $\overline{x}$ and $\overline{y}$ are the respective sample means, n is the length of the time series, $Cxy(k)$ is the cross-covariance function between the input and output sequences, $Cx(0)$ and $Cy(0)$ are the variances of the input and output sequences, and $r_{xy}(k)$ is the cross-correlation function between the input and output sequences.

3.3. Evaluation of Groundwater Dynamic Impact Factors by GIS-Based Multi-Factor Weighted Comprehensive Evaluation System (MFWCES)

The groundwater response to precipitation is, to some extent, a reflection of fluctuations caused by net recharge. Therefore, the evaluation system considers both recharge and discharge factors. Factors such as hydraulic conductivity, soil media, aquifer thickness, groundwater depth, and land use type govern the rate and volume of groundwater recharge. On the discharge side, the primary factors include extraction intensity and phreatic evaporation. The MFWCES selects seven influencing factors, as outlined above, and utilizes GIS to conduct independent assessments of each factor. The vector maps are then converted into raster data, and the independent evaluation factors are superimposed using map algebra and the raster calculator function, ultimately yielding a comprehensive evaluation result for groundwater dynamics [22]. See the formula below for details.

$$MFWCES_{index} = W_C L_C + W_S L_S + W_A L_A + W_D L_D + W_{LUT} L_{LUT} - W_{EI} L_{EI} - W_E L_E \tag{9}$$

where W is the weight of factor, and L is the assessment levels.

The factors in formula (9) include: hydraulic conductivity (C), soil media (S), aquifer thickness (A), groundwater depth (D), land use type (LUT), groundwater extraction intensity (EI), and groundwater evaporation(E).

Regional groundwater storage serves as the "reservoir" for groundwater dynamics, with its variations influencing dynamic characteristics through hydrogeological conditions and human activities. Tangshan City, situated in the northeastern part of the densely populated North China Plain, suffers from severe groundwater overexploitation. Under the influence of the South-to-North Water Diversion Project, groundwater extraction emerges as the primary driver of groundwater level fluctuations [25]. Wang et al., using expert scoring methods, identified key factors affecting groundwater dynamics in the North China Plain, including topography and groundwater depth [26]. Permeability and soil lithology accurately reflect variations in regional rock types and pore structures. Additionally, Etuk et al. employed a multi-factor weighted overlay approach to delineate groundwater potential in regions characterized by complex geology, high population density, and water scarcity. Utilizing the Satty Analytic Hierarchy Process (AHP) in combination with expert judgment, weights were assigned to various factors, revealing that geological factors exert the greatest influence on groundwater dynamics (25%), while land use type contribute the least (5%), and soil media account for approximately 15% [27]. Based on the above research and the *Guidelines for Groundwater Dynamics Analysis and Evaluation* published by the Hydrology Society in 2023 [28]. The evaluation system filtrates seven independent factors from the numerous variables that influence groundwater dynamics for assessment, and the weighting assignments for each factor in our evaluation framework are presented in Table 1.

Table 1. Introduction to the MFWCES and the weights assigned to them. Table 1 presents the seven indicators incorporated into the evaluation system. The second column provides a concise description of each indicator, while the third column lists the respective weights assigned in the evaluation framework.

Factors	Introduction	Weight
Hydraulic conductivity (C)	Measures the aquifer's permeability to determine the flow rate within the aquifer.	6
Soil media (S)	Soil particle size is related to infiltration rate.	5
Aquifer thickness (A)	Aquifer thickness is obtained by combining borehole data with groundwater depth.	4
Groundwater depth (D)	Groundwater depth is the distance from the surface to the phreatic water table, which influences the migration time of precipitation.	6
Land use type (LUT)	Land use type represent the natural and artificial distribution of features on the Earth's surface. It affects the spatial and temporal dynamics of groundwater systems.	2
Extraction intensity of groundwater (EI)	Groundwater extraction intensity is used to measure the rationality of groundwater development and utilization, as it can weaken the recharge effect of precipitation on groundwater.	5
Groundwater evaporation (E)	An increase in phreatic evaporation depletes groundwater, leading to a decline in groundwater levels. There exists a critical depth beyond which evaporation effectively ceases and can be considered negligible.	3

Each of the seven influencing factors mentioned above will be assigned an evaluation level, with scores ranging from 0.1 to 0.8, as detailed in Table 2. The score reflects the frequency of groundwater level fluctuations: the higher the score, the greater its impact on the comprehensive evaluation value, indicating a faster response of groundwater dynamics. On the contrary, a low score reflects a slow response of the factor to groundwater dynamics. The evaluation level of each factor increases with an increase in infiltration coefficient, particle size of soil media, groundwater extraction intensity, and groundwater evaporation and decreases with an increase in aquifer thickness and groundwater depth.

Table 2. Factors in the MFWCES and their assessment levels. Each indicator in the evaluation system is assigned a distinct evaluation level, with a corresponding score ranging from 0.1 to 0.8 that reflects the magnitude of its impact on groundwater dynamics.

Factors	Assessment Levels							
	0.1	0.2	0.3	0.4	0.5	0.6	0.7	0.8
C (m/d)	[0, 12]	(12, 20]	(20, 30]	(30, 35]	(35, 40]	(40, 60]	(60, 80]	>80
S	clay loam	silt loam	loam	sandy loam	swelling or condensing clay	silt-sand/fine sand	gravel-cobble/medium sand/coarse sand	thin layer or missing
A (m)	>40	(35, 40]	(30, 35]	(25, 30]	(20, 25]	(15, 20]	(10, 15]	≤10
D (m)	>25	(20, 25]	(15, 20]	(10, 15]	(6, 10]	(4, 6]	(2, 4]	≤2
LUT	artificial surfaces	forest	wetland	water bodies	grassland	cultivated land	bare land	
EI(10^4 m^3/km^2 a)	≤2	(2, 4]	(4, 6]	(6, 8]	(8, 10]	(10, 12]	(12, 15]	>15
E(10^4 m^3)	≤1000	(1000, 8000]	(8000, 12,000]	(12,000, 20,000]	(20,000, 90,000]	(90,000, 110,000]	(110,000, 160,000]	>160,000

4. Results and Discussion

4.1. Results of the Correlation Analysis

Due to vegetation cover, surface soil moisture content, artificial influence, and other factors, groundwater buried in each region will be affected by precipitation factors to different degrees. To clarify the relationship between groundwater level response and precipitation in Tangshan, the correlation between observation wells and precipitation was analyzed. The correlation coefficient between each observation well and rainfall was calculated, and the cross-correlation between precipitation and groundwater level lag time, as well as the auto-correlation of groundwater level lag time, was plotted.

The degree of persistence and variability of groundwater systems can be characterized by auto-correlation analysis. Correlation analysis is widely used to assess the average response time between precipitation and groundwater. As a stochastic process, the stochastic dependence between groundwater levels and precipitation can be revealed by the cross-correlation function [29,30]. By using these two methods, we can more clearly understand how precipitation inputs affect groundwater dynamics at various timescales, thereby providing a critical basis for subsequent classification of groundwater dynamic types and investigation of their underlying mechanisms.

The auto-correlation function diagram of groundwater level is demonstrated in Figure 3a,b. When the lag time of groundwater level is 1 month, the auto-correlation of each point is relatively high, and the correlation average is more than 0.75, with a strong positive correlation. The correlation is generally relatively low at a lag time of 2 months, approximately 0.45, indicating a weak positive correlation. The rate of decline of the auto-correlation coefficient can be expressed in terms of the number of days it takes for it to fall from 0.8 to 0.2 [31]. It takes an average of 60–75 days for the auto-correlation coefficient of the water level at each point to fall to 0.2, with an overall trend of the correlation decreasing slowly with increasing lag time. This means that the groundwater level has a strong memory effect, as well as a strong ability to continuously respond to past precipitation, human activities, etc. This lag is attributable to prolonged groundwater extraction, which has increased the thickness of the vadose zone and thus extended the recharge delay [32]. In ZHS, the considerable groundwater depth and low aquifer permeability lead to longer recharge pathways and reduced percolation rates, resulting in a pronounced lag response of groundwater levels to precipitation.

The cross-correlation functions between groundwater levels and precipitation are displayed in Figure 3c,d. Negative correlations peak at lag times of 0–2 months, with the correlation approaching zero around 3–4 months after precipitation. In contrast, positive correlations reach a maximum at a lag of 5–6 months, indicating a significant manifestation of precipitation-induced groundwater recharge. This is because, in the initial period following precipitation, groundwater levels do not exhibit significant changes. Over time, precipitation gradually infiltrates and recharges groundwater, causing groundwater levels to begin rising. Hydrogeological conditions, human activities, meteorological conditions, and other elements will affect the correlation between groundwater level and precipitation. This shows that the response of groundwater level to precipitation in the plain and hilly areas of Tangshan City had a certain complexity.

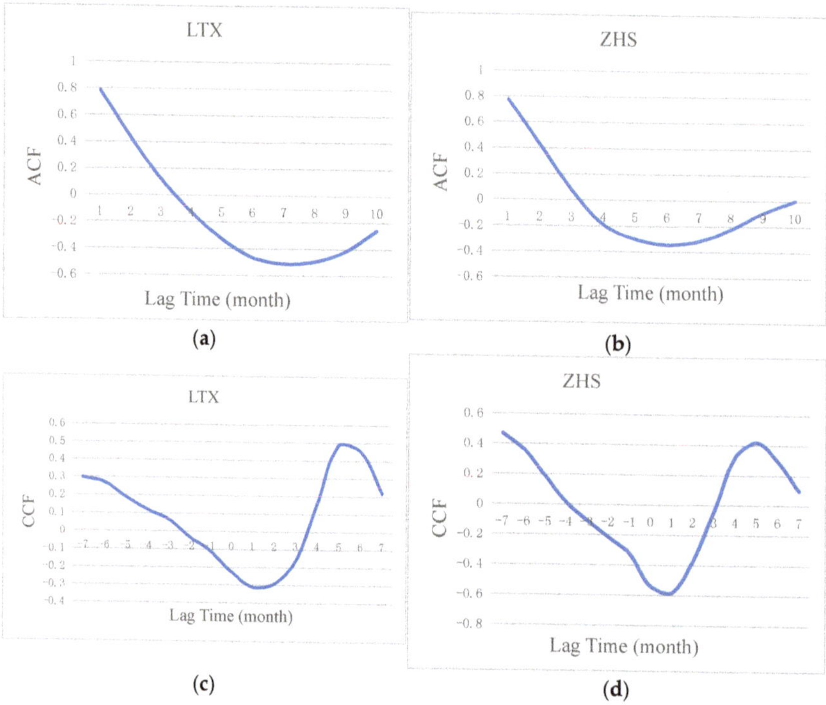

Figure 3. Correlation analysis diagram. Correlation maps for monitoring wells at LTX and ZHS were generated. Figures (**a**,**b**) illustrate the groundwater level autocorrelation functions, whereas Figures (**c**,**d**) present the correlation functions between groundwater level and precipitation.

4.2. Results of the Groundwater Dynamics

Based on the selection and data processing from monitoring wells and rainfall stations, monthly average values of groundwater levels and precipitation were obtained from January 2018 to August 2020 (the specific data sources are shown in Table 3). Groundwater dynamic curves were plotted and classified according to fundamental theories of groundwater dynamics in "hydrogeology" and the aforementioned correlation analysis results. According to analysis of the curve characteristics, the main types of groundwater dynamics in the study area can be categorized as follows.

Table 3. Sources of primary data and their time span.

Data	Source of Data	Time Span
groundwater level	Groundwater data were obtained from actual monitoring wells, the locations of which have been projected in Figure 1.	January 2018 to August 2020
precipitation	Precipitation data were obtained from three precipitation gauge stations located in Zunhua, Tangshan, and Leting.	January 2018 to August 2020

Precipitation-infiltration type: This type is widely distributed across the study area, primarily in regions where the aquifer is deeply buried and the vadose zone has good lithological permeability. The groundwater level varies with changes in precipitation, with its peak often coinciding with or slightly lagging behind peak precipitation. There is considerable intra-annual variation in groundwater levels, as illustrated in Figure 4a,b.

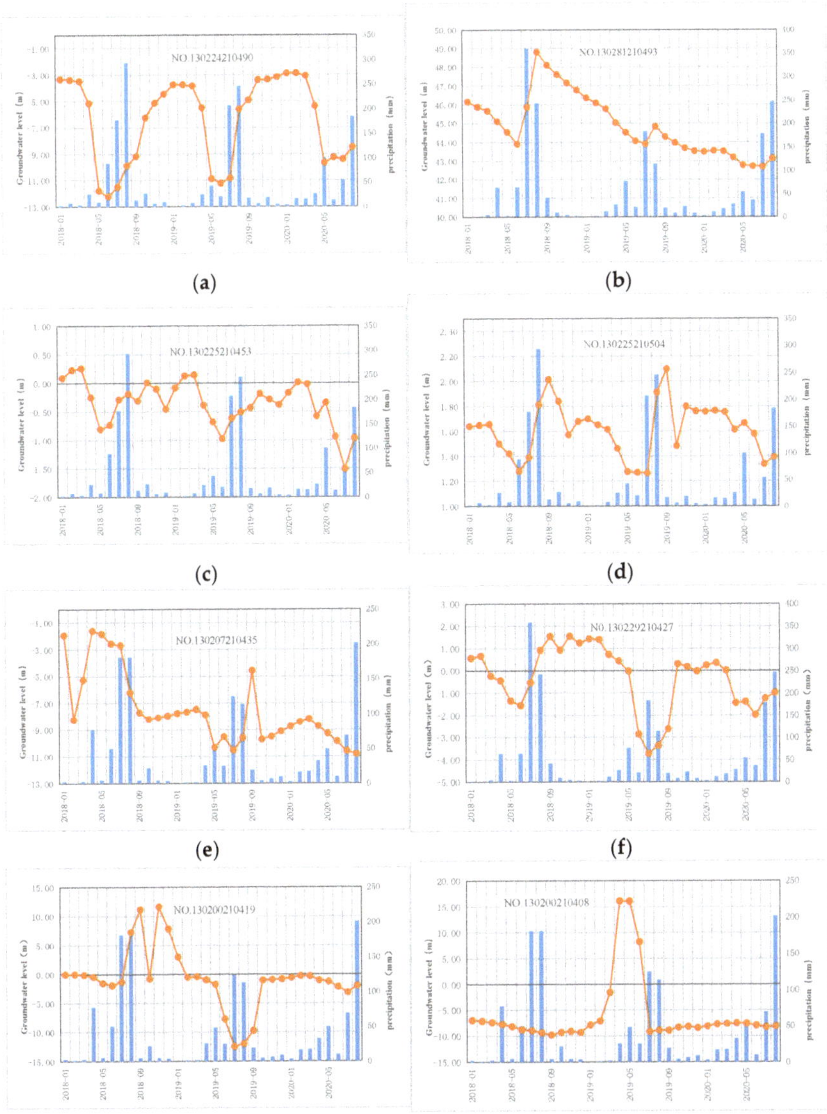

Figure 4. Maps of the basic types of groundwater dynamics. Temporal variations in groundwater levels and precipitation were mapped for representative wells during the period from January 2018 to August 2020. (**a**,**b**) Precipitation-infiltration type (**c**,**d**) Runoff type (**e**,**f**) Irrigation-infiltration type (**g**,**h**) Artificial exploitation type. In the above picture, blue color represents precipitation, and orange color represents the depth of groundwater.

Runoff type: It is mainly distributed in regions with favorable groundwater runoff conditions and extensive recharge areas and where groundwater is deeply buried or the aquifer is covered by an aquitard. The groundwater shows gentle intra-annual variations, with a small annual fluctuation range, and the peak groundwater level generally lags behind peak precipitation, as illustrated in Figure 4c,d.

Irrigation-infiltration type: This type is distributed in irrigation areas with surface water irrigation, where the vadose zone soil has a certain degree of permeability, and the groundwater depth is not very large. During the agricultural irrigation period, groundwater levels show a clear upward trend, with the high water level period often extending for a prolonged duration within the year, as illustrated in Figure 4e,f.

Artificial exploitation type: It is mainly distributed in areas with intensive groundwater extraction, where groundwater dynamics vary significantly with changes in extraction activities. During peak precipitation seasons, the groundwater level either shows a minimal increase or even decreases. When the extraction volume exceeds the annual recharge of groundwater, a trend of continuous annual decline in groundwater levels is observed, as illustrated in Figure 4g,h.

An understanding of the sources of groundwater recharge, as well as seasonal or interannual flow variations characteristics, can be obtained through analysis of the above characteristics of groundwater dynamics. The future trend of groundwater level is reasonably predicted based on exploring the factors influencing groundwater dynamics under different hydrogeological, climatic, and anthropogenic influences. It provides a reference for groundwater monitoring, protection, development, and utilization and is of great significance for the timely detection of adverse changes such as saltwater intrusion into freshwater aquifers and groundwater pollution.

4.3. *Dynamics Discussion of Impact Evaluation for Groundwater Dynamic Characterization*

4.3.1. Factors Influencing the Dynamic Characterization of Groundwater

After analyzing the natural dynamic patterns of groundwater levels and precipitation, we can discern the seasonal and interannual variability of groundwater in the study area. This analysis underpins the subsequent investigation of groundwater dynamics in response to multiple influencing factors and serves as the basis for classifying groundwater response types. This section examines the primary driving factors influencing groundwater dynamics in the study area. Based on analysis of natural groundwater dynamics, it is evident that the groundwater dynamics in the study area, as a typical coastal region, are subject to the combined stress of hydrological conditions, geological factors, and human activities [33–35]. The MFWCES selects five influencing factors, including hydraulic conductivity, soil media, aquifer thickness, groundwater depth, and groundwater evaporation, from the natural influencing factors, as well as land use type and groundwater extraction intensity as human factors. Through GIS mapping, the scores for each influencing factor under individual effects are obtained. Subsequently, the natural and human factors are combined according to the weights pre-defined in Table 1 to investigate the groundwater dynamics under the stress of composite factors.

1. Hydraulic Conductivity

Hydraulic conductivity is an important geological and geotechnical property [36], which is a parameter to measure the strength of aquifer permeability, and the magnitude of the permeability coefficient mainly depends on the scale of connected voids in the aquifer [37]. When the hydraulic gradient is the same, the size of the hydraulic conductivity determines the size of the seepage rate of the aquifer. Typically, aquifers with larger voids have larger hydraulic conductivity and relatively faster seepage rates, with faster response to groundwater dynamics.

In the Tangshan plain area, aquifers are primarily composed of loose sediments with relatively uniform pore structures. Therefore, the collected borehole data are processed using interpolation, and the hydraulic conductivity is determined by combining the results with empirical values. In contrast, aquifers in the hilly areas are mainly fissured bedrock aquifers, where the degree of fissure development is uneven, making determination of the hydraulic conductivity more complex. First, the hilly aquifers are classified based on lithology and pore type. Then, pumping tests conducted in bedrock boreholes are used to determine the hydraulic conductivity by analyzing the relationship between water level drawdown and discharge. Finally, the hydraulic conductivity for different regions is imported into GIS, where the plain and hilly areas are stitched together, and values are assigned to each region according to their assessment levels.

The hydraulic conductivity in the study area was divided into three regions according to the topography, namely, the northern hilly region, the central pre-mountainous region, and the southern coastal region. The hydraulic conductivity of QXX, located in the northern mountainous area, is 0–10 m/d in most of the areas. From Figure 5a, it can be seen that the hydraulic conductivity of the central premontane plains shows a clear banded distribution with distinct levels, and the values are generally large, mostly higher than 20 m/d. Among them, the highest score of 0.8 is found at the junction of FRQ and KPQ, in the eastern part of GZQ and in the southern part of LZS, while the lowest value is concentrated within ZHS and QAS, with a rating of 0.4. The hydraulic conductivity in the southern marine plains is small, with most areas scoring 0.1–0.3.

2. Soil Media

Infiltration of precipitation into the subsurface requires a certain response time, a process that depends heavily on the hydraulic conductivity of the soil [36], while the infiltration rate is controlled by the physical properties of the soil [38]. Therefore, soil media assessment levels are divided according to soil particle size as well as infiltration capacity. Specifically, soils with larger particle sizes are given higher recharge scores because of their high porosity and fast infiltration rate, which significantly accelerate precipitation infiltration and thus are more likely to raise the groundwater level. Soils with smaller particle sizes, such as silt or clay, are given lower recharge scores because of their high viscosity and low porosity, which prevent infiltration of precipitation to recharge groundwater.

The distribution of various soil media types in Tangshan City is relatively concentrated, among which sandy loam and clay loam are more widely distributed, as shown in Figure 5b. In the northern hilly areas, the topsoil layer is very thin, and most of the soil layer in QXX is missing, with a rating of 0.8. In the central part of the city, including LZS, LNX, LBQ, LNQ, and FNQ, most of the areas have a rating of 0.4 to 0.6, indicating that sandy loam and silt-sand are concentrated in these areas. The distribution in the coastal areas is dominated by clay loam and silt loam.

3. Aquifer Thickness

For aquifers with relatively uniform water-bearing properties, their thickness is determined based on groundwater levels, the particle composition of loose rock formations revealed by drilling, and the lithological structure [39]. The thickness of bedrock aquifers should be determined by analyzing rock fractures and karst development revealed through drilling, combined with hydrogeological observations.

Based on the foregoing analysis of how aquifer thickness influences groundwater dynamics, this study further examines the spatial distribution and regional variability of aquifer thickness within the study area to elucidate the controlling role of thickness heterogeneity on groundwater behavior.

Aquifer thickness is one of the key factors influencing groundwater dynamics. Generally, thicker aquifers exhibit greater water storage capacity and support longer flow paths. With sufficient precipitation, the slower and more uniform seepage within thicker aquifers results in smaller water level fluctuations. This, in turn, helps to regulate water levels and minimizes seasonal variations. In contrast, thinner aquifers have limited recharge capacity and rely more heavily on precipitation. When precipitation occurs in concentrated bursts, flow velocities increase, resulting in more sensitive groundwater level changes.

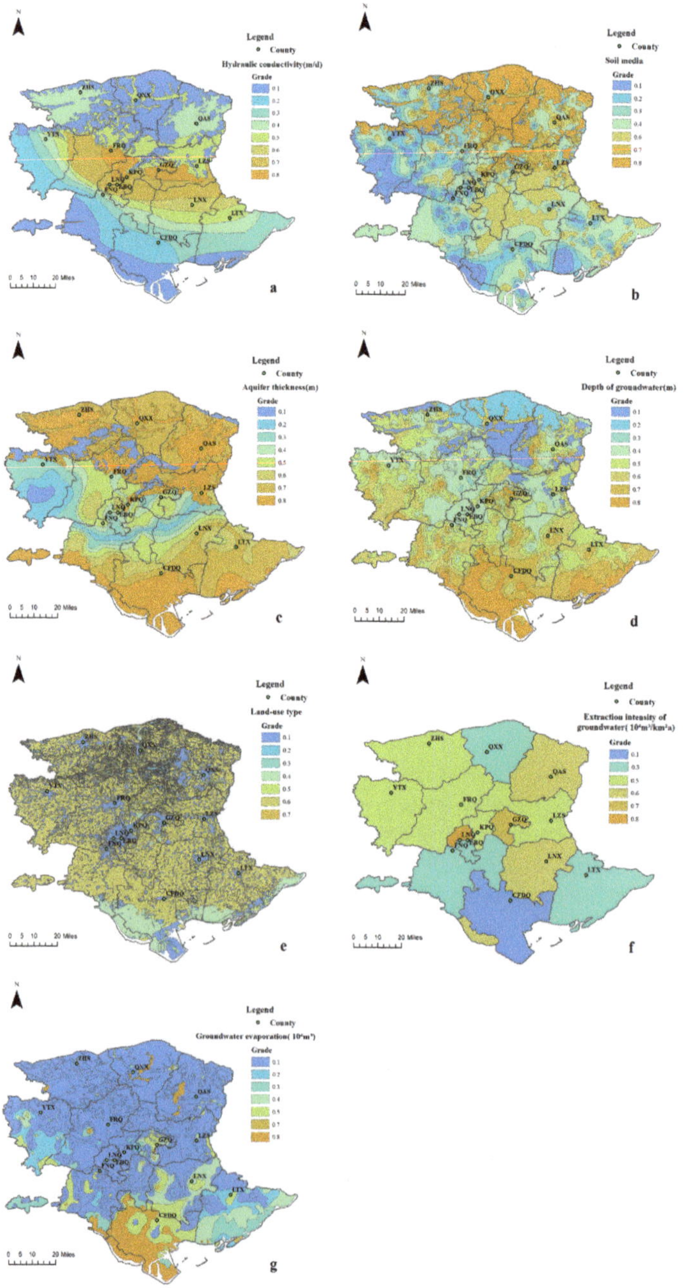

Figure 5. Results from the single-factor evaluation. Figures (**a**–**g**) present the evaluation results of groundwater dynamics under the isolated influence of each of the seven factors (hydraulic conductivity, soil media, aquifer thickness, depth of groundwater, land use type, groundwater extraction intensity, groundwater evaporation) incorporated into the MFWCES.

The distribution of aquifer thickness across the study area is significantly influenced by sedimentary environments, geological structures, and fluvial deposition, resulting in considerable spatial variability. In the northern region near river channels, such as QAS and QXX, the aquifer thickness is generally less than 10 m, due to the effects of fluvial erosion and sedimentation. In the central alluvial plain, which is dominated by silt and fine sand deposits, aquifer thickness ranges from 25 to 35 m, corresponding to evaluation grade levels between 0.2 and 0.5. In the southern coastal region, aquifer thickness remains below 15 m. Across the entire study area, a few zones of relatively higher aquifer thickness are concentrated in ZHS, YTX, FRQ, and LZS.

4. Depth of Groundwater

Generally, changes in groundwater depth affect the thickness of the vadose zone. If the vadose zone thickens, it delays and alters the process of precipitation infiltration recharging groundwater, causing the response rate of groundwater dynamics to lag [40].

The study area is located in the estuarine coastal zone, a transition zone connecting land and coastal ocean. The dynamics of groundwater depth in this area are influenced by the interaction between land and ocean [41]. Thus, different methods are applied to determine groundwater depth in Tangshan City's plains and hilly areas. In plains and valleys, the depth of groundwater is determined by the Kriging interpolation method based on data collected from the survey with reference to the geological profile. For hilly areas with complex topography, groundwater depth is related to aquifer thickness. According to the range of buried depth, the buried depth parameter of groundwater in Tangshan City is divided into eight grades; the greater the depth of the groundwater is, the lower the rating. As can be seen from the Figure 5d, the shallow groundwater depth in Tangshan City presents an overall state of deepness in the north and shallowness in the south. In the northern hilly area, due to the higher topography and the thickness of the vadose zone, the depth of the groundwater is approximately 20 m, with assessment ratings ranging from 0.1 to 0.3, and locally the groundwater level exceeds 25 m. The average depth of groundwater in KPQ, LNQ, LBQ, FRQ, and northern FNQ in the central part of the city is 6–15 m. Most of the groundwater depths in the southern alluvial and marine plains are less than 2 m.

5. Land use type

Groundwater recharge is regulated by the combined effects of climate change and anthropogenic activities [42], with land use type (LUT) serving as a critical intervention factor in the hydrological cycle [43]. Different land use types influence groundwater dynamics by modifying processes such as interception, infiltration, evapotranspiration, and surface runoff [44].

In this study, the LUT classification for the research area was derived from GlobeLand30 raster data through mosaicking and masking procedures. The study area encompasses seven land use types, among which cultivated land is the most widespread, followed by impervious surfaces, while bare land occupies the smallest area. In the early stages of agricultural cultivation, frequent mechanical tillage increases the proportion of large soil pores, temporarily enhancing soil permeability. However, as precipitation and freeze–thaw cycles act on the soil, pore heterogeneity increases, leading to spatial variability in the groundwater system's response to precipitation [45]. With continued agricultural activities, degradation of the soil structure diminishes the soil's water transport capacity, thereby weakening the groundwater response. In each administrative region, impervious surfaces are not only concentrated in certain areas, but also widely distributed in a fragmented pattern; for instance, in LBQ, impervious surfaces cover over 90% of the area. These artificial, impermeable surfaces predominantly generate surface runoff, reducing infiltration and resulting in the most delayed response to precipitation. Conversely, water bodies—primarily located in CFDQ—maintain a strong hydraulic connection with the groundwater system. Although fluctuations in the water levels of these water bodies can directly reflect the amount of precipitation received, the recharge process associated with them differs from the direct infiltration observed with bare land.

6. Groundwater Extraction Intensity

Exploitation and utilization of groundwater alter its natural equilibrium, causing changes in permeability processes. Continuous groundwater extraction leads to declining water levels and also affects the response rate and effectiveness of groundwater dynamics in response to precipitation [46,47]. While extreme rainfall events may cause a rapid rise in groundwater levels, overextraction can offset such dynamic trends.

Groundwater extraction intensity is calculated as the ratio of the groundwater extracted in each county or district to the respective administrative area. The area of each administrative unit is determined using calculated fields in the attribute table of the geographic layers. Extraction intensity values are then assigned to administrative regions for evaluation. Analysis reveals that the northern and central parts of the region, characterized as alluvial plains, have better groundwater recharge conditions and higher water resource availability. As a result, extraction intensity in this area is higher than in the southern coastal zones. Within the study area, the highest groundwater extraction intensity is found in LBQ, due to its high population density and the concentration of industrial parks, which together result in significant water demand for both industrial production and residential use.

In regions suffering from groundwater overexploitation, water stress is primarily mitigated through strategies such as zoned management, water source substitution projects, agricultural water conservation, and land subsidence monitoring. In industrial clusters and coal-mining subsidence zones (e.g., FNQ and GZQ), groundwater extraction is regulated by designating prohibition and restriction zones, thereby banning or limiting the installation of new extraction wells. In agricultural irrigation areas, the promotion of canal-based water diversion facilitates partial replacement of groundwater with surface water. Moreover, in agricultural regions like YTX, the implementation of high-efficiency drip irrigation systems is encouraged to further conserve water resources.

7. Groundwater Evaporation

Phreatic evaporation is evaluated assuming a critical depth of 6 m, with evaporation considered negligible for regions where the groundwater depth exceeds this threshold. The geographical coordinates of seven evaporation stations—Fengnan, Fengrun, Luan County, Tanghai, Laoting, Yutian, and Tangshan—were first imported into GIS to create an evaporation station layer. Then, a Thiessen polygon network covering the study area was generated, with the environment parameters predefined. This process identified the effective control area for each evaporation station. The phreatic evaporation coefficients were determined for each region based on the vadose zone media type and empirical values, ultimately yielding the phreatic evaporation volume.

Figure 5g illustrates that phreatic evaporation varies significantly across different areas of Tangshan. In the southern coastal plain, where the average groundwater depth is less than 4 m, evaporation is more sensitive to temperature variations. During summer, higher temperatures accelerate molecular motion, resulting in substantially increased phreatic evaporation. In contrast, the piedmont alluvial plains exhibit relatively deeper groundwater levels, leading to lower evaporation volumes.

According to the findings of the team's previous studies on groundwater vulnerability in the region, groundwater vulnerability directly reflected the degree of groundwater susceptibility to external pollutants [22]; that is, it indirectly reflected the degree of change in groundwater dynamics, so there is a certain degree of duplication in the two studies in the selection of indicators.

4.3.2. Results of the MFWCES-Based Evaluation of Groundwater Dynamics

Based on GIS and the study area profile, following the weighting requirements in the multi-factor weighted comprehensive evaluation system, the seven single-factor evaluation results were superimposed using the raster calculator function to obtain a comprehensive evaluation value of groundwater dynamics, with values ranging from 2.4 to 12.7. The obtained evaluation values are categorized into four types of response mechanisms, namely: rapid response (10.3–12.7), dual response to precipitation and anthropogenic extraction (9.6–10.3), delayed response (7.6–9.6), and strong superimposed response to human activities (2.4–7.6). Larger values represent a faster response of groundwater dynamics; conversely, smaller values indicates a relatively delayed response of groundwater dynamics. From Figure 6, the following conclusions can be obtained.

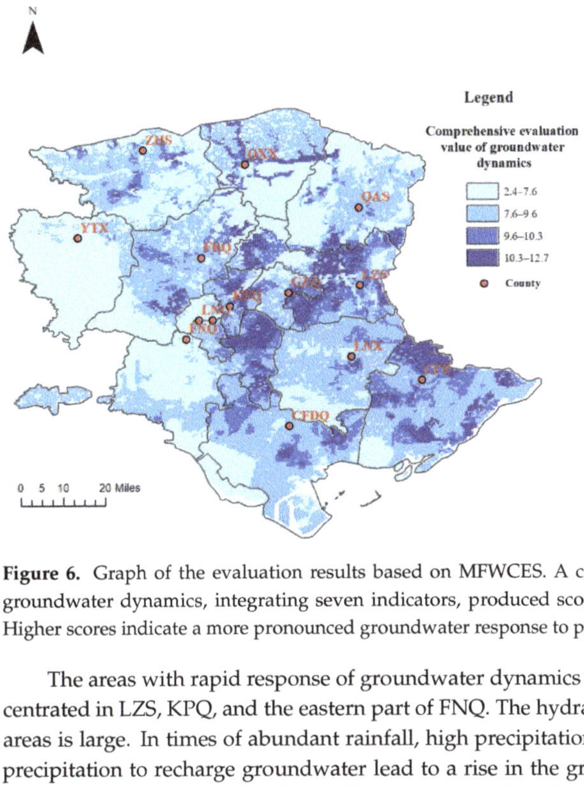

Figure 6. Graph of the evaluation results based on MFWCES. A comprehensive evaluation of groundwater dynamics, integrating seven indicators, produced scores ranging from 2.4 to 12.7. Higher scores indicate a more pronounced groundwater response to precipitation.

The areas with rapid response of groundwater dynamics to rainfall are mainly concentrated in LZS, KPQ, and the eastern part of FNQ. The hydraulic conductivity in these areas is large. In times of abundant rainfall, high precipitation and rapid infiltration of precipitation to recharge groundwater lead to a rise in the groundwater level. During the dry season, the amount of rainfall decreases significantly, so groundwater dynamics respond more quickly. Similarly, the valley plain area located in QXX responds quickly to rainfall because of the uniform and single lithological composition and rapid infiltration rate. In addition, there are other areas where such response mechanisms are distributed in small portions, e.g., the northern part of LTX. Regions of dual response to precipitation and anthropogenic extraction are scattered across the alluvial plain. The dynamic genesis of this type of groundwater is more complex, characterized by the interaction of natural recharge and anthropogenic discharge. Concentrated recharge from precipitation during the rainy season may temporarily mask the effects of mining; when precipitation is low during the dry season, extraction leads to an accelerated decline in water levels, and the combination of the two leads to fluctuations in water levels that depend on the dominant factors at different times of the year. Densely populated plains areas and agricultural cultivation lead to frequent tillage of the soil, which accelerates rainfall infiltration. To some extent, heavy groundwater extraction can also reduce the recharge efficiency of precipitation. In the southern regions of LNX and CFDQ, delayed-response mechanisms are predominantly distributed. The soils in these areas are primarily composed of loam and silt, characterized by low hydraulic conductivity and poor water transmissivity, which hinder recharge. CFDQ, situated near the coast, features low-lying terrain that facilitates moisture accumulation. The high local temperatures in this environment enhance evaporation, consuming part of the precipitation and causing a relatively delayed groundwater response to rainfall. The strong anthropogenic overlay response is characterized by groundwater level variations closely tied to human activities. This type of response is mainly observed in FNQ and YTX. In FNQ and YTX, where agriculture is prevalent, groundwater withdrawal for agricultural irrigation is the main reason for the lag in groundwater dynamics in response to precipitation.

4.4. Response Mechanisms and Validation of Groundwater Dynamic Characterization

4.4.1. Impact of Hydraulic Engineering and Human Activities on Groundwater Dynamics

The spatial distribution characteristics of the impacts of the seven factors on the groundwater dynamics in this study area have been derived from previous studies. As a typical coastal area with dense population and rapid industrial development, the construction of water conservancy projects and the impacts of human activities also affect the dynamic changes of groundwater.

Niu et al. [48] explored the interactions among climate, human activities, and groundwater depth. It was concluded that anthropogenic variables were the main factors affecting shallow groundwater in urban areas with intensive human activities, while the influence of climate was gradually increasing in suburban areas. Li et al. [49] studied the response patterns of groundwater to its driving forces under the influence of human activities and found that areas with higher urbanization are more prone to groundwater depletion. Ahmed et al. [50] found that the trends in groundwater levels in five basins in Morocco were mainly related to groundwater extraction rates and rainfall. Lee et al. [51] assessed groundwater sustainability based on natural and anthropogenic factors and their spatio-temporal changes. They found that downstream groundwater wells were influenced by local human activities and recharged from upstream, causing a blurred relationship between precipitation and water levels.

Standard deviation, as the most commonly used indicator of the degree of statistical distribution in probability statistics, responds to the degree of dispersion of data within a group. In this study, the standard deviation in groundwater level was used to characterize the degree of response of groundwater dynamics to rainfall, which was more intuitive than exploring fluctuations in groundwater level with cumulative precipitation, cumulative groundwater level rise, and groundwater level variance. The data of groundwater depth of 24 wells for 32 months (January 2018 to August 2020) were counted, and the data for each well for 32 months were taken as a group to calculate the standard deviation of the groundwater level in each well, and the results were obtained: the standard deviation ranged from 0.21 to 6.51. The obtained results were divided into four parts, marked with dots according to a certain size and scale and differentiated by color, and plotted into the groundwater dynamic evaluation results of MFWCES and the two base maps of rivers and water conservancy projects.

The larger the standard deviation (the darker the color and larger the diameter of the dots), the faster the response of groundwater dynamics to precipitation, and vice versa: the lighter the color and smaller the diameter of the dots, the more lagging the response to precipitation. As can be seen in Figure 7, the red and orange dots are located in areas where water conservancy construction and human activities are more intensive. The northern mountainous areas has relatively little water conservancy construction, and the dots are lighter in color and smaller in diameter.

Figure 7. Map of major watersheds and hydraulic engineering. Bold blue lines delineate the study area's coastline, dark blue lines represent the major watersheds, pink lines indicate significant hydraulic projects, and circles denote the standard deviation in groundwater levels.

4.4.2. Response Mechanisms for Characterizing Groundwater Dynamics

Based on the correspondence between peak precipitation and the month of occurrence of the peak groundwater level, as well as the length of time that the high groundwater level continues to occur, we calculated dynamic characteristics, such as the amplitude of the water level variation during the year. The response relationship of groundwater dynamics to precipitation were categorized into four types: rapid response, precipitation–artificial mining dual response, delayed response, and strong superimposed response to human activities.

1. Rapid response

A rapid response is characterized by dynamics, and groundwater variability within the year is small, ranging from 1 to 5 m. Generally, peak precipitation occurred in August of each year, and the peak in groundwater levels occurred in September of each year; the peak in groundwater levels would have a rapid response to peak precipitation. The lowest value of precipitation occurred in January of each year, and the lowest value for groundwater levels occurred in February-May of each year. Influenced by the characteristics of groundwater runoff and drainage of different geomorphic types and the regulating effect of the vadose zone on groundwater evaporation, the response of low groundwater values to low precipitation values in different regions presented different characteristics.

2. Precipitation–artificial mining dual response

The dynamic of the precipitation–artificial mining dual response is characterized by the fact that the general groundwater level does not show a significant response relationship with precipitation. The groundwater level will drop significantly during the period of artificial mining and will rise rapidly after the end of mining. When the amount of exploitation is greater than the annual recharge of groundwater, negative equilibrium of groundwater occurs, and the groundwater level decreases year by year.

3. Delayed response

The dynamic characteristics of delayed response show that the variation of groundwater within the year was large, ranging from 2 to 10 m. Generally, peak precipitation occurs in August of each year, while the peak in the groundwater level occurs in February of the following year; thus, there is a six-month response period for the peak groundwater level to follow peak precipitation. The lowest value of precipitation occurs in January of each year, and the lowest value of groundwater level occurs in May–July of each year; the response period of the low values of both is also about half a year. In different areas, the groundwater level will start to fall during the two agricultural irrigation periods of March-May and May-July, followed by a period of high groundwater levels for about four consecutive months.

4. Strong superimposed response to human activities

The dynamics of strong superimposed response to human activities characterize a groundwater level that is closely related to human activities. Human activities change the natural dynamics of groundwater by adding new sources of recharge or new destinations for discharge [49]. During the dry season, April-June, when precipitation is low and recharge from precipitation is insufficient, groundwater needs to be extracted for agricultural irrigation, resulting in lowering of the groundwater level. During the rainy season, July-September, when precipitation increases and less water is used for agriculture, the depth of the groundwater rises.

4.4.3. Response Mechanism and Validation of Groundwater Dynamic Characterization

Natural factors (including precipitation and depth of groundwater), as well as land use type and other human activities, had a significant impact on groundwater dynamics [49]. Hydraulic conductivity, soil media type, etc. fluctuate groundwater levels by affecting the infiltration capacity and rate of rainfall. Topography and depth of groundwater reflect some of the topographic and geomorphological features, which can have an impact on groundwater dynamics. In addition, human activities will also cause some damage to the balance of natural groundwater and disturb the groundwater dynamics. MFWCES comprehensively considered seven major factors, such as aquifer thickness, assigned different weights to them according to their impacts on changes in the groundwater level, and superimposed the evaluation scores of the seven factors, so as to obtain a comprehensive evaluation value that can objectively respond to the characteristics of changes in groundwater dynamics. Based on the comprehensive evaluation values, combined with the temporal relationship between peak precipitation and peak groundwater levels, groundwater level variation patterns, human activities, and hydraulic engineering construction, the response of groundwater dynamics to precipitation is classified into four types.

In Figure 8, the color and diameter of the dots represent the standard deviation of the groundwater level; the larger the standard deviation is, the more significant the change in groundwater dynamics throughout the year. The standard deviation threshold of groundwater in this region ranged from 0.21 to 6.51 m. From the overall spatial distribution characteristics, most of the points with large standard deviation fall in the areas with a high MFWCES index, and the correlation can reach more than 60%. The correlation is more ambiguous in other areas due to climate and other unconsidered factors. After verification, the data are accurate, and the method is reasonable.

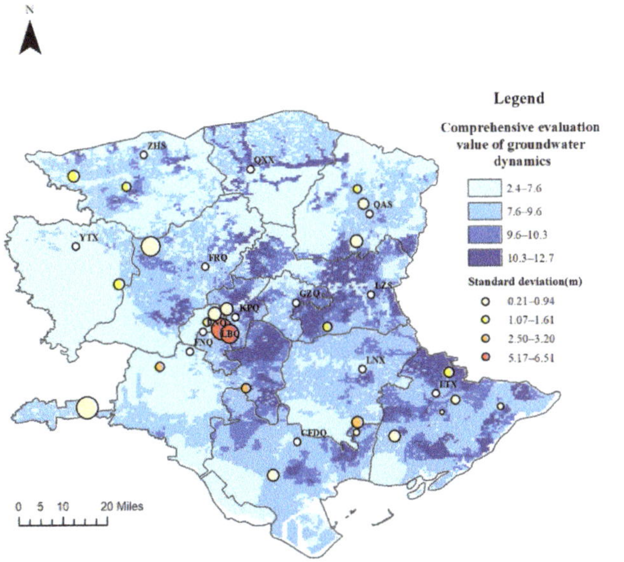

Figure 8. Distribution of groundwater level standard deviations and MFWCES index values. The figure combines the results of the comprehensive evaluation of groundwater dynamics and the standard deviation of groundwater levels.

4.5. Future Monitoring and the Long-Term Sustainability of Groundwater Supplies

Sustainable management of groundwater resources is of paramount importance; however, current management measures may significantly lag behind the rate of aquifer depletion, especially in the severely over-exploited North China Plain. The MFWCES model developed in this study does not incorporate key factors such as drainage line density, indicating that the evaluation framework requires further refinement. In addressing complex nonlinear groundwater dynamics, machine learning techniques—particularly ensemble methods like random forest—show substantial potential [52]. An integrated approach that combines the evaluation framework, machine learning-based hydrological models, and both in situ and remote sensing data can estimate parameters that are otherwise difficult to measure directly, thereby providing reliable information for sustainable economic and hydrological management [53]. Currently, methodologies for studying changes in groundwater storage are constrained by limitations in data resolution, temporal span, and model uncertainty; as a result, the overall complexity, local precision, and long-term applicability of the conclusions require further verification [54]. Future research should integrate higher-resolution remote sensing data, field monitoring, and policy analysis to support the development of more precise sustainable management strategies.

5. Conclusions

In this study, a typical hilly plain area is used as a case study to investigate the dynamic characteristics and response mechanisms of groundwater under compound stress. Supported by well observation data and other geological datasets, the analysis—which integrates mathematical statistics, correlation analysis, and a GIS-based multi-factor weighted comprehensive evaluation system (MFWCES)—yields the following conclusions:

(1) The auto-correlation coefficient of the groundwater level decreased from 0.8 to 0.2 in 60–75 days on average, and the correlation decreased slowly with an increase in lag time. The cross-correlation analysis of groundwater level and precipitation showed that, when the lag time was between 0 and 7 months, the cross-correlation coefficient changed from a weak negative correlation to no correlation to a weak positive correlation.

(2) According to the characteristics of the groundwater dynamic curve and the results of the correlation analysis, the dynamic type of groundwater in Tangshan City was classified into four categories: precipitation-infiltration type, runoff type, irrigation-infiltration type, and artificial exploitation type.

Precipitation-infiltration type: Strong seasonal variation, with groundwater peaks aligned with or lagging 1–3 months behind precipitation peaks.

Runoff type: The changes throughout the year are relatively stable. The peak of the groundwater level often occurs later than the peak of rainfall.

Irrigation-infiltration type: Clear response during irrigation periods, reflecting agricultural water use.

Artificial exploitation type: Significant fluctuations driven by intensive groundwater extraction in plain areas.

(3) The results of the GIS-based evaluation of MFWCES groundwater dynamics had a score size ranging from 2.4 to 12.7. Combining the two main categories of influencing factors of groundwater dynamics—natural factors and human activities—the response relationship of groundwater dynamics to precipitation was classified into four categories: rapid response (10.3–12.7), dual response to precipitation and anthropogenic extraction (9.6–10.3), delayed response (7.6–9.6), and strong superimposed response to human activities (2.4–7.6). Finally, the evaluation results were validated using the standard deviation of the groundwater level. The standard deviation was calculated to be in the range of 0.21 to 6.51. After validation, it was found that wells with relatively large standard deviations fell in areas with a high MFWCES evaluation index value. Moreover, the impact of water conservancy construction and human activities is particularly significant in LBQ, KPQ, LNX, and FNQ, while tides increase the periodicity of the dynamics. This study's classification of groundwater dynamic response types, together with its quantitative assessment of groundwater reactions to precipitation, offers a novel perspective for regional water resource planning, ecological protection, and the development of differentiated groundwater management strategies under compound stress.

Author Contributions: Q.Z.: writing—original draft; M.Z. (Meng Zhang): writing—review and editing; W.J.: writing—review and editing; Y.H.: data; F.C.: software; M.Z. (Mucheng Zhang): investigation. All authors have read and agreed to the published version of the manuscript.

Funding: This study was supported by the National Key Research and Development Program of China (No. 2023YFD1900802-01) and the National Natural Science Foundation of China (No. 41907149).

Data Availability Statement: All the data in this manuscript are derived from field surveys, reports, and references.

Conflicts of Interest: The authors declare no conflicts of interest.

Appendix A

Table A1. Abbreviations and full names of administrative region.

Abbreviation of Administrative Region	Full Name of Administrative Region
ZHS	Zunhua Shi
QXX	Qianxi Xian
QAS	Qianan Shi
YTX	Yutian Xian
FRQ	Fengrun Qu
KPQ	Kaiping Qu
GZQ	Guzhi Qu
LZS	Luanzhou Shi
LBQ	Lubei Qu
LNQ	Lunan Qu
FNQ	Fengnan Qu
LNX	Luannan Xian
CFDQ	Caofeidian Qu
LTX	Leting Xian

References

1. Satoh, Y.; Kahil, T.; Byers, E.; Burek, P.; Fischer, G.; Tramberend, S.; Greve, P.; Flörke, M.; Eisner, S.; Hanasaki, N.; et al. Multi-model and multi-scenario assessments of Asian water futures: The Water Futures and Solutions (WFaS) initiative. *Earth's Future* **2017**, *5*, 823–852. [CrossRef]
2. Van Camp, M.; Mtoni, Y.; Mjemah, I.C.; Bakundukize, C.; Walraevens, K. Investigating seawater intrusion due to groundwater pumping with schematic model simulations: The example of the Dar es Salaam coastal aquifer in Tanzania. *J. Afr. Earth Sci.* **2014**, *96*, 71–78. [CrossRef]
3. Sun, Y.; Xu, S.; Wang, Q.; Hu, S.; Qin, G.; Yu, H. Response of a Coastal Groundwater System to Natural and Anthropogenic Factors: Case Study on East Coast of Laizhou Bay, China. *Int. J. Environ. Res. Public Health* **2020**, *17*, 5204. [CrossRef]
4. Schreiner-McGraw, A.P.; Ajami, H. Delayed response of groundwater to multi-year meteorological droughts in the absence of anthropogenic management. *J. Hydrol.* **2021**, *603*, 126917. [CrossRef]
5. Sheng, Y.; Wang, G.; Zhao, D.; Hao, C.; Liu, C.; Cui, L.; Zhang, G. Groundwater Microbial Communities Along a Generalized Flowpath in Nomhon Area, Qaidam Basin, China. *Groundwater* **2018**, *56*, 719–731. [CrossRef]
6. Giordano, M. Global groundwater? Issues and solutions. *Annu. Rev. Environ. Resour.* **2009**, *34*, 153–178. [CrossRef]
7. Ahmad, S.; Liu, H.; Günther, A.; Couwenberg, J.; Lennartz, B. Long-term rewetting of degraded peatlands restores hydrological buffer function. *Sci. Total Environ.* **2020**, *749*, 141571. [CrossRef]
8. Alfio, M.R.; Pisinaras, V.; Panagopoulos, A.; Balacco, G. Groundwater level response to precipitation at the hydrological observatory of Pinios (central Greece). *Groundw. Sustain. Dev.* **2024**, *24*, 101081. [CrossRef]
9. Cui, Y.; Liao, Z.; Wei, Y.; Xu, X.; Song, Y.; Liu, H. The Response of Groundwater Level to Climate Change and Human Activities in Baotou City, China. *Water* **2020**, *12*, 1078. [CrossRef]
10. Balacco, G.; Alfio, M.R.; Fidelibus, M.D. Groundwater Drought Analysis under Data Scarcity: The Case of the Salento Aquifer (Italy). *Sustainability* **2022**, *14*, 707. [CrossRef]
11. Jin, G.; Shimizu, Y.; Onodera, S.; Saito, M.; Matsumori, K. Evaluation of drought impact on groundwater recharge rate using SWAT and Hydrus models on an agricultural island in western Japan. *Proc. IAHS* **2015**, *371*, 143–148. [CrossRef]
12. Yin, X.; Shu, L.; Wang, Z.; Lu, C.; Liu, B. Time-frequency analysis of groundwater depth variation based on the ICA-WTC composite method. *J. Hydrol.* **2023**, *617*, 128914. [CrossRef]
13. Russo, T.A.; Lall, U. Depletion and response of deep groundwater to climate-induced pumping variability. *Nat. Geosci.* **2017**, *10*, 105–108. [CrossRef]
14. Wang, D.; Li, P.; He, X.; He, S. Exploring the response of shallow groundwater to precipitation in the northern piedmont of the Qinling Mountains, China. *Urban Clim.* **2023**, *47*, 101379. [CrossRef]
15. Qi, P.; Zhang, G.; Xu, Y.J.; Wang, L.; Ding, C.; Cheng, C. Assessing the Influence of Precipitation on Shallow Groundwater Table Response Using a Combination of Singular Value Decomposition and Cross-Wavelet Approaches. *Water* **2018**, *10*, 598. [CrossRef]
16. Kellner, E.; Hubbart, J.A. Land use impacts on floodplain water table response to precipitation events. *Ecohydrology* **2018**, *11*, e1913. [CrossRef]
17. Jiang, W.; Sheng, Y.; Shi, Z.; Guo, H.; Chen, X.; Mao, H.; Liu, F.; Ning, H.; Liu, N.; Wang, G. Hydrogeochemical characteristics and evolution of formation water in the continental sedimentary basin: A case study in the Qaidam Basin, China. *Sci. Total Environ.* **2024**, *957*, 177672. [CrossRef]

18. Schiavo, M. Probabilistic delineation of subsurface connected pathways in alluvial aquifers under geological uncertainty. *J. Hydrol.* **2022**, *615*, 128674. [CrossRef]
19. Meng, Q.; Xing, L.; Liu, L.; Xing, X.; Zhao, Z.; Zhang, F.; Li, C. Time-lag characteristics of the response of karst springs to precipitation in the northern China. *Environ. Earth Sci.* **2021**, *80*, 348. [CrossRef]
20. Fronzi, D.; Mirabella, F.; Cardellini, C.; Caliro, S.; Palpacelli, S.; Cambi, C.; Valigi, D.; Tazioli, A. The Role of Faults in Groundwater Circulation before and after Seismic Events: Insights from Tracers, Water Isotopes and Geochemistry. *Water* **2021**, *13*, 1499. [CrossRef]
21. Tian, Y.; Yang, Y.; Bao, Z.; Song, X.; Wang, G.; Liu, C.; Wu, H.; Mo, Y. An Analysis of the Impact of Groundwater Overdraft on Runoff Generation in the North China Plain with a Hydrological Modeling Framework. *Water* **2022**, *14*, 1758. [CrossRef]
22. Zhang, Q.; Shan, Q.; Chen, F.; Liu, J.; Yuan, Y. Groundwater Vulnerability Assessment and Protection Strategy in the Coastal Area of China: A GIS-Based DRASTIC Model Approach. *Appl. Sci.* **2023**, *13*, 10781. [CrossRef]
23. Dai, X.; Xie, Y.; Liao, A.; Wang, C.; Lin, J.; Wu, J. Quantifying hydrological responses to monsoon-controlled precipitation across the soil-groundwater-stream continuum with long-term high-frequency hydrometric monitoring. *Hydrol. Process.* **2024**, *38*, e15089. [CrossRef]
24. Royer-Carenzi, M.; Hassani, H. Deviations from Normality in Autocorrelation Functions and Their Implications for MA(q) Modeling. *Stats* **2025**, *8*, 19. [CrossRef]
25. Li, W.; Bao, L.; Yao, G.; Wang, F.; Guo, Q.; Zhu, J.; Zhu, J.; Wang, Z.; Bi, J.; Zhu, C.; et al. The analysis on groundwater storage variations from GRACE/GRACE-FO in recent 20 years driven by influencing factors and prediction in Shandong Province, China. *Sci. Rep.* **2024**, *14*, 5819. [CrossRef] [PubMed]
26. Wang, S.; Song, X.; Wang, Q.; Xiao, G.; Liu, C.; Liu, J. Shallow groundwater dynamics in North China Plain. *J. Geogr. Sci.* **2009**, *19*, 175–188. [CrossRef]
27. Etuk, M.N.; Igwe, O.; Egbueri, J.C. An integrated geoinformatics and hydrogeological approach to delineating groundwater potential zones in the complex geological terrain of Abuja, Nigeria. *Model. Earth Syst. Environ.* **2023**, *9*, 285–311. [CrossRef]
28. China Hydrology Society. *Guidelines for Groundwater Dynamics Analysis and Evaluation*; China Institute of Water Resources and Hydropower: Beijing, China, 2023.
29. Zhong, X.; Gong, H.; Chen, B.; Zhou, C.; Xu, M. Study on the evolution of shallow groundwater levels and its spatiotemporal response to precipitation in the Beijing Plain of China based on variation points. *Ecol. Indic.* **2024**, *166*, 112466. [CrossRef]
30. Duvert, C.; Jourde, H.; Raiber, M.; Cox, M.E. Correlation and spectral analyses to assess the response of a shallow aquifer to low and high frequency rainfall fluctuations. *J. Hydrol.* **2015**, *527*, 894–907. [CrossRef]
31. Tirogo, J.; Jost, A.; Biaou, A.; Valdes-Lao, D.; Koussoubé, Y.; Ribstein, P. Climate Variability and Groundwater Response: A Case Study in Burkina Faso (West Africa). *Water* **2016**, *8*, 171. [CrossRef]
32. Cao, G.; Scanlon, B.R.; Han, D.; Zheng, C. Impacts of thickening unsaturated zone on groundwater recharge in the North China Plain. *J. Hydrol.* **2016**, *537*, 260–270. [CrossRef]
33. Richardson, C.M.; Davis, K.L.; Ruiz-González, C.; Guimond, J.A.; Michael, H.A.; Paldor, A.; Moosdorf, N.; Paytan, A. The impacts of climate change on coastal groundwater. *Nat. Rev. Earth Environ.* **2024**, *5*, 100–119. [CrossRef]
34. Chala, D.C.; Quiñones-Bolaños, E.; Mehrvar, M. An integrated framework to model salinity intrusion in coastal unconfined aquifers considering intrinsic vulnerability factors, driving forces, and land subsidence. *J. Environ. Chem. Eng.* **2022**, *10*, 106873. [CrossRef]
35. Zuecco, G.; Rinderer, M.; Penna, D.; Borga, M.; van Meerveld, H.J. Quantification of subsurface hydrologic connectivity in four headwater catchments using graph theory. *Sci. Total Environ.* **2019**, *646*, 1265–1280. [CrossRef]
36. Kim, B.; Roh, G.; Lee, J.; Yoon, J.; Lee, J. Characterizing the hydraulic conductivity of soil based on the moving average of precipitation and groundwater level using a regional database. *AQUA–Water Infrastruct. Ecosyst. Soc.* **2023**, *72*, 1459–1473. [CrossRef]
37. Lorenzo-Lacruz, J.; Garcia, C.; Morán-Tejeda, E. Groundwater level responses to precipitation variability in Mediterranean insular aquifers. *J. Hydrol.* **2017**, *552*, 516–531. [CrossRef]
38. Corona, C.R.; Ge, S.; Anderson, S.P. Water-table response to extreme precipitation events. *J. Hydrol.* **2023**, *618*, 129140. [CrossRef]
39. Nair, A.M.; Prasad, K.R.; Srinivas, R. Groundwater vulnerability assessment of an urban coastal phreatic aquifer in India using GIS-based DRASTIC model. *Groundw. Sustain. Dev.* **2022**, *19*, 100810.
40. Cai, Y.; Huang, R.; Xu, J.; Xing, J.; Yi, D. Dynamic Response Characteristics of Shallow Groundwater Level to Hydro-Meteorological Factors and Well Irrigation Water Withdrawals under Different Conditions of Groundwater Buried Depth. *Water* **2022**, *14*, 3937. [CrossRef]
41. Luo, F.; Wang, D.; Tian, X.; Bi, X.; Zheng, Q.; Zhou, Z.; Tang, Z. Estuarine groundwater level response to and recovery from extreme precipitation events: Typhoon Lekima in the Yellow River Delta. *J. Hydrol.* **2024**, *632*, 130918. [CrossRef]
42. Siddik, M.S.; Tulip, S.S.; Rahman, A.; Islam, M.N.; Haghighi, A.T.; Mustafa, S.M.T. The impact of land use and land cover change on groundwater recharge in northwestern Bangladesh. *J. Environ. Manag.* **2022**, *315*, 115130. [CrossRef] [PubMed]

43. Pinsri, P.; Shrestha, S.; Kc, S.; Mohanasundaram, S.; Virdis, S.G.P.; Nguyen, T.P.L.; Chaowiwat, W. Assessing the future climate change, land use change, and abstraction impacts on groundwater resources in the Tak Special Economic Zone, Thailand. *Environ. Res.* **2022**, *211*, 113026. [CrossRef]
44. Yifru, B.A.; Chung, I.-M.; Kim, M.-G.; Chang, S.W. Assessing the Effect of Land/Use Land Cover and Climate Change on Water Yield and Groundwater Recharge in East African Rift Valley using Integrated Model. *J. Hydrol. Reg. Stud.* **2021**, *37*, 100926. [CrossRef]
45. Chandrasekhar, P.; Kreiselmeier, J.; Schwen, A.; Weninger, T.; Julich, S.; Feger, K.-H.; Schwärzel, K. Why We Should Include Soil Structural Dynamics of Agricultural Soils in Hydrological Models. *Water* **2018**, *10*, 1862. [CrossRef]
46. Wang, L.; Jia, B.; Xie, Z. How climate change and groundwater extraction will shape the global terrestrial water cycle pattern. *J. Hydrol.* **2024**, *641*, 131780. [CrossRef]
47. Wakode, H.B.; Baier, K.; Jha, R.; Azzam, R. Impact of urbanization on groundwater recharge and urban water balance for the city of Hyderabad, India. *Int. Soil Water Conserv. Res.* **2018**, *6*, 51–62. [CrossRef]
48. Niu, X.Y.; Lu, C.P.; Zhang, Y.; Zhang, Y.; Wu, C.C.; Saidy, E.; Liu, B.; Shu, L.C. Hysteresis response of groundwater depth on the influencing factors using an explainable learning model framework with Shapley values. *Sci. Total Environ.* **2023**, *904*, 166662. [CrossRef]
49. Li, H.; Hou, E. Groundwater dynamic response mechanism and quantity vulnerability assessment under the influence of human activities. *Environ. Sci. Pollut. Res.* **2020**, *27*, 22046–22064. [CrossRef]
50. Ahmed, M.; Aqnouy, M.; El Messari, J.S. Sustainability of Morocco's groundwater resources in response to natural and anthropogenic forces. *J. Hydrol.* **2021**, *603*, 126866. [CrossRef]
51. Lee, J.M.; Kwon, E.H.; Woo, N.C. Natural and Human-Induced Drivers of Groundwater Sustainability: A Case Study of the Mangyeong River Basin in Korea. *Sustainability* **2019**, *11*, 1486. [CrossRef]
52. Agarwal, V.; Akyilmaz, O.; Shum, C.K.; Feng, W.; Yang, T.-Y.; Forootan, E.; Syed, T.H.; Haritashya, U.K.; Uz, M. Machine learning based downscaling of GRACE-estimated groundwater in Central Valley, California. *Sci. Total Environ.* **2023**, *865*, 161138. [CrossRef] [PubMed]
53. Deines, J.M.; Kendall, A.D.; Butler, J.J.; Basso, B.; Hyndman, D.W. Combining Remote Sensing and Crop Models to Assess the Sustainability of Stakeholder-Driven Groundwater Management in the US High Plains Aquifer. *Water Resour. Res.* **2021**, *57*, e2020WR027756. [CrossRef]
54. Rodell, M.; Velicogna, I.; Famiglietti, J.S. Satellite-based estimates of groundwater depletion in India. *Nature* **2009**, *460*, 999–1002. [CrossRef] [PubMed]

water

MDPI

Article

Study on the Factors Affecting the Drainage Efficiency of New Integrated Irrigation and Drainage Networks and Network Optimization Based on Annual Cost System

Zhiwei Zheng [1], Mingrui Li [1], Tianzhi Wang [2] and Hejing Ren [3,*]

1 College of Water Conservancy Engineering, Tianjin Agricultural University, Tianjin 300392, China; zhengzhiwei@tjau.edu.cn (Z.Z.); mingrui86717@163.com (M.L.)
2 School of Environmental Science and Engineering, Tianjin University, Tianjin 300350, China; wangtianzhi@tju.edu.cn
3 Department of Irrigation and Drainage, China Institute of Water Resources and Hydropower Research, Beijing 100048, China
* Correspondence: renhj@iwhr.com

Abstract

With the frequent occurrence of extreme weather events worldwide, the compound frequency of drought and flood events has significantly increased, imposing multidimensional pressures on agricultural water resource management. Agricultural water consumption accounts for approximately 70%, with severe waste, as a large amount of water is lost during transmission and distribution. Faced with increasingly severe and frequent extreme weather, traditional drainage systems may become unsustainable. Identifying the factors influencing drainage time is crucial for efficient drainage. The MIKE URBAN model has significant potential in farmland waterlogging simulation and drainage network optimization. This study validated the model's accuracy based on infiltration well overflow capacity experiments, with Average Relative Error (ARE) values of 2.29%, 6.52%, 4.41%, 3.17%, 4.37%, and 5.69%, demonstrating good simulation accuracy. The MIKE URBAN model was used to simulate drainage networks, explore factors affecting drainage time, establish an annual cost system for the drainage network, and optimize the network using a genetic algorithm with the objective of minimizing annual costs. Research findings: There is a clear negative correlation between the maximum inflow of infiltration wells and drainage time. As inflow increases, drainage becomes faster, but beyond 0.0075 m^3/s (27 m^3/h), the efficiency gains level off. This indicates that selecting infiltration wells with at least a 20% opening ratio is essential. Similarly, increasing the collector's diameter enhances drainage efficiency significantly, though the effect follows a diminishing return pattern. While smaller lateral spacing improves local water collection, it may lead to flow congestion if the collector is undersized; conversely, larger spacing increases drainage paths and delays, even if the collector is large. An optimal spacing range of 100–150 m is suggested alongside the collector diameter. Lateral diameter also affects performance: increasing it reduces drainage time, but the benefit plateaus around 200 mm, making further enlargement cost-ineffective. The genetic algorithm helped to optimize the drainage network design. Utilizing the genetic algorithm, the drainage network was optimized in just 15 iterations. The fitness function value rapidly decreased from 351,000 CNY to 55,000 CNY and then stabilized, reducing the annual cost from 59,640.67 CNY to 45,337.86 CNY—a 24% savings—highlighting the approach's effectiveness in designing efficient and economical farmland drainage systems. This study has shown that the simulation-based optimization of drainage networks provides a more rational and cost-effective approach to planning drainage infrastructure.

Water **2025**, *17*, 1201 https://doi.org/10.3390/w17081201

Keywords: MIKE URBAN; drainage; pipeline network optimization; genetic algorithm; network annual cost

1. Introduction

Against the backdrop of increasing extreme weather events globally, the frequency of compound drought and flooding events has significantly risen [1], and the problem of waterlogging is becoming more severe. Without effective drainage measures, crops under flooding stress are directly threatened by yield reduction [2], and agricultural water resources management is facing multidimensional pressures. Groundwater is generated from large paleolake deposits controlled by multiple tectonic movements [3]. However, due to excessive groundwater extraction, many areas have experienced phenomena such as land subsidence and surface cracks, leading to severe consequences. With the acceleration of urban modernization, natural ground surfaces are gradually being destroyed and replaced by hard, impermeable surfaces. This severely hinders precipitation infiltration, increases surface runoff, and consequently leads to flooding [4]. According to the Food and Agriculture Organization (FAO) of the United Nations, about 33% of irrigated farmland globally faces the dual threat of seasonal water scarcity and inadequate drainage capacity due to climate change [5], while only 25–30% of existing irrigation facilities are resilient enough to adapt to enhanced rainfall variability [6]. Studies have shown that traditional irrigation and drainage systems exhibit significant flaws in extreme hydrological events. For instance, during the 2015 drought in the Ganges Plain of India, the irrigation network had a 37% water supply gap [7]. In China's North China Plain, the lack of drainage infrastructure led to crop yield losses. Additionally, the separation of irrigation and drainage functions can lead to soil salinization [8]. These challenges urgently require systematic optimization to enhance the resilience of water infrastructure. As early as the mid-19th century, European countries used tiles to build irrigation and drainage systems, and with the improvement in research and technology, the area of drained farmland nationwide steadily increased [9]. In India, the over-extraction of groundwater has become a serious issue, and to address the land–water–salt balance, research into underground drainage systems was initiated relatively early. Some studies show that although farmers need to invest in drainage system construction upfront to address waterlogging issues, once the system is built, it can effectively mitigate the yield and production losses caused by internal flooding and ensure economic benefits [10]. However, faced with increasingly severe and frequent extreme weather events, traditional irrigation and drainage systems may no longer be safe.

Traditional drainage methods occupy a large amount of arable land, significantly increase conveyance losses, result in a low irrigation water utilization coefficient, and fail to adequately guarantee drainage performance—potentially even affecting crop yields. As research progressed, the concept of dual-purpose irrigation and drainage channels emerged, which saved some farmland areas and reduced water head loss. Some scholars have pointed out that the height of dual-purpose channels should be between that of irrigation and drainage channels, but the drop is minimal, making it suitable only for flat areas; it is not suitable for hilly regions [11].

Due to the limitations of dual-purpose irrigation and drainage channels, such as occupying a large area of farmland and having poor compatibility with new water-saving irrigation methods, scholars began researching irrigation and drainage pipelines. Pipeline irrigation and drainage involves irrigation and drainage through pipelines, which operates quite differently from surface open channel irrigation and drainage. By studying the soil moisture movement mechanism, researchers have explored the differences between pipeline irrigation and drainage models and traditional open channel systems. For example, in the salinized farmland of Wuyuan County, the migration and distribution of soil salts were compared under two models: "drip irrigation + open drainage" and "flood irrigation + buried pipe drainage". The study showed that both models resulted in salt removal from the soil, with the "flood irrigation + buried pipe drainage" model having a relatively better desalting effect [12]. Using the HYDRUS-1D model, a one-dimensional numerical simulation was conducted to evaluate the salinity improvement effect on saline–alkali land under water-saving irrigation, conventional irrigation, buried pipe drainage, and no drainage. The results indicated that buried pipe drainage improved leaching efficiency and effectively suppressed the process of soil salinization [13]. Furthermore, simulations using the Vedernikov infiltration equation and the Van der Molen leaching desalination equation studied the improvement process of coastal saline soil irrigation and drainage. The study pointed out that uniform flooding and leaching across the entire region could result in excessive leaching near the buried pipes, wasting water resources, while areas farther from the pipes experienced insufficient leaching, thus reducing leaching efficiency [14]. A double-layer buried pipe drainage experiment with alternating shallow and deep pipes was conducted, and the results showed that double-layer buried pipe drainage was more effective in flood control and salinity reduction than single-layer buried pipe drainage [15]. A comparison of surface drip irrigation and underground straw composite pipeline irrigation indicated that the latter helped promote winter wheat growth after the jointing stage, improving yield and water use efficiency. It also showed higher economic benefits and promising prospects for field crops with high planting density in supplemental irrigation areas [16].

In addressing the optimization of pressurized water distribution networks, several scholars have proposed various methods. One proposed an optimal path identification method based on genetic algorithms [17]; another utilized a GIS system to optimize the irrigation network in eastern Arkansas, USA [18]. A three-stage optimization model was introduced, where the first stage determines the lowest cost layout and pipe diameter, the second stage identifies the annual investment in pump stations and energy costs to meet crop water demand, and the third stage minimizes the total cost [19]. Some scholars introduced reliability as a constraint in the optimization model for the first time, integrating it with cost considerations and analyzing the relationship between the two [20]. An optimized pipe diameter model for agricultural irrigation systems was designed and applied in the Sinwankim area of Korea, saving 11% of costs compared to the original plan [21]. A three-level irrigation network planning was proposed for an irrigation district in India, aiming to determine the optimal pipe diameter at the lowest cost. The study showed that compared to traditional distribution methods, this approach significantly reduced losses and saved costs [22]. Genetic algorithms were used to optimize the network layout in agricultural irrigation systems, reducing construction and operational costs while ensuring water supply demands [23]. The PHSM algorithm was proposed to improve the efficiency of evolutionary algorithms in optimizing distribution systems [24]. For pressurized irrigation networks, an optimization program was written using Excel VBA, revealing that even minor changes in the network could lead to significant changes in optimization outcomes [25]. Complex network theory (CNT)was applied to replace traditional methods, reducing uncertainties in the distribution network [26]. These studies primarily focus on water distribution networks, but the optimization of complex networks remains a challenging problem that still needs to be addressed.

Integrated irrigation and drainage networks provide a technical solution to address the aforementioned contradictions by combining water delivery and drainage functions.

The core design concept lies in using a modular network structure to dynamically switch between irrigation and drainage modes. Systematic studies have shown that such systems can significantly improve water resource utilization efficiency and reduce the impact of groundwater level fluctuations on crop growth by integrating both irrigation and drainage functions [27]. The integrated irrigation and drainage pipeline, designed for dual use, is especially crucial in low-lying, flood-prone areas, playing a vital role in ensuring crop safety during floods and supporting normal growth. This technology is of great significance for reforming the old canal systems and management models in farmland [28]. However, existing research primarily focuses on optimizing hydraulic performance or the cost optimization of single-function networks, and there is still a lack of universal models for the quantitative assessment of annual costs for complex networks (including construction investment, maintenance costs, etc.).

In the field of agricultural water management, models like DRAINMOD, SWAP, HYDRUS-1D/2D/3D are widely used [29], along with other open-source software [30,31]. Some scholars have simulated the dynamic water and nitrogen balance in paddy fields under traditional irrigation and drainage modes versus controlled irrigation and drainage modes using the DRAINMOD model. The results showed that the controlled irrigation and drainage mode achieved better water-saving and emission–reduction effects in paddy fields [32]. Other scholars have introduced the latest research and application progress of the HYDRUS model in China within the field of hydrological and hydraulic modeling, demonstrating its widespread application in soil moisture, salinity, heat, and nutrient transport [33]. Compared to other models, the MIKE URBAN model has the advantage of multi-module coupling capability, an easy-to-use operating system, and excellent visualization. However, it is primarily used for simulation analysis in municipal and river network systems, with relatively few applications in agricultural water management (DHI, 2021).

In a drainage network, infiltration wells serve as critical nodes, and their overflow capacity directly impacts both the drainage efficiency and the uniformity of irrigation. Physical experiments that measure the overflow capacity of infiltration wells are conducted to establish a hydraulic parameter database applicable to the MIKE URBAN model. This database is then used to validate the MIKE URBAN model's simulation accuracy in farmland drainage and to develop a hydraulic network model that simulates drainage performance under different topological configurations and pipe diameter selections, thereby exploring the factors affecting drainage speed. Finally, a genetic algorithm is employed—with the objective of minimizing annual construction and annual operation and maintenance (O and M) costs—to optimize the network design. Specifically, our goals are the following:

(1) Study on infiltration well overflow capacity:

Drainage experiments were conducted to investigate the practical overflow capacity (i.e., drainage speed) of infiltration wells. The experiments considered four factors: infiltration well diameter, connection pipe diameter, water depth, and opening rate. A total of 10 experimental schemes were designed, with each scheme repeated 24 times (resulting in 240 tests covering all variable combinations), to analyze the effects of these factors on the drainage efficiency of the wells.

(2) Simulation of drainage network layout based on the MIKE URBAN model:

An experimental field in Yanggu County, Shandong, was selected as the simulation area to study the drainage performance and marginal effects of different network layouts. Multiple simulations using the MIKE URBAN model were performed for various drainage network layouts. Each layout was evaluated based on drainage speed, and the study investigated the factors influencing drainage speed as well as the marginal impact of different network configurations.

(3) Optimization of the drainage network:

In designing and optimizing the drainage network, annual cost is a key indicator of the system's economic efficiency. With the objective of minimizing annual cost—and under constraints such as pipe diameter, pipe length, and pipe spacing—a fitness function was constructed. The algorithm structure was then optimized using tournament selection to enhance convergence and identify the optimal network layout.

2. Materials and Methods

2.1. Experimental Background and Objectives

In the integrated irrigation and drainage network, infiltration wells, as key nodes, directly affect the drainage efficiency of the system. This experiment quantifies the flow capacity of infiltration wells through physical testing, thereby establishing a hydraulic parameter database suitable for the MIKE model.

The experimental area is located in Yanggu County, Liaocheng City, Shandong Province, at 36°11′ N and 115°79′ E, with an elevation of 39.62 m. The long-term average annual rainfall is 560.2 mm, and the average temperature is 13.6 °C, falling within a semi-humid continental temperate monsoon climate. Yanggu County is situated on the north bank of the Yellow River in the western Shandong Plain, with a gently sloping terrain from southwest to northeast. It borders Dong'a County to the east, Xin County to the west, Taixian County of Henan Province to the south, and Dongchangfu District to the north; across the Yellow River to the southeast, it faces Dongping County.

2.2. Experimental Design

This experiment focuses on the flow capacity of infiltration wells. Through controlled head, variable head, and float valve comparative tests on infiltration wells of different specifications, the study observes their drainage performance under various flooding scenarios. The objective is to identify the factors influencing drainage efficiency, quantify the infiltration wells' flow capacity, and provide data support for subsequent model simulations.

The experimental site is a 7.1 m × 6.1 m test pool, as shown in Figure 1, designed to simulate field irrigation and drainage processes effectively. The perimeter of the site has been treated with anti-infiltration measures or enclosures to ensure that test water does not leak or overflow, thereby ensuring the accuracy of measurement data. The core components include the following:

(1) Infiltration Well System: The infiltration wells are installed within the test pool to simulate the function of field drainage wells or irrigation–drainage piles. When the water level in the test pool reaches a certain height, water flows into the well through perforations in the well cover and is then discharged or collected through pipes connected to the drainage network.

Infiltration wells with diameters of dn200, dn250, and dn315 are used, all with a uniform height of 51 cm. The variation in well diameter is intended to investigate its effect on drainage capacity and provide reference parameters for subsequent drainage network optimization.

The well cover, as shown in Figure 2, is made of PVC material and features 10 mm diameter perforations with a 16 mm spacing. The perforation rates are 5%, 10%, 15%, 20%, 25%, and 30%. By adjusting the number of perforations, the experiment explores the flow capacity of infiltration wells under different perforation rates. Since the perforation rate significantly influences water inflow and drainage efficiency, comparative analysis of different rates helps determine an optimal perforation configuration.

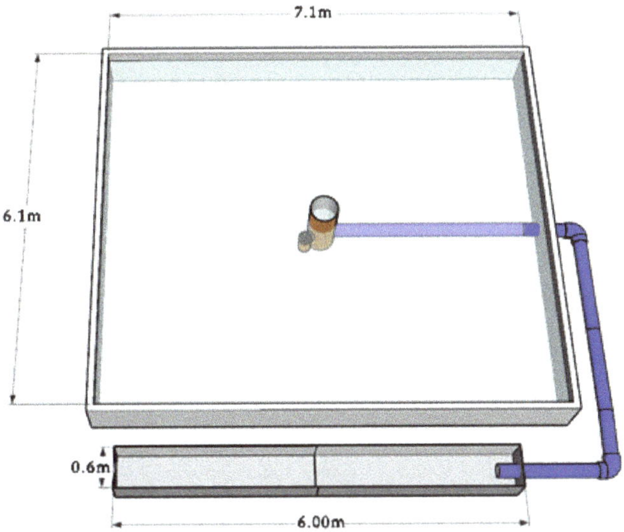

Figure 1. Test field.

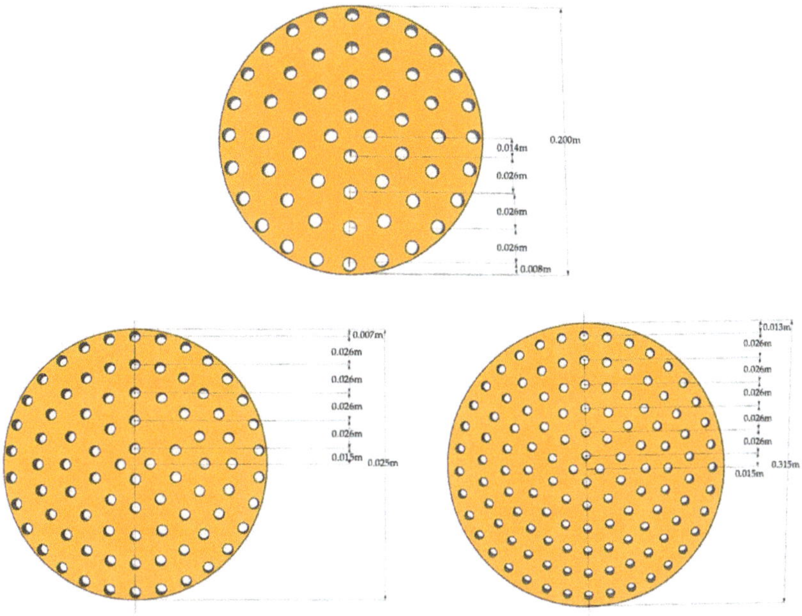

Figure 2. The top cover of an infiltration well with different specifications.

(2) Connecting Pipes: The connecting pipes are attached to the infiltration wells using sealed joints to prevent leakage while allowing easy assembly and adjustment. These pipes simulate the water transport function of various levels of field drainage networks (e.g., collectors, laterals). By testing different pipe diameter combinations, the impact of pipe diameter on drainage efficiency can be analyzed.

The connecting pipes are made of PVC material with diameters of 160 mm, 110 mm, and 90 mm, each corresponding to different drainage capacities and hydraulic conditions.

(3) Float Valve: Made of PVC material, the float valve is used in comparative experiments by installing it on the connecting pipes. It automatically opens or closes based on water levels, controlling inflow and outflow speeds. This setup allows observation of the float valve's buffering effect in response to water level fluctuations and sudden flow changes.

(4) Triangular Weir: The triangular weir is positioned at the drainage outlet of the test pool. By recording the height of the water level, it is possible to obtain the flow rate through the triangular weir. The drainage flow and process can be accurately monitored, providing essential data for subsequent analysis and model validation.

As shown in Figure 3, the triangular weir is constructed from welded steel plates, approximately 6 m in length and 0.6 m in width, with a relatively smooth inner surface. The overflow notch is designed according to ISO 1438 [34] standards to ensure accurate flow measurement under different discharge conditions. When water flows through the triangular weir, a high-precision liquid level sensor (sampling frequency: 10 Hz) continuously records water level variations in real time. The collected water level data are then converted using the corresponding weir height-flow rate reference table.

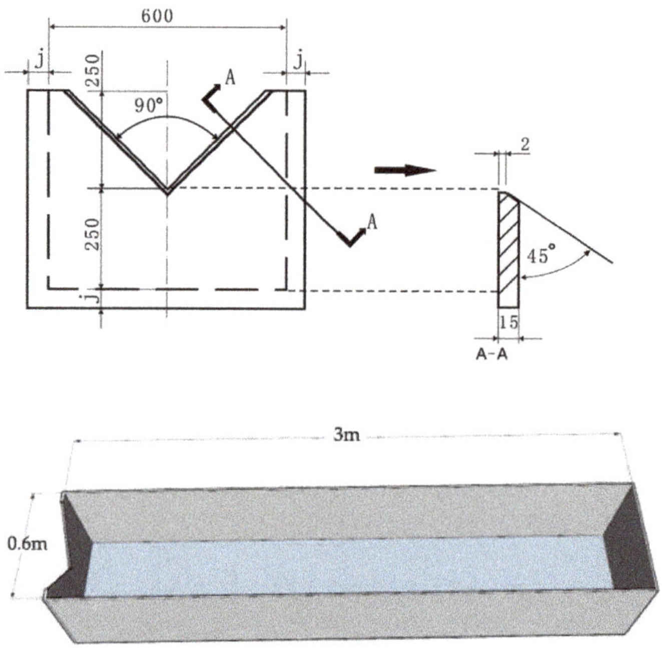

Figure 3. A triangular weir.

(5) Water Level Control System: Water supply valves are installed in the test pool, allowing manual or automatic water flow control to simulate rainfall or irrigation-induced ponding. Based on experimental requirements, different target water levels (e.g., 5 cm, 10 cm, 20 cm) can be set to observe drainage performance under varying ponding depths.

For constant water head tests, the inflow is finely adjusted using a manual valve to maintain a stable water level in the test pool (accuracy: ±1 cm). The experiment monitors flow rate variations at the triangular weir, the infiltration capacity of the wells, and the efficiency of water transport through the connecting pipes under stable water level conditions.

A measuring scale is placed inside the test pool to monitor water levels in real time. The water level in the test pool is manually recorded by reading the scale measurements at regular intervals.

The experiment was designed with the following four factors (Table 1):

Table 1. Experimental Schemes.

Scheme Serial Number	Main Observed Content	Infiltration Well Diameter (mm)	Connecting Pipe Diameter (mm)	Water Depth (cm)	Hole Ratio and Hole Position	Number of Trials
1	Triangular Weir Water Surface Elevation	200	160	20, 10, 5, and 20 cm Variable Water Head	Top Cover Hole Ratios: 30%, 25%, 20%, 15%, 10%, 5%	24
2		200	110			24
3		200	90			24
4		250	160			24
5		250	110			24
6		250	90			24
7		315	160			24
8		315	110			24
9		315	90			24
10		200	160 (with Float Valve)			24

(1) Infiltration well diameter: 200 mm, 250 mm, 315 mm.

(2) Connecting pipe diameter: 160 mm, 110 mm, 90 mm.

(3) Ponding depth: Constant water head (5 cm, 10 cm, 20 cm) and variable water head (initial depth of 20 cm).

(4) Perforation ratio: 5% to 30% (adjusted by the number of perforations on the well cover).

These four factors (infiltration well diameter, connecting pipe diameter, ponding depth, and perforation ratio) are mutually independent. Each factor is assigned different values, forming distinct experimental schemes through their combinations. Each infiltration well diameter, connecting pipe diameter, ponding depth, and perforation ratio combination results in a unique experimental setup.

A special control scheme (Scheme 10) was included, where a float valve was installed in the connecting pipe to analyze its buffering effect during sudden water level changes or high-flow impact. This control scheme also covered 4 ponding depths $\times$ 6 perforation ratios, totaling 24 tests.

In summary, the experiment consisted of 240 tests in total.

The experimental procedure is divided into the following steps:

(1) Constant head test:

Water is added to reach the target depth, maintaining the water level at 5 cm, 10 cm, and 20 cm within the test pool using dynamic valve adjustments. The flow rate over the triangular weir and water level changes in the test pool are recorded. Once the flow over the triangular weir stabilizes, a high-precision liquid level sensor (accuracy $\pm$0.1 mm) is used to record the water surface height.

(2) Variable head test:

The test pool is filled to an initial ponding depth of 20 cm via an irrigation–drainage pile, and then inflow is stopped. Water drains naturally through the infiltration well, causing the water level to drop. Once the triangular weir flow stabilizes, the water level decline rate and triangular weir flow rate are monitored in real time, and the water surface height is recorded.

(3) Float valve control test:

A float valve is installed in the connecting pipe of the infiltration well system as a control mechanism to analyze its buffering effect on sudden flow fluctuations.

2.3. Experimental Measurement Items and Methods

2.3.1. Water Level Measurement

The measurement of water levels in the experiment is mainly divided into two parts: the water level in the test tank and the upstream water level of the triangular weir.

(1) Test Tank Water Level: The dynamic changes in the water level within the test tank are recorded in real time using a graduated scale inside the tank. The measurements include both constant water head conditions (5 cm, 10 cm, 20 cm) and variable water head conditions (initial depth of 20 cm).

(2) Upstream Water Level of the Triangular Weir: The water level changes are continuously collected in real time using a high-precision liquid level sensor (sampling frequency 10 Hz), whose measuring range is 0~500 mm; the resolution is 0.1 mm; accuracy ±0.1 mm; supply voltage is 12 V, with the data transmitted to the computer for analysis.

2.3.2. Drainage Time Measurement

For the determination of drainage time, a high-precision liquid level sensor is used, with a time step set to 3 s. The sensor continuously collects the triangular weir water level data for each time step and records them in the computer. In the non-fixed head test, after stopping the water inflow, the duration is recorded for the water level to drop from the initial value (20 cm) to the point where there is no obvious water accumulation. The recording stops when the water level in the experimental pool reaches the point where no significant accumulation remains. In the fixed head test, the time from the start of water inflow to the stabilization of flow is recorded as an indicator of the system's response time.

2.3.3. Triangular Weir Flow Calculation Method

The triangular weir water level height is measured in real time using a high-precision liquid level sensor. The calculation formula for the flow rate of the ISO standard triangular weir is as follows:

$$Q = C \times \frac{8}{15}\sqrt{2g} \times H^{\frac{5}{2}} \quad (1)$$

In the formula:

Q—Flow rate, m^3/h;

C—Flow coefficient (related to the weir angle and weir characteristics);

H—Water depth in front of the weir, m.

Based on the above formula, the flow rate of the infiltration well can be calculated according to the water depth in front of the weir.

3. MIKE URBAN Model Data Processing

3.1. Infiltration Well (Manhole) Parameters

In the MIKE URBAN model, the infiltration well (manhole) is a key node that connects adjacent pipes and serves as the infiltration point for rainwater during drainage. Through the infiltration well flow capacity test conducted in Yanggu County, the flow capacity of the infiltration well was quantified as the drainage flow rate per hour, as shown in Table 2.

Table 2. Overflow capacity (drainage flow rate) of infiltration wells of different specifications.

Serial Number	Diameter of Infiltration Well (mm)	Connecting Pipe Diameter (mm)	Drainage Flow (m^3/h)					
			5%	10%	15%	20%	25%	30%
1	200	160	6	30	33	35	37	40
2		110	6	17	25	30	36	38
3		90	6	16	20	27	31	33
4	250	160	26	35	42	43	44	45
5		110	20	26	35	38	41	42
6		90	16	18	34	36	40	41
7	315	160	26	38	43	44	45	48
8		110	23	36	41	42	43	46
9		90	20	30	38	40	41	43
10	200 (with Float Valve)	160	6	16	28	31	35	37

Based on the experimental results, the flow capacity of infiltration wells with various specifications, composed of different components, can be simplified into a single value, greatly reducing the difficulty of parameter settings in the MIKE URBAN model. Additionally, based on these results, the influence of various factors on the drainage rate of infiltration wells can be analyzed:

(1) Effect of opening rate on drainage flow: As the opening rate increases, the drainage flow increases. In all combinations, as the opening rate increases from 5% to 30%, the drainage flow shows an upward trend. There is a diminishing marginal benefit, with some combinations showing a slowdown in the flow rate increase when the opening rate exceeds 20%. In most cases, the flow rate approaches its maximum when the opening rate is between 20% and 25%, and further increasing the opening rate has a limited impact on the flow rate. Therefore, the optimal range for the opening rate is 20% to 25%.

(2) Effect of infiltration well diameter on drainage flow: The larger the diameter of the infiltration well, the higher the drainage flow. A 315 mm infiltration well shows a 20% higher flow rate than a 200 mm infiltration well, and this growth trend follows the principle of increased overflow area in fluid dynamics.

(3) Effect of connection pipe diameter on drainage flow: As the pipe diameter increases, the drainage flow increases significantly.

(4) Effect of float ball device on drainage flow: The float ball causes a decrease in flow because it increases local resistance and limits the flow rate. However, the float ball can prevent backflow, and its use can be considered depending on the actual situation.

3.2. *Pipe Diameter and Length Parameters*

3.2.1. Drainage Parameter Design

In farmland drainage, the main hydraulic parameters affecting the pipeline are the drainage modulus or drainage flow rate.

The design drainage modulus (drainage flow per unit area) is typically calculated based on the design runoff depth R, and the calculation formula is as follows:

$$q = \frac{R}{86.4t} \tag{2}$$

In the formula, q is the design drainage modulus in $m^3/(s \cdot km^2)$; R is the regional design runoff depth in mm; and t is the drainage duration, taken as the crop's allowable waterlogging duration in days.

The surface runoff coefficient of Yanggu County is higher than the average value of other counties in Liaocheng City by 0.18–0.32. The western and northwestern counties have values lower than the city's average by 0.14–0.44. Based on this, the surface runoff in Yanggu County ranges from 0.32 to 0.44, with an average value of 0.38. Therefore, the design runoff depth is 46.70 mm.

Most of the experimental area is under dryland farming. According to the China High-Standard Farmland Construction Plan (2021–2030) [35] and the Technical Specifications for Farmland Drainage Engineering [36], the drainage standard for dryland farming is to drain the rainfall within 1 to 3 days, starting from the moment crops are submerged until there is no water accumulation on the field. In this study, the drainage standard used is to drain the 1-day rainfall within 1 day until there is no water accumulation on the field. Using Formula (2), the design drainage modulus is calculated to be 0.54 $m^3/(s \cdot km^2)$.

The design of the water conveyance pipeline should be based on the drainage capacity it is intended to carry. The design cross-section is calculated using the following formula:

$$Q_{dr} = \frac{1}{n} {R_h}^{2/3} i^{1/2} \times \pi r^2 \tag{3}$$

In the formula, Q_{dr} represents the drainage capacity to be carried by the pipe; i is the longitudinal slope of the pipe; n is the pipe's roughness coefficient; R_h is the hydraulic radius of the pipe; and r is the radius of the pipe.

For full pipe flow, $R_h = r/2$; hence, Formula (3) can be rewritten as follows:

$$r = (\frac{Q_{dr} n}{\pi i^{1/2}} \times 2^{2/3})^{3/8} \tag{4}$$

3.2.2. Impact Mechanism of Drainage Pipe Diameter and Length

The drainage pipe diameter is calculated based on the drainage modulus, which is 0.54 $m^3/(s \cdot km^2)$. The analysis of the main drainage influencing parameters is conducted based on the basic parameter settings, with additional analysis for 25%, 50%, 100%, and 200% variations.

The basic parameters are set as follows (Table 3):

Table 3. Basic values of parameters affecting pipe diameter.

Parameter	Unit	Numerical Value
Lateral length	m	60
Collector length	m	150

(1) Impact of lateral length:

Figure 4 shows the effect of lateral length on both the lateral and collector diameters under drainage conditions. It can be observed that lateral length significantly affects the drainage pipe diameters, exhibiting a quadratic parabolic increase. When the lateral length is between 20 m and 180 m, the required diameter for the lateral ranges from 100 mm to 250 mm, while the collector diameter ranges from 200 mm to 450 mm. It is evident that the influence of lateral length on the main drainage pipe diameter is greater than the influence on the lateral diameter. The primary factor is that the area controlled by the collector is more significantly impacted by the lateral length, leading to a greater influence on the pipe diameter.

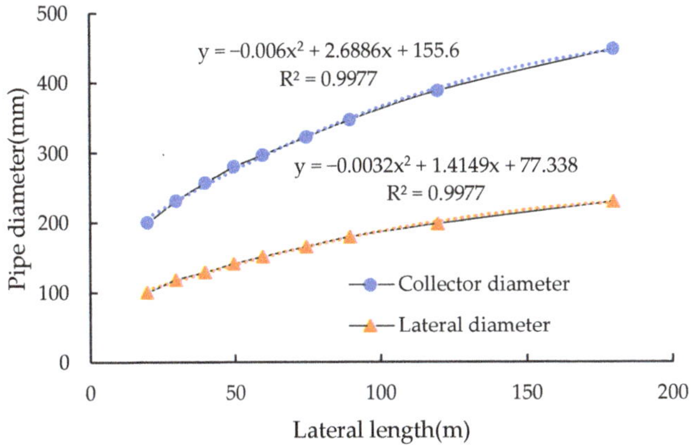

Figure 4. The influence of lateral length on the diameter of lateral and collectors.

(2) Impact of collector length:

Figure 5 illustrates the effect of collector length on both the lateral and collector diameters under drainage conditions. It can be observed that the collector length still has a significant impact on the drainage pipe diameters, exhibiting a quadratic parabolic increase. When the collector length is between 50 m and 450 m, the required lateral diameter ranges from 100 mm to 230 mm, while the collector diameter ranges from 200 mm to 450 mm. The trend of its influence is quite similar to that of the lateral length.

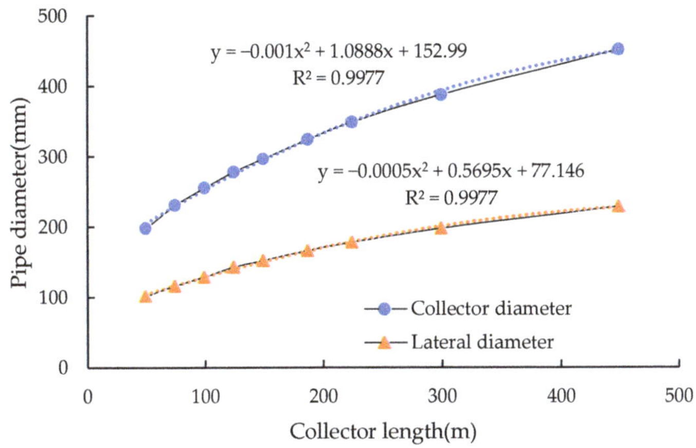

Figure 5. The influence of the length of the collector on the diameter of laterals and collectors.

3.2.3. Selection of Pipe Diameter and Length

In conclusion, under the drainage modulus of 0.54 $m^3/(s{\cdot}km^2)$, it is recommended to control the collector diameter between 200 mm and 450 mm and the lateral diameter between 100 mm and 250 mm. The collector length should be between 50 m and 450 m, while the lateral length should be between 20 m and 180 m.

3.3. *Rainfall Boundary Condition Data*

In the MIKE URBAN model, rainfall boundary data are one of the necessary input parameters. Rainfall pattern design typically refers to the setting or selection of the time-based rainfall distribution process for a design storm in drainage, flood control, or water resources planning in order to simulate or evaluate the rainfall intensity variation over time.

Using the rainfall data from the Cha Cheng meteorological monitoring station near the test area as the basis for the design rainfall, rainfall data from 1957 to 2013 were collected. The results yielded a 10-year return period rainfall of 122.92 mm, a 5-year return period rainfall of 102.29 mm, and a 20-year return period rainfall of 142.11 mm.

According to the "Design Standard for Irrigation and Drainage Engineering" (GB50288-2018) [37], the general drainage standard is typically based on a return period of 5 to 10 years. The "National High-Standard Farmland Construction Plan (2021-2030) [35]" specifies that the drainage design for dryland farming areas should have a storm recurrence period of 5–10 years. In Shandong province, a 10-year recurrence period is used. Therefore, in accordance with the above construction requirements, the preliminary drainage standard is set to a 10-year recurrence period, with the design rainfall of 122.92 mm.

The Chicago rainfall pattern is selected as the synthetic rainfall boundary data, with the rainfall design process curve shown in Figure 6.

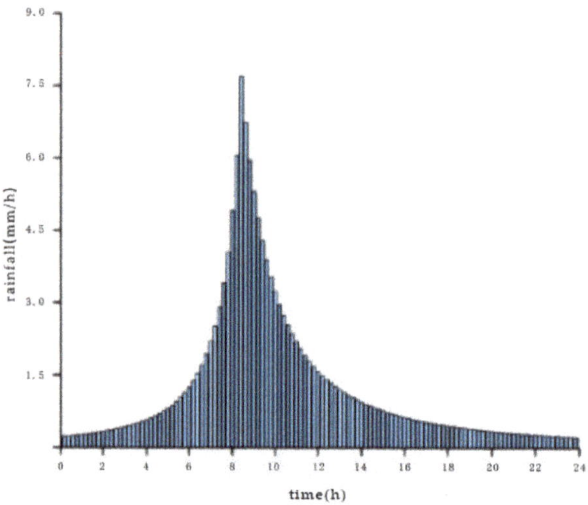

Figure 6. p = 10a Rainfall design process line.

3.4. *Subdivision of Sub-Catchment Areas*

A sub-catchment area refers to a smaller unit within a larger watershed or catchment area that is divided for hydrological or drainage system analysis. In each unit, precipitation or surface runoff is collected into a specific outflow point, and these smaller areas are collectively referred to as sub-catchment areas.

In the MIKE URBAN model, sub-catchment areas can be automatically delineated based on factors such as the layout of the pipeline network, topography, and rainfall conditions.

3.5. *Model Accuracy Validation*

To verify the application effectiveness of the MIKE URBAN model in drainage systems, this study uses the infiltration well overflow capacity test as a basis. The MIKE URBAN model is used to simulate the same conditions as the experiment, and a comparative analysis is conducted between the simulated and measured values to test the model's accuracy. The drainage process of the experimental pool provides real hydraulic conditions, and by comparing the simulation results with the actual measured data, the reliability of the model in practical operation can be evaluated.

This study verifies the model's accuracy using the experimental data from infiltration wells with three different diameters and six different opening rates in the unsteady head and 5 cm steady head tests. The model simulation accuracy is evaluated using the Average Relative Error (ARE). ARE < 10%, 10% < ARE < 20%, and 20% < ARE < 30% represent very good, average, and poor simulation performance, respectively.

The model verification results are shown in Figures 7 and 8. Under the unsteady head condition, the ARE between the simulated and measured values for the dn200 infiltration well is 2.29%, for the dn250 infiltration well is 6.52%, and for the dn315 infiltration well is 4.41%. Under the 5 cm steady head condition, the ARE for the dn200 infiltration well is 3.17%, for the dn250 infiltration well is 4.37%, and for the dn315 infiltration well is 5.69%. The verification results indicate that the model performs well in simulating flooding and pipe drainage, with good agreement with the measured values. It can accurately simulate farmland drainage.

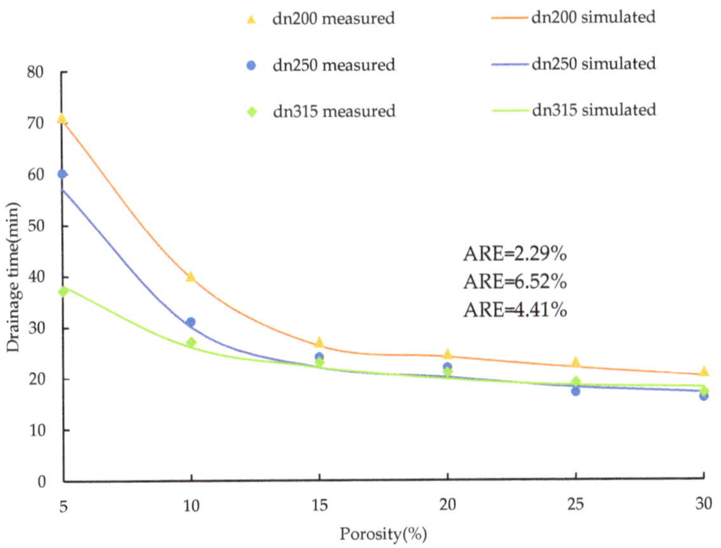

Figure 7. Uncertain head test values and simulated values.

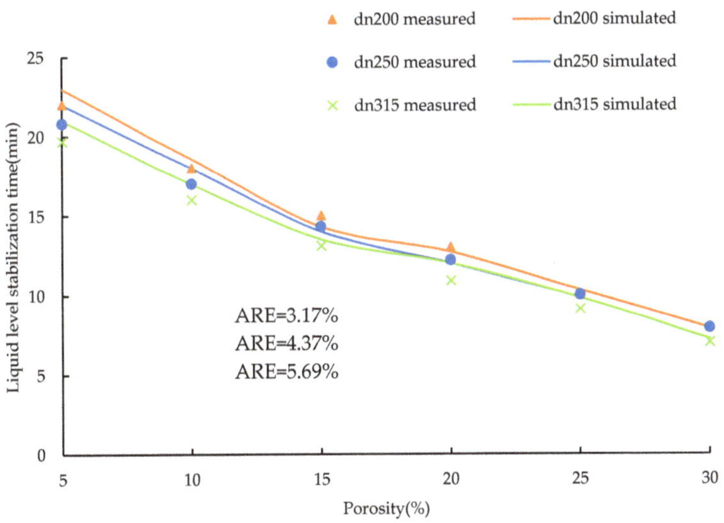

Figure 8. 5 cm constant head test value and simulated value.

4. Flooding Simulation of Farmland Based on the MIKE URBAN Model

Farmland flooding refers to the phenomenon where excess water in the field cannot be drained in a timely manner, leading to excessively high soil moisture content, which in turn affects crop growth and may even result in reduced yield or crop failure. It is usually caused by multiple factors, including both natural and human factors. From a natural perspective, intense and concentrated rainfall generates a large amount of runoff in a short period of time. If the area is low-lying or the soil has poor permeability (such as clay layers or high groundwater levels), the excess rainwater will be difficult to infiltrate and drain quickly. From a human perspective, if the agricultural water infrastructure is incomplete or in poor condition, such as insufficient cross-sections of drainage ditches, inadequate pump station capacity, or lack of drainage pipe systems, and coupled with irrational land planning and blind reclamation that leads to the disappearance of natural water storage areas, these factors significantly increase the risk of flooding. Additionally, excessive groundwater extraction causes land subsidence, soil compaction, and crop root sensitivity to excess water, which further raises the probability of flooding.

In agricultural water projects, flooding not only leads to crop yield loss but can also cause a series of problems, such as soil degradation and infrastructure damage. With the increasing frequency of extreme weather events and the growing complexity of hydrological conditions, how to accurately simulate farmland flooding processes and optimize irrigation and drainage systems based on this has become an important issue to improve agricultural water resource utilization and ensure food security.

MIKE URBAN is characterized by its multi-module coupling ability, excellent visualization, and user-friendly operation. In this chapter, the MIKE URBAN model is used to simulate flooding in an experimental field in Yanggu County, Liaocheng City, Shandong Province. To consider the worst-case drainage scenario, soil infiltration was ignored in this simulation to ensure that the drainage system design can handle the maximum possible water accumulation. By simulating the drainage process in detail, we assess the drainage effectiveness of different pipeline layouts and analyze the factors influencing drainage speed. By adjusting key parameters and observing the simulation results, we further explore the "boundary effects" of these parameters, that is, the impact on drainage efficiency and system stability when parameters approach or exceed certain critical values. The above content will provide the necessary data support and theoretical basis for subsequent network economic analysis and comprehensive optimization.

4.1. Overview of the Simulation Area

In order to study the drainage performance of the irrigation and drainage network, this study selected an experimental field in Yanggu County, Liaocheng City, Shandong Province, as the simulation area, with the layout diagram shown in Figure 9. The area of the simulation block is approximately 550 m $\times$ 550 m, representing a typical farmland drainage area. The recurrence period for heavy rainfall is chosen as 10 years. The drainage requirements in this area are high. Therefore, a reasonable pipeline layout is needed to ensure that the drainage system can discharge water in a timely and effective manner to meet the farmland drainage needs.

The original layout of the area was as follows: a total of 66 infiltration wells, with a central vertical collector diameter of 315 mm, a horizontal lateral diameter of 200 mm, a lateral spacing of 100 m, a horizontal layout spacing of infiltration wells of 50 m, and a slope of 1‰.

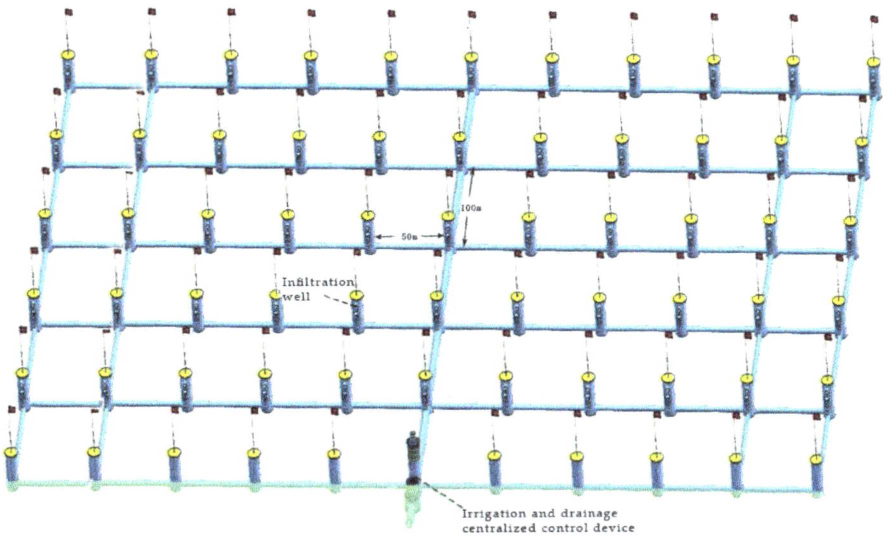

Figure 9. Schematic diagram of simulated block layout.

4.2. Simulation Objectives and Parameter Settings

The main purpose of this simulation is to assess whether the original layout meets the drainage standards and to analyze the differences in drainage speed across different pipeline layouts. It aims to explore the factors affecting drainage time and their marginal effects, providing a theoretical basis for subsequent pipeline optimization. To achieve this, the simulation considers the following key variable parameters:

Collector diameter: The diameter of the collector directly affects the water flow capacity. Therefore, different collector diameters are simulated to evaluate their impact on drainage speed. The collector diameter is selected within the range of 200 mm to 450 mm.

Maximum inflow rate of the infiltration wells: The inflow rate of the infiltration wells determines the upper limit of the drainage capacity. Thus, it is adjusted to simulate the drainage effect under different maximum inflow rates. The hole opening rate is selected at 20% and above (with an infiltration well discharge flow > 27 m^3/h).

Lateral spacing: The spacing of laterals affects the distribution of water flow. Appropriate lateral spacing can improve drainage efficiency. By adjusting the lateral spacing, its impact on drainage speed is explored.

Lateral diameter: The diameter of the lateral affects the flow velocity within the lateral, thus influencing the overall flow distribution of the drainage system. By adjusting the lateral diameter, the impact on water flow and drainage efficiency is further evaluated. The lateral diameter is selected within the range of 100 mm to 250 mm.

4.3. Simulation Results

By using the MIKE URBAN model, simulations were conducted for multiple different pipeline layout configurations, and each configuration was evaluated based on drainage speed. The simulation results are presented in tabular form, which lists the drainage speed under different parameter settings.

According to the Irrigation and Drainage Engineering Design Standard (GB 50288-2018) [37], the drainage system is considered adequate if floodwater is removed from the field surface within 24 h after crop submersion.

The original layout of the block included 66 infiltration wells, while the hydrological module of MIKE URBAN automatically divided the catchment area into 56 sub-catchments, resulting in the use of 56 infiltration wells.

4.4. Results Analysis

Based on the simulation results (Table 4), we can analyze the drainage performance of each layout. The following is the main analysis of the results.

(1) Impact of maximum inflow of infiltration wells:

The simulation results indicate that when blocking is not considered, the maximum inflow of the infiltration well seems to have little effect on the drainage speed. Instead of using the inflow values obtained from previous experiments, a set of values increasing from small to large was artificially set for further simulation. The study found that the maximum inflow of the infiltration well has a negative correlation with the drainage time (Figure 10). Specifically, as the maximum inflow increases, the drainage time gradually decreases. However, when the maximum inflow exceeds 0.0075 m^3/s (27 m^3/h), the drainage speed stabilizes, and further increases in inflow no longer have a significant impact on the drainage speed. This suggests that when selecting the specifications for the infiltration well, an opening rate of at least 20% should be chosen.

(2) Impact of collector diameter:

To investigate the effect of collector diameter on drainage time, a strategy was adopted to modify the collector diameter based on the original layout, with the addition of both a larger diameter and a smaller diameter control group, as shown in Table 5.

Table 4. Simulation results.

Serial Number	Diameter of Infiltration Well (mm)	Collector Diameter (mm)	Porosity (%)	Lateral Spacing (m)	Lateral Diameter (mm)	Whether it Meets the Requirements of Drainage	Maximum Drainage Time (h)
1	200	315	20	100 (66)	100	No	48.74
2					125	No	32.87
3					150	No	26.85
4					180	No	25.61
5					200	No	25.31
6					225	No	25.11
7					250	No	25
8					400	No	24.59
9			25	50 (56)	180	No	27.5
10					200	No	27.4
11					225	No	27.26
12					250	No	27.13
13				100 (56)	180	No	25.61
14					200	No	25.31
15					225	No	25.11
16					250	No	25
17				100 (56)	180	No	27.88
18					200	No	27.74
19					225	No	27.62
20					250	No	27.56

Table 4. *Cont.*

Serial Number	Diameter of Infiltration Well (mm)	Collector Diameter (mm)	Porosity (%)	Lateral Spacing (m)	Lateral Diameter (mm)	Whether it Meets the Requirements of Drainage	Maximum Drainage Time (h)
21	200	315	25	150 (56)	180	No	26.14
22					200	No	25.94
23					225	No	25.83
24					250	No	25.76
25				200 (56)	180	No	26.46
26					200	No	25.99
27					225	No	25.93
28					250	No	25.85
29			30	100 (66)	200	No	25.31
30		400	25	50 (56)	180	Yes	18.23
31					200	Yes	18.19
32					225	Yes	18.17
33					250	Yes	18.15
34				100 (66)	180	Yes	15.04
35					200	Yes	14.82
36					225	Yes	14.61
37					250	Yes	14.49
38				100 (56)	180	Yes	19.04
39					200	Yes	18.87
40					225	Yes	18.73
41					250	Yes	18.60
42				150 (56)	180	Yes	20.31
43					200	Yes	20.01
44					225	Yes	19.63
45					250	Yes	19.51
46				200 (56)	180	Yes	21
47					200	Yes	20.14
48					225	Yes	19.75
49					250	Yes	19.65

Note: The numbers in parentheses in the table represent the number of infiltration wells. "66" refers to the simulation based on the original layout of the block, while "56" corresponds to the simulation where the catchment areas were automatically divided by the MIKE URBAN model.

Table 5. Comparison of collector diameter.

Diameter of Infiltration Well (mm)	Collector Diameter (mm)	Porosity (%)	Lateral Spacing (m)	Lateral Diameter (mm)	Whether It Meets the Requirements of Drainage	Maximum Drainage Time (h)
200	250	20	100	200	No	44.83
	315				Yes	25.31
	400				Yes	14.82
	450				Yes	12.22
	500				Yes	8.43

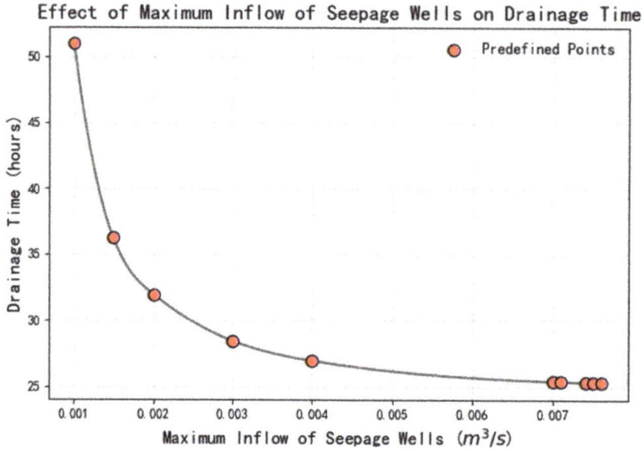

Figure 10. The impact of the maximum inflow of permeable wells on drainage time.

As illustrated in Figure 11, the impact of collector diameter on drainage time follows a quadratic decreasing trend. As the collector diameter increases, the drainage speed significantly improves. This is because a larger collector diameter can better accommodate and transport water flow, thereby enhancing the overall drainage efficiency of the system.

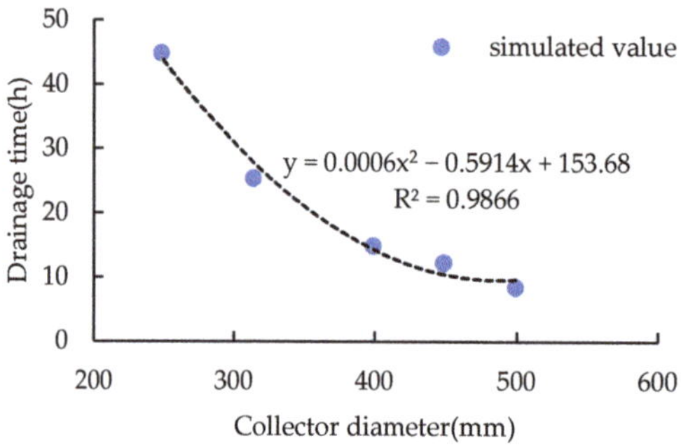

Figure 11. The influence of collector diameter on drainage time.

In the simulation, with the lateral diameter and spacing kept constant, the collector diameter exhibited a negative correlation with drainage time. Therefore, the selection of the collector diameter directly influences the overall drainage capacity of the system.

(3) Impact of lateral spacing:

As shown in Figures 12 and 13, the effect of lateral spacing on drainage performance is relatively complex and exhibits a synergistic interaction with the collector diameter.

When the lateral spacing is small, the high coverage density allows water to be quickly diverted into the drainage network. However, if the collector diameter is small, the inflow from multiple laterals can cause uneven flow distribution, which slows down the drainage process. Increasing the collector diameter can effectively mitigate this issue, thereby improving drainage speed.

Conversely, when the lateral spacing is large, each lateral covers a significantly larger area. Although the localized runoff can still enter the collector, the accumulation and transport distance of water at the lateral ends increases, leading to longer local drainage paths and a slower overall drainage process. Even with a larger collector diameter, this adverse effect cannot be completely offset.

Therefore, an optimal lateral spacing should be within the range of 100 m to 150 m, and its selection should be considered in conjunction with the collector diameter.

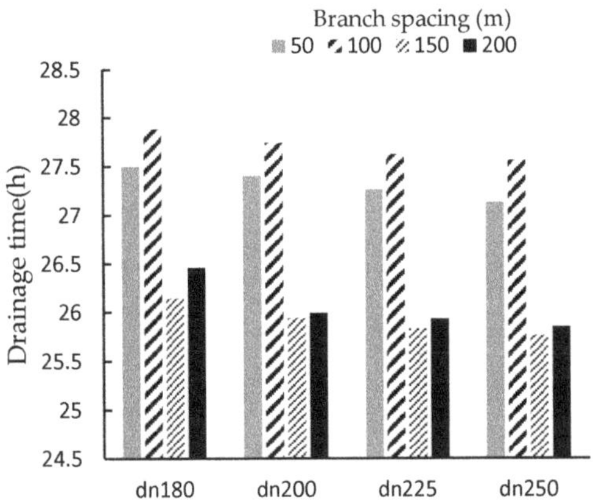

Figure 12. Comparison of drainage time for different lateral diameters under four different lateral spacings when the collector diameter is 315 mm.

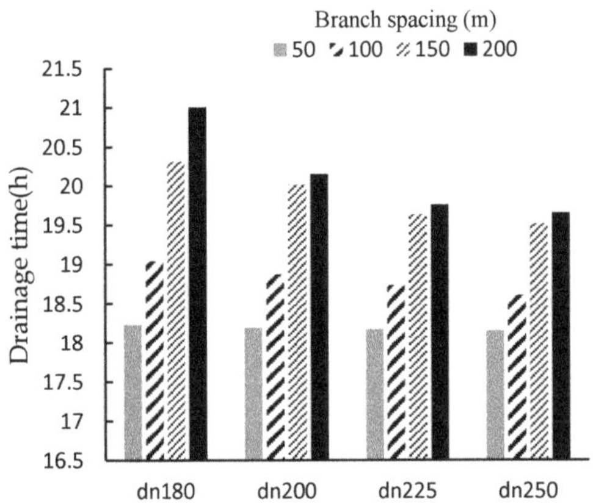

Figure 13. Comparison of drainage time for different lateral diameters under four different lateral spacings when the collector diameter is 400 mm.

(4) Impact of lateral diameter:

To investigate the impact of lateral diameter on drainage time, a strategy was implemented to adjust the lateral diameter while maintaining a collector diameter of 315 mm and a lateral spacing of 100 m.

Drainage time decreases significantly as the lateral diameter increases. However, once the pipe diameter reaches a certain threshold, the rate of improvement gradually slows down. When the lateral diameter increases from 100 mm to 150 mm, the drainage time is drastically reduced from 48.74 h to 26.85 h, indicating that increasing the pipe diameter within the smaller size range has a substantial effect on improving drainage capacity. Primarily, this is because smaller pipe diameters often become a "bottleneck" in the system, and enlarging them can significantly alleviate drainage inefficiencies.

However, when the pipe diameter exceeds 180 mm, the reduction in drainage time becomes much less pronounced. For example, increasing the lateral diameter from 200 mm (25.31 h) to 250 mm (25.00 h) only shortens the drainage time by 0.31 h. This suggests that once the pipe diameter reaches a certain level, the water conveyance capacity is already sufficient to meet drainage demands. Further increasing the diameter yields only marginal efficiency gains, demonstrating a classic case of diminishing marginal returns.

Simulation data indicate that when the lateral diameter reaches approximately 200 mm, the reduction in drainage time levels off. In this range, the relationship between drainage efficiency and pipe diameter enters a "plateau phase", where further increasing the pipe diameter offers little benefit but leads to higher investment and maintenance costs for the drainage network.

4.5. Summary

This chapter conducted a simulation study on the drainage network layout of the irrigation area in Yanggu County, Shandong Province, using the MIKE URBAN model. The study explored the impact of different network parameters (collector diameter, maximum infiltration capacity of the infiltration wells, lateral spacing, and lateral diameter) on drainage efficiency. Based on the analysis of the results, the following key findings regarding factors influencing drainage time were obtained:

(1) Impact of maximum infiltration capacity of infiltration wells:

As the maximum inflow capacity of infiltration wells increases, drainage time gradually decreases. However, when the inflow capacity exceeds 0.0075 m^3/s (27 m^3/h), the drainage speed stabilizes, and further increasing the inflow capacity no longer significantly improves the drainage rate. This indicates that a minimum perforation rate of 20% or higher should be considered when selecting infiltration well specifications.

(2) Impact of collector diameter:

The effect of collector diameter on drainage time follows a parabolic decreasing trend—as the collector diameter increases, the drainage speed significantly improves. This is because larger collectors can better accommodate and transport water flow, thereby enhancing the overall drainage efficiency. In the simulations, given constant lateral diameters and spacing, collector diameter was negatively correlated with drainage time. Therefore, the selection of an appropriate collector diameter is crucial for optimizing the overall system's drainage capacity.

(3) Synergistic effect between lateral spacing and collector diameter:

The interaction between lateral spacing and collector diameter plays a complex role in determining drainage efficiency. Excessively small lateral spacing allows for rapid water diversion, but if the collector diameter is insufficient, multiple laterals discharging simultaneously may cause flow congestion. When collector diameter increases, congestion can be alleviated, resulting in faster drainage. Conversely, excessively large lateral spacing significantly increases the drainage area per lateral and extends the water transmission path at the lateral ends. In such cases, even if the collector diameter is expanded, it may not fully offset the inefficiencies caused by the prolonged drainage pathway. Considering both drainage efficiency and collector diameter, the optimal lateral spacing should be controlled within the range of 100 m to 150 m.

(4) Impact of lateral diameter:

As the lateral diameter increases, drainage time significantly decreases. However, after reaching a certain diameter threshold, the rate of improvement slows down. When the pipe diameter exceeds a certain level, the water conveyance capacity is already sufficient, and further enlarging the diameter only provides marginal efficiency improvements, exhibiting a typical "diminishing marginal returns" effect. Simulation results indicate that when the lateral diameter reaches approximately 200 mm, the reduction in drainage time flattens out, and in this range, the relationship between drainage efficiency and pipe diameter enters a "plateau phase".

5. Optimization of Drainage Networks

This study is based on a 550 m × 550 m experimental field in Yanggu County, Liaocheng City, Shandong Province, where five key variables—collector diameter, lateral diameter, collector length, lateral length, and the number of infiltration wells—were set. Using a genetic algorithm, the drainage network was simulated and optimized, employing tournament selection to retain the best-performing individuals. The optimal network layout was determined with the objective of minimizing the annual cost of the drainage system.

5.1. The Composition of the Annual Cost System

In the process of pipeline network design and optimization, the annual cost system is an important indicator for evaluating the system's economic efficiency. The annual cost is mainly composed of two parts: annual construction costs and annual operation and maintenance (O and M) costs. These two components cover various investments during the pipeline network construction process and various maintenance and operational expenditures during the network operation.

5.1.1. Annual Construction Costs (C_1)

The annual construction cost refers to the various expenditures incurred during the pipeline network construction process, including pipe laying, equipment installation, and all related construction costs. The formula for calculating the annual construction cost is as follows:

$$C_1 = \xi(B + A) \tag{5}$$

where:

ξ—capital recovery factor, $\xi = \frac{(1+r)^y}{(1+r)^y - 1}$, where y is the depreciation period, taken as 30 years; r is the annual interest rate, taken as 6%.

B—Construction and installation costs, which include the following subitems:

(1) Drainage well costs (drainage well unit price × number of drainage wells: $C_m \times n_2$);
(2) Pipe material cost (price per meter of pipe × length: $\Sigma\, C_i \times l_i$, $i = 1, 2\ldots\ldots$);
(3) Pipe installation cost (typically 10% of the material cost) (installation price per meter of pipe * length: $\Sigma\, C_k \times l_i$, $k = 1, 2\ldots\ldots$);
(4) Trench excavation cost: $w = \sum\left(2dh + D_i h + \frac{h^2}{p}\right) \times l_i \times y_w$;
and backfilling costs: $t = \left(\sum\left(2dh + D_i h + \frac{h^2}{p} - \frac{\pi D_i^2}{4}\right) \times l_i \times y_t\right)$, where:
D_i is the pipe diameter in meters, $i = 1, 2\ldots\ldots$; y_w is the excavation unit price in yuan; y_t is the backfilling unit price in yuan; d is the excavation width on both sides of the pipeline; h is the pipeline burial depth; p is the excavation side slope.
(5) Other expenses x (e.g., road crossing facilities, anti-freeze facilities, etc.).

A—Independent fee, which can be calculated as A = (B + E) × β, where β is the proportion of the total engineering investment cost and can be taken as 8%.

5.1.2. Annual Operation and Maintenance Costs (C_2)

The annual operation and maintenance costs mainly include pipeline maintenance and pipeline dredging costs (*M*). The calculation formula for annual operation and maintenance costs is as follows:

$$C_2 = M \tag{6}$$

where:

M—Annual pipeline maintenance and dredging costs, M (per meter pipeline maintenance and dredging cost C_p × required maintenance and dredging pipeline length l_p), which can be taken as 0.5% of the material cost.

5.2. Objective Function

The minimization of annual costs is the core objective of this system. The annual cost F is the sum of the annual construction cost C_1 and the annual operation and maintenance cost C_2. By optimizing the network layout, equipment configuration, and system parameters. The annual cost can be minimized while meeting the drainage standards, thus achieving the most economically optimal network layout.

The objective function is as follows:

$$\min F = \xi \times \left(C_m \times n_2 + \sum C_i \times l_i + w + t + \sum C_k \times l_i + x\right) + C_p \times l_p \tag{7}$$

5.3. Constraint Condition

5.3.1. Pipe Diameter Constraints

Due to the requirements of engineering construction, the pipe diameter should follow standard specifications. Additionally, based on the simulation results above, the constrained range for the collector diameter is determined as follows:

$$200 \leq D_i \leq 450 \tag{8}$$

In the equation:
D_i—Diameter of the i-th collector, mm.
The constraint for the lateral diameter is as follows:

$$100 \leq d_i \leq 250 \tag{9}$$

d_i—Diameter of the i-th lateral, mm.

5.3.2. Pipe Length Constraints

In practical engineering, the length of pipelines cannot be increased or decreased without limitations. Therefore, constraints should be applied to pipeline length. Within the set of pipe diameters, the constraint for the collector length is as follows:

$$50 \leq L_i \leq 450 \tag{10}$$

In the equation:
L_i—The length of the i-th collector, mm.
The constraint for the lateral length is as follows:

$$20 \leq l_i \leq 180 \tag{11}$$

l_i—The length of the i-th lateral, mm.

5.3.3. Time Constraints of Drainage

According to the "Irrigation and Drainage Engineering Design Standard" (GB 50288-2018) [37], to meet the drainage requirements, the drainage time should ensure that the water is drained to the field surface within 24 h after the crop is flooded. Therefore, the drainage time must be constrained. The constraint on drainage time is as follows:

$$0 < t \leq 24 \tag{12}$$

In the equation:
t—Drainage time, h.

5.3.4. Constraints on Lateral Spacing

If the lateral spacing is too large, the coverage area of the infiltration wells will increase, leading to poor drainage in localized areas, which may cause water accumulation problems and fail to meet drainage requirements. When the lateral spacing is smaller, although the lateral coverage density is higher, water can be diverted more quickly into the network. However, if the collector diameter is too small, multiple laterals will merge into the collector, causing flow blockages or uneven flow distribution, which slows down the drainage speed. Additionally, the construction cost of the pipes will increase significantly, reducing the economic feasibility. Based on the above, the constraint on lateral spacing is as follows:

$$100 \leq p \leq 150 \tag{13}$$

In the equation:
p—Lateral spacing, m.

5.3.5. Flow Velocity Constraint

When the flow velocity is too low, it may result in a low drainage efficiency of the system, causing water to remain in the field for more than 24 h, failing to meet the design requirements. Additionally, sediments, suspended particles, and impurities carried by the water may accumulate at the bottom of the pipes, gradually forming sedimentation, leading to blockages and increased maintenance costs. On the other hand, when the flow velocity is too high, the fast-moving water increases the scouring force on the pipes, leading to pipe erosion. Therefore, the constraint on flow velocity is as follows:

$$0.6 \leq v \leq 2.5 \tag{14}$$

In the equation:
v—Flow velocity, m/s.

5.4. Fitness Function

In the genetic algorithm, the fitness function is used to measure the performance of an individual in the population, with the goal of evaluating and ranking individuals to ensure that better individuals can be selected and evolved. To achieve the objective of transforming the constraints into a part of the objective function during the optimization process, penalty terms are introduced in the construction of the fitness function. This means that if a solution violates the pre-set constraints, an additional penalty is added to its objective function value, which worsens its "fitness" and makes it less likely to be selected during the selection operation. The fitness function in this study can be expressed as follows:

$$F = C + P \tag{15}$$

In the equation:
C—Annual cost.
P—Penalty term.

5.5. Optimization Analysis of Pipeline Network

5.5.1. Algorithm Analysis

The optimization control has many parameters, with a population size of 500, a maximum of 400 iterations, and crossover and mutation probabilities set at 70% and

50%, respectively. A higher level of mandatory constraint is applied to the pipe diameter classification, with its penalty coefficient set to 10^5 to quickly eliminate invalid individuals, while the penalty coefficient for other constraints is set to 10^4.

In the previous iteration, the minimum fitness value significantly decreased, fully demonstrating the advantages of tournament selection. At the same time, the convergence curve showed a monotonic decrease, indicating that the offspring retained the advantages of the parents during the genetic process.

5.5.2. Annual Cost Analysis of Pipeline Network

Through the iteration of this population, as shown in Figure 14, the optimal solution for the network's annual cost objective function is obtained after 15 genetic iterations. The fitness function rapidly decreased from an initial value of 351,000 yuan to 55,000 yuan and stabilized at convergence.

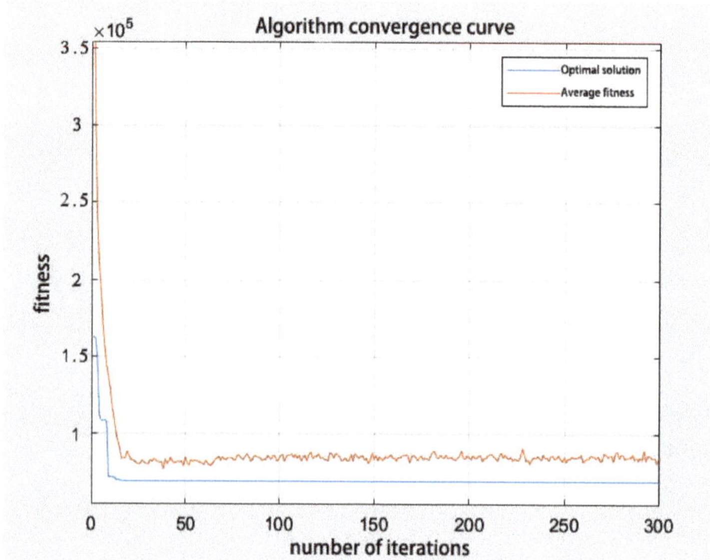

Figure 14. Optimization process based on genetic algorithm.

From the simulation results in the previous and current chapters, it can be seen that compared to the original layout (Scheme 35), although the number of infiltration wells, lateral diameter, and total lateral length have been reduced, the drainage time condition can still be met. According to Table 6, after optimization by the genetic algorithm, the new layout consists of 56 infiltration wells, a collector diameter of 400 mm, a lateral diameter of 160 mm, a total collector length of 500 m, and a total lateral length of 2120 m. Under this layout, the drainage time still meets the requirements, and the annual cost is 45,337.86 yuan. Compared to the original layout of the plot, not only has the layout been optimized, but the annual cost has also been reduced by 24%.

Table 6. Comparison between simulation results and original layout.

	Number of Infiltration Wells	Collector Diameter (mm)	Lateral Diameter (mm)	Collector Total Length (m)	Lateral Total Length (m)	Annual Cost (CNY)
Original layout	66	400	200	500	3000	59,640.67
Simulated layout	56	400	160	500	2120	45,337.86

5.6. Summary

This section systematically analyzes the two main components of annual costs for irrigation and drainage networks, annual construction costs and annual operation and maintenance costs, and provides an economic evaluation basis for network optimization. During the network design process, annual construction costs include pipeline installation, equipment installation, and related construction expenses, while annual operation and maintenance costs cover pipeline maintenance, dredging costs, and other expenses. By quantifying these costs, the contribution of each cost component to the total annual cost can be clarified, providing a scientific basis for optimizing the network layout. In this chapter, the genetic algorithm is applied to the case study of the experimental fields in Yanggu County, Liaocheng City, Shandong Province. The objective is to minimize the annual network cost, subject to constraints on pipe diameter, pipe length, lateral spacing, drainage time, and flow velocity. Penalty terms are introduced, and the fitness function for the genetic algorithm is constructed. With a population size of 500, a maximum of 400 iterations, and crossover and mutation probabilities of 70% and 50%, respectively, the algorithm is iterated, and after 15 iterations, the results converge. The optimized layout results in an annual cost of 45,337.86 yuan, compared to the original layout's cost of 59,640.67 yuan, a 24% savings in annual costs, demonstrating that network simulation and optimization can provide more rational and economical guidance for actual agricultural field layouts.

6. Conclusions

This study, based on the infiltration well overflow capacity test, quantifies the overflow capacity of infiltration wells, enabling the validation of the MIKE URBAN model. The validation results show that the Average Relative Errors (ARE) for different scenarios were 2.29%, 6.52%, 4.41%, 3.17%, 4.37%, and 5.69%, demonstrating the good simulation accuracy of the model. Furthermore, a hydraulic database for MIKE URBAN was constructed. The MIKE URBAN model was used to simulate different drainage network layouts and analyze the influence of different network parameters (e.g., maximum inflow of infiltration wells, collector diameter, lateral spacing, and lateral diameter) on drainage time. The simulation results indicate that the maximum inflow of infiltration wells, collector diameter, lateral spacing, and lateral diameter significantly affect drainage time.

The maximum inflow of infiltration wells shows a negative correlation with drainage time. As the maximum inflow increases, the drainage time decreases. However, when the maximum inflow exceeds 0.0075 m^3/s (27 m^3/h), the drainage speed stabilizes, and further increases in inflow have no significant impact. This suggests that when selecting the specifications for infiltration wells, an opening ratio of at least 20% should be chosen.

The influence of collector diameter on drainage time follows a quadratic decreasing pattern. As the diameter of the collector increases, the drainage speed significantly improves. This is because larger collectors can better carry and transmit water, improving the overall drainage efficiency of the system. In the simulations, with lateral diameter and spacing remaining the same, there was a negative correlation between the collector diameter and drainage time. Therefore, the choice of collector diameter directly impacts the overall drainage capacity of the system.

The effect of lateral spacing on drainage efficiency is more complex and shows a synergistic effect with the collector diameter. When the lateral spacing is small, the coverage density of the laterals is high, enabling faster diversion of water. However, when the collector diameter is small, multiple laterals converging into the collector lead to uneven flow distribution, slowing down the drainage speed. As the collector diameter increases, this congestion is alleviated, improving drainage speed. On the other hand, larger lateral spacing significantly increases the coverage area of each lateral. While the water collected from these larger areas may flow smoothly into the collector, the water flow accumulates, and its transmission distance increases, leading to longer drainage paths and slower drainage. Even if the collector diameter increases, this negative effect cannot be fully offset. Therefore, lateral spacing should be between 100 m and 150 m and should be considered in conjunction with the collector diameter.

As the lateral diameter increases, the drainage time decreases significantly, but after the diameter reaches a certain level, the rate of improvement in drainage time slows down. When the lateral diameter increases from 100 mm to 150 mm, the drainage time is reduced significantly from 48.74 h to 26.85 h. This shows that in the smaller pipe diameter range, increasing the diameter greatly improves the drainage capacity, as small diameters tend to create a "bottleneck" in the system. Expanding the diameter alleviates drainage issues significantly. When the diameter exceeds 180 mm, the reduction in drainage time continues but at a much slower rate (from 25.31 h with 200 mm to 25 h with 250 mm, only a 0.31 h decrease), suggesting that after a certain pipe diameter is reached, the water flow capacity is sufficient to meet drainage requirements. Further increases in pipe diameter only result in minimal improvements, showing typical "diminishing marginal returns." Simulation data show that when the lateral diameter reaches about 200 mm, the drainage time reduction stabilizes. Within this range, the relationship between drainage efficiency and pipe diameter enters a "plateau phase," and further increases in pipe diameter yield negligible benefits while increasing investment and maintenance costs.

This study, based on the 550 m × 550 m experimental field in Yanggu County, Liaocheng City, Shandong Province, sets five variables for the optimization of the irrigation and drainage network layout: collector diameter, lateral diameter, collector length, lateral length, and the number of infiltration wells. The study sets five constraints: pipe diameter, pipe length, drainage time, lateral spacing, and flow velocity, and uses a genetic algorithm to optimize the network layout. Different penalty coefficients were applied for each constraint, with a population size of 500, a maximum of 400 iterations, and crossover and mutation probabilities of 70% and 50%, respectively. After 15 iterations, the results stabilized. The optimized layout was determined with 56 infiltration wells, a collector diameter of 400 mm, a lateral diameter of 200 mm, a total collector length of 500 m, and a total lateral length of 2120 m, resulting in an annual cost of 45,337.86 yuan. This layout saved 24% in annual costs compared to the original design, demonstrating that the simulation optimization of the network can provide more rational and cost-effective guidance for actual agricultural field layouts.

Author Contributions: M.L. , H.R. and Z.Z. designed the review. M.L. analyzed the data and wrote the manuscript. M.L., H.R. and Z.Z. improved the figures and manuscript writing. Z.Z., H.R. and T.W. read and approved the contents. All authors have read and agreed to the published version of the manuscript.

Funding: This study was funded by the Science and Technology Major Project of the Ministry of Water Resources of China (SKS-2022050), National Key Research and Development Program of

China (NO. 2023YFD1900802-01), The Gansu Province Science and Technology Program—East–West Cooperation Special Project (23CXNA0025).

Data Availability Statement: The data presented in this study are available upon request from the corresponding author. The data are not publicly available due to institutional policies regarding sensitive information.

Conflicts of Interest: The authors declare that the review was written in the absence of any commercial or financial relationships that could be construed as a potential conflict of interest.

References

1. IPCC. *Climate Change 2021: The Physical Science Basis. Contribution of Working Group i to the Sixth Assessment Report of the Intergovernmental Panel on Climate Change*; Cambridge University Press: Cambridge, UK, 2021.
2. Narayan, B.; Seok, K.H.; Su-Gon, H.; Tae-Myung, Y. Photosynthetic traits and plantâ-water relations of two apple cultivars grown as bi-leader trees under long-term waterlogging conditions. *Environ. Exp. Bot.* **2020**, *176*, 104111.
3. Jiang, W.; Sheng, Y.; Shi, Z.; Guo, H.; Chen, X.; Mao, H.; Liu, F.; Ning, H.; Liu, N.; Wang, G. Hydrogeochemical characteristics and evolution of formation water in the continental sedimentary basin: A case study in the qaidam basin, china. *Sci. Total Environ.* **2024**, *957*, 177672. [CrossRef] [PubMed]
4. Yuan, Y.; Zhang, Q.; Chen, S.; Li, Y. Evaluation of comprehensive benefits of sponge cities using meta-analysis in different geographical environments in china. *Sci. Total Environ.* **2022**, *836*, 155755. [CrossRef] [PubMed]
5. FAO. *Water for Sustainable Food and Agriculture*; Food and Agriculture Organization of the United Nations: Rome, Italy, 2017.
6. Elliott, J.; Deryng, D.; Muller, C.; Frieler, K.; Konzmann, M.; Gerten, D.; Glotter, M.; Florke, M.; Wada, Y.; Best, N.; et al. Constraints and potentials of future irrigation water availability on agricultural production under climate change. *Proc. Natl. Acad. Sci. USA* **2014**, *111*, 3239–3244. [CrossRef]
7. Fishman, R.; Devineni, N.; Raman, S. Can improved agricultural water use efficiency save india's groundwater? *Environ. Res. Lett.* **2015**, *10*, 84022. [CrossRef]
8. Rengasamy, P. Soil processes affecting crop production in salt-affected soils. *Funct. Plant Biol.* **2010**, *37*, 613–620. [CrossRef]
9. Kulhavý, Z.; Doležal, F.; Fučík, P.; Kulhavý, F.; Kvítek, T.; Muzikář, R.; Soukup, M.; Švihla, V. Management of agricultural drainage systems in the czech republic. *Irrig. Drain.* **2007**, *56*, S141–S149. [CrossRef]
10. Satyanarayana, T.V.; Boonstra, J. Subsurface drainage pilot area experiences in three irrigated project commands of andhra pradesh in india. *Irrig. Drain.* **2007**, *56*, S245–S252.
11. Wang, S. Application of Integrated Drainage and Irrigation Channels in the Basic Farmland Protection Demonstration Area of Mudanjiang. *Water Resour. Plan. Des.* **2013**, *2013*, 14–15. [CrossRef]
12. Gao, H.; Liu, X.; Gao, X.; Qu, Z. Salt Content and Ion Migration Patterns of Saline-Alkali Soils under Different Irrigation and Drainage Modes and Their Equilibrium Analysis. *J. Soil Water Conserv.* **2023**, *37*, 361–370. [CrossRef]
13. Sun, J.; Yu, M. Dynamic Simulation and Evaluation of Soil Salinity under Different Irrigation and Drainage Modes. *Agric. Res. Arid Areas* **2011**, *29*, 157–163.
14. Zhang, J.; Liu, M.; Qian, H.; Zhang, Q.; Wang, Z. Characteristics of Water and Salt Spatiotemporal Variation in Coastal Saline Soil Improved by Furrow Irrigation, Leaching, and Subsurface Drainage Synergy. *Trans. Chin. Soc. Agric. Eng.* **2018**, *34*, 98–103. [CrossRef]
15. Hornbuckle, J.W.; Christen, E.W.; Faulkner, R.D. Evaluating a multi-level subsurface drainage system for improved drainage water quality. *Agric. Water Manag.* **2007**, *89*, 208–216.
16. Wu, F.; Liu, L.; Zai, S.; Wei, H.; Li, X.; Liu, D. Effects of Straw Composite Pipe Subsurface Irrigation on Winter Wheat Growth and Water Use Efficiency. *Water-Sav. Irrig.* **2024**, *67*, 50–55. [CrossRef]
17. Smith, D.K.; Walters, G.A. An evolutionary approach for finding optimal trees in undirected networks. *Eur. J. Oper. Res.* **2000**, *120*, 593–602. [CrossRef]
18. Lin, L.Y.; James, T.; Fletcher, L. Irrigation pipe network design using geographic information system. In *Environmental and Pipeline Engineering 2000: Proceedings of the ASCE National Conference on Environmental and Pipeline Engineering, July 23–26, 2000, Kansas City, MO, USA*; American Society of Civil Engineers: Reston, VA, USA, 2000. [CrossRef]
19. Alandi, P.P.; Martin-Benito, J.M.T.; Alvarez, J.F.O. Optimization of irrigation water distribution networks, layout included. *Agric. Water Manag.* **2007**, *88*, 110–118. [CrossRef]
20. Ostfeld, A.; Oliker, N.; Salomons, E. Multiobjective optimization for least cost design and resiliency of water distribution systems. *J. Water Resour. Plan. Manag.-Asce* **2014**, *140*, 12. [CrossRef]
21. Yoo, D.G.; Kim, Y.H.; Kim, Y.D.; Cho, J.; Kim, J.H. Development of optimal pipe size design tool for irrigation systems and its application to saemangeum reclamation area. *Irrig. Drain.* **2016**, *65*, 58–68. [CrossRef]

22. Gajghate, P.W.; Mirajkar, A. Irrigation pipe network planning at tertiary level: An indian case study. *Ksce J. Civ. Eng.* **2020**, *24*, 322–335. [CrossRef]
23. Srivastava, S.K.; Kumar, R.; Dikshit, P.K.S. Optimization of irrigation pipe network design using genetic algorithms. *I-Manag. J. Civ. Eng.* **2015**, *5*, 31–38.
24. Bi, W.; Dandy, G.C.; Maier, H.R. Improved genetic algorithm optimization of water distribution system design by incorporating domain knowledge. *Environ. Model. Softw. Environ. Data News* **2015**, *69*, 370–381. [CrossRef]
25. Wang, K.; Dal, M.R.M. Optimization and modelling of pressurized irrigation networks. *Turk. J. Water Sci. Manag.* **2017**, *1*, 62–80.
26. Alaggio, J.; Barros, D.; Brentan, B.; Carpitella, S.; Herrera, M.; Izquierdo, J. Assessment of criticality in water distribution networks via complex network theory. *Processes* **2025**, *13*, 408. [CrossRef]
27. Wang, T.; Wang, Z.; Chen, L.; Zhang, J.; Li, W. A Review of the Effects of Integrated Irrigation and Drainage Projects on Groundwater Depth and Crop Growth. *J. Water Resour. Water Eng.* **2020**, *31*, 174–180. [CrossRef]
28. Sun, X. A Comprehensive Review of Low-Pressure Pipe Irrigation and Drainage Integration Technology. *Heilongjiang Water Resour. Sci. Technol.* **2012**, *40*, 191. [CrossRef]
29. Liang, C.; Zhou, B.; Chen, X.; Li, S. Application and Progress of the Drainmod Model in Agricultural Drainage Systems. *J. Irrig. Drain.* **2020**, *39*, 101–106. [CrossRef]
30. Planchon, O.; Delestre, O.; Cordier, S.; Lucas, C.; Du, M.; Laguerre, C.; Darboux, F.; James, F. *Fullswof: A Software for Overland Flow Simulation*; Gourbesville, P., Caignaert, G., Cunge, J., Eds.; Springer Singapore Pte. Limited: Singapore, 2014; pp. 221–231.
31. Ion, S.; Marinescu, D.; Cruceanu, S.G. Asterix: Module for modelling the water flow on vegetated hillslopes. *Environ. Model. Softw.* **2025**, *186*, 106336. [CrossRef]
32. Xie, Y.; Xu, M.; Gao, S.; Shen, L. Simulation of Water and Nitrogen Transport in Paddy Fields under Different Irrigation and Drainage Modes Based on the Drainmod Model. *J. Irrig. Drain.* **2021**, *40*, 37–44. [CrossRef]
33. Gao, Z.; Zhong, R.; Yang, S.; Li, X.; Yang, X. Recent Research and Application Progress of the Hydrus Model in China. *Soil* **2022**, *54*, 219–231. [CrossRef]
34. *ISO 1438*; Hydrometry—Open Channel Flow Measurement Using Thin-plate Weirs. British Standards Institute: London, UK, 2017.
35. *China High-Standard Farmland Construction Plan (2021–2030)*; Ministry of Agriculture and Rural Affairs of the People's Republic of China: Beijing, China, 2021.
36. *Technical Specifications for Farmland Drainage Engineering*; Ministry of Water Resources of the People's Republic of China: Beijing, China, 2020.
37. *GB50288-2018*; Design Standard for Irrigation and Drainage Engineering. The Ministry of Housing and Urban Rural Development of the People's Republic of China and the General Administration of Quality Supervision, Inspection and Quarantine of the People's Republic of China: Beijing, China, 2018.

MDPI AG
Grosspeteranlage 5
4052 Basel
Switzerland
Tel.: +41 61 683 77 34

Water Editorial Office
E-mail: water@mdpi.com
www.mdpi.com/journal/water

www.ingramcontent.com/pod-product-compliance
Lightning Source LLC
Chambersburg PA
CBHW061518210926
51850CB00008B/1966
* 9 7 8 3 7 2 5 8 6 6 5 0 2 *